MATERIALS SCIENCE AND TECHNOLOGIES

PROCEEDINGS OF THE 2018 INTERNATIONAL CONFERENCE ON "PHYSICS, MECHANICS OF NEW MATERIALS AND THEIR APPLICATIONS"

MATERIALS SCIENCE AND TECHNOLOGIES

Additional books and e-books in this series can be found on Nova's website under the Series tab.

MATERIALS SCIENCE AND TECHNOLOGIES

PROCEEDINGS OF THE 2018 INTERNATIONAL CONFERENCE ON "PHYSICS, MECHANICS OF NEW MATERIALS AND THEIR APPLICATIONS"

IVAN A. PARINOV
SHUN-HSYUNG CHANG
AND
YUN-HAE KIM
EDITORS

nova
science publishers
New York

Library of Congress Cataloging-in-Publication Data

ISBN: 978-1-53615-862-5
Library of Congress Control Number:2019944362

Published by Nova Science Publishers, Inc. † New York

CONTENTS

PREFACE

Novel materials are very important for a whole set of prospective directions of modern science and techniques, defining the most interesting areas of modern technologies and industry. The developed conceptions, approaches and methods to study of advanced materials and composites initiate the main achievements and define possibilities of modern natural and technical sciences. Direct improvements of the characteristics of materials and devices are based on different chemical, physical and mechanical studies, methods of mathematical modeling and physical experiment. Broad set of applications determine constant and tremendous interest to these studies. Rapid growth of new knowledge and R&D of theoretical, experimental and numerical methods and approaches state the problem of continuous improvement of experimental facilities, and development of original computer hard- and software of higher level. Obviously, new scientific knowledge is based on the results of these researches, which give a possibility to understand and estimate very fine technological processes and transformations of structural-sensitive properties, taking place in manufacture of novel developed materials, composites and goods. These discoveries define optimal physical and mechanical loads and operation of devices in the conditions of critical impacts and fields.

This collection of 50 papers presents selected reports of the 2018 International Conference on “Physics, Mechanics of New Materials and Their Applications” (PHENMA-2018), which has been taken place in Busan, South Korea, 9-11 August, 2018 (http://phenma2018.math.sfedu.ru) The conference was sponsored by the Busan Tourism Organization (South Korea), Ministry of Education and Science of Russian Federation, South Scientific Center of Russian Academy of Science, Russian Foundation for Basic Research, Ministry of Science and Technology of Taiwan, The Korean Society of Ocean Engineering (South Korea), New Century Education Foundation (Taiwan), Ocean & Underwater Technology Association (Taiwan), Unity Opto Technology Co. (Taiwan), Fair Well Fishery Co. (Taiwan), Woen Jinn Harbor Engineering Co. (Taiwan), Lorom Group (Taiwan), Longwell Co. (Taiwan), University of 17 Agustus 1945 Surabaya

(Indonesia), University of 45, Surabaya (Indonesia), University of Islam Kadiri (Indonesia), University of Darul Ulum, Jombang (Indonesia), University of Maarif Hasyim Latif, Sidoarjo (Indonesia), PDPM Indian Institute of Information Technology, Design and Manufacturing (India), Don State Technical University (Russia), South Russian Regional Centre for Preparation and Implementation of International Projects.

The topics of the PHENMA-2018 continued ideas of previous international symposia and conferences: PMNM-2012 (http://pmnm.math.rsu.ru), PHENMA-2013 (http://phenma.math.sfedu.ru), PHENMA-2014 (http://phenma2014.math.sfedu.ru), PHENMA-2015 (http://phenma2015.math.sfedu.ru), PHENMA-2016 (http://phenma2016.math.sfedu.ru) and PHENMA-2017 (http://phenma2017.math.sfedu.ru), whose results have been published in the following edited books "Physics and Mechanics of New Materials and Their Applications", Ivan A. Parinov, Shun Hsyung-Chang (Eds.), Nova Science Publishers, New York, 2013, 444 p. ISBN: 978-1-62618-535-7; "Advanced Materials – Physics, Mechanics and Applications", Springer Proceedings in Physics. Vol. 152. Shun-Hsyung Chang, Ivan A. Parinov, Vitaly Yu. Topolov (Eds.), Springer, Heidelberg, New York, Dordrecht, London, 2014, 380 p. ISBN: 978-3319037486; "Advanced Materials – Studies and Applications", Ivan A. Parinov, Shun-Hsyung Chang, Somnuk Theerakulpisut (Eds.), Nova Science Publishers, New York, 2015, 527 p. ISBN: 978-1-63463-749-7; Proceedings of the 2015 International Conference on Physics and Mechanics of New Materials and Their Applications, devoted to 100-th Anniversary of the Southern Federal University, Ivan A. Parinov, Shun-Hsyung Chang, Vitaly Yu. Topolov (Eds.). Nova Science Publishers, New York, 2016, 582 p. ISBN: 978-1-63484-577-9, Proceedings of the 2016 International Conference on Physics and Mechanics of New Materials and Their Applications, Ivan A. Parinov, Shun-Hsyung Chang, Muaffaq A. Jani (Eds.). Nova Science Publishers, New York. 2017, 794 p. ISBN: 978-1-53611-033-3 and Proceedings of the 2017 International Conference on Physics and Mechanics of New Materials and Their Applications, Ivan A. Parinov, Shun-Hsyung Chang, Vijay K. Gupta (Eds.). Nova Science Publishers, New York. 2018, 619 p., ISBN: 978-1-53614-083-5, respectively.

The presented reports are divided into five scientific directions: (i) processing techniques, (ii) physics, (iii) mechanics, (iv) applications, and (v) industry and management.

These PHENMA 2018 Proceedings discuss numerous and promising problems, arising in studying advanced materials, composites with specific properties and also devices, manufactured on their base. The proposed theoretical and experimental approaches cover new processing methods of nanomaterials, piezoelectrics (including environmentally-friendly) and other advanced materials and composites. In particular, this book presents numerous results of theoretical and experimental investigations of advanced materials and devices with before-given and improved structure-sensitive properties, developed on the base of methods of inorganic and organic chemistry, modern medicine, electric elasticity and physics of condensed matter. The obtained results also include computational algorithms and original software, used in realization of numerical methods (in particular, finite-element modeling), showing new fascinating results for wide spectrum of advanced materials (which, in particular, could be obtained due to reprocessing natural materials, wasters, fruits and plants) and devices. The scientific developments, new goods and obtained results, presented in the book, demonstrate higher and improved comparative characteristics. They give a new knowledge, which is necessary for numerous applications and industrial developments. The original theoretical, numerical and experimental methods, fabricated devices and set-ups show significant possibilities in expanding the research spectrum of various physical processes and phenomena. They provide different improvements in investigation of various structure-sensitive characteristics of solids and media.

The results of the book are related to the mathematical modeling and experimental studies of advanced devices, in particular: energy-harvesters, piezogenerators, piezoelectric transducers and sensors, medical technologies etc., based on developed nano-structures, ferro-piezoelectrics and other materials with specific characteristics. The book studies very interesting modern nano- and microstructure techniques for manufacture of modern materials (for example, ZnO nanostructures), which present great interest for educational process and industrial purposes, unification and modernization of various expertise, design and analysis. Many achievements discussed in the book are linked with advanced devices, demonstrating high exactness, longevity, durability and extended opportunities to work under critical conditions. The improved properties, shown by developed materials and structures, open new possibilities to investigate numerous physical-mechanical processes and phenomena.

The book will be useful to students, post-graduate students, scientists and engineers, investigating and developing a new generation of nano-materials and nano-structures, piezo-ferroelectrics, other advanced materials with special properties, and also different devices, manufactured on their base and used in various areas of science, technique and technology. The book presents new research methods and scientific results in the Condensed Matter Physics, Materials Science, Physical and Mechanical Experiment, Processing Techniques and Engineering of Nanomaterials, Piezoelectrics, and other Advanced Materials and Composites, Numerical Methods, and also different applications and developed devices.

Ivan A. Parinov (Ed.)
Shun-Hsyung Chang (Ed.)
Yun Hae Kim (Ed.)

December, 2018

PART 1. PROCESSING TECHNIQUES

In: Proceedings of the 2018 International Conference … ISBN: 978-1-53615-862-5
Editors: Ivan A. Parinov et al.

Chapter 1

STRUCTURAL FEATURES OF NANOCOMPOSITE ION-PLASMA DEPOSITED COATINGS CONTAINING ALUMINUM

V. N. Varavka, O. V. Kudryakov, I. Yu. Zabiyaka and N. I. Bronnikova
Department of Material Science, Don State Technical University, Rostov-on-Don, Russia

ABSTRACT

The nanocomposite coatings of the Ti/AlSi system obtained by ion-plasma technology using magnetron evaporation are investigated. Experimental data on the structure and elemental composition of coatings are presented. The emphasis is on two features of coatings: the formation of 3D-morphology of the structure instead of the expected layered 2D-morphology and low Si content in the coating. A physical concept for the formation of a 3D-nanocomposite in Ti/AlSi coatings is proposed. Its key points are the influence of a low-melting element (Al), the possibility of a significant decrease in the temperature of structure formation under vacuum conditions, the occurrence of nonvariant transformations involving the liquid phase.

Keywords: ion-plasma deposition, composite coatings, nanostructure, scanning electron microscopy, spectral analysis

1. INTRODUCTION

The structure and properties of nanocomposite coatings obtained by ion-plasma deposition technology have been studied in a number of our previous papers [1 – 4]. Among the coatings studied there were systems including Al and Si. The results of the studies showed significant differences in the structure and properties of such coatings in comparison with coatings of other cermet systems. We attributed these differences to the influence of silicon, which causes amorphization of the coatings, which was confirmed by the results of qualitative X-ray diffraction analysis. However, strictly speaking, this hypothesis needed a broader argumentation – it required deeper instrumental diagnostics and theoretical justification. It was this task that was posed in the present paper.

2. SAMPLES AND EXPERIMENTAL METHODS

The Ti/AlSi coatings obtained using ion-plasma technology on samples of steel 20Cr13 (AISI 420) were used as an object of research. The base metal of the samples had a sorbite structure formed by standard bulk hardening with high tempering. The purity of the surface of the base metal before the coating was $R_Z = 0.08$ μm.

Coatings were obtained using an ion-plasma deposition unit with an unbalanced magnetron evaporation system. The technological task of coatings was formulated as the preparation of a multilayer composite coating (2D-nanocomposite) with a total thickness of 3 – 15 μm (optimally 6 – 8 μm) at a thickness of individual monolayers of 5 – 30 nm.

The structural MDT-diagram (the Movchan-Demchishin-Thornton diagram) was used as the theoretical base for obtaining defect-free nanocrystalline coatings [5, 6]. The usual technology of applying such coatings involves, as an initial operation, ion implantation (assisting) to a depth of about 10 nm by ions of one of the metals that make up the composite. Further, to ensure maximum adhesion of the coating to the surface of the base metal, a base layer is applied and only then the coating itself is formed. Due to the high adhesion to the steel surface, pure Ti is often used as the base layer (especially when nitride coatings are applied). If Ti is part of the composite, as in the Ti/AlSi system, then its layer is applied first, plays the role of the base layer and has a somewhat larger thickness than in the composite.

The structure of the coatings was studied using a high-resolution scanning electron microscope (SEM) “Mira Tescan 3 LMU”. It provides a study of the structure of the material in the nanoscale, and is also equipped with a probe energy-dispersive microanalyzer (EDMA) of the chemical composition “X-Max EDAX” (Oxford Instruments, UK). In addition, a high-precision GD-PROFILER 2 glow discharge spectrometer (HORIBA Jobin Yvon, France) was also used to determine the elemental composition of the coating. It allows to obtain and analyze the total emission spectrum of the material using the built-in library of spectral lines, has a spectral range of 120 – 700 nm, the resolution of the polychromator is 0.007 nm, the resolution of the monochromator is 0.015 nm, the sensitivity for the detectable elements is in the range from 1 to 10 ppm.

3. Results and Discussion

The typical structure of a 2D-nanocomposite coating obtained by ion-plasma technology is a dense stacking of several hundred layers of nanometric thickness (Figure 1 a). Such a structure had coatings of many systems studied by us (Ti/C, Ti/Mo, TiN/MoN, Ti/Zr(Nb), TiN/Zr(Nb)N, etc.). Exceptions were coatings, which included Al and Si (Ti/AlSi, TiN/AlSiN). The structure of these coatings can not be attributed to 2D-nanocomposites, since it is not layered (Figure 1b).

Figure 1b clearly shows that the initial stages of coating formation have a clear correlation with the ion-plasma technology used: a base Ti layer (a light horizontal interlayer in the central part of Figure 1b) is formed on the surface of the base metal, about two dozen layers of nanometric thickness are located on it. But with further deposition, the layered 2D-morphology of the coating gets broken and becomes 3D-morphology (a matrix with dispersed weakly ordered particles of compact form).

Table 1. Data of statistical metallographic measurements and chemical analysis (EDMA) of Ti/AlSi coating samples

Samples of coatings	Dark phase area, %		Average size of particles (dark phase), nm		Elemental composition of coating (relative units)
	in different zones	averaged	in different zones	averaged: value/range	
1 (Figure 1b)	21.3	21.25	56.6	49.8 / 10 – 60	Al-40 Ti-16 Si-6
	21.85		48.75		
	20.6		44.17		
2	9.5	10.33	22.0	21.94 / 10 – 40	
	9.83		22.6		
	11.66		21,22		
3 (Figure 2a)	14.3	16.62	50.7	51.7 / 10 – 50	Al-100 Ti-40 Si-15
	19.45		49.6		
	16.11		54.87		
Average statistical values		**19.5**	**48.5***		–

*Computer analysis did not take into account the elongated shape of the particles.

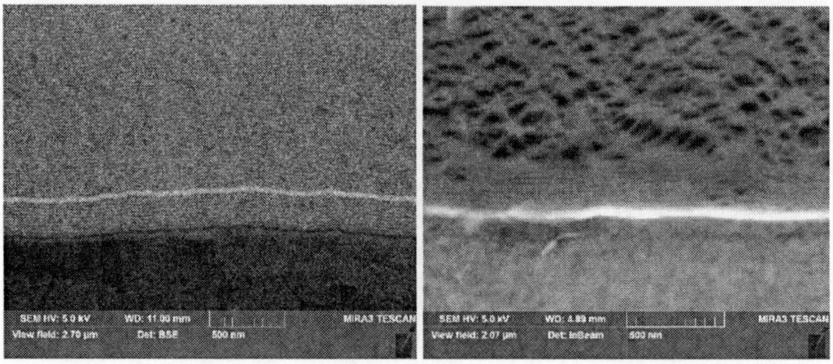

(a) (b)

Figure 1. Structure of nanocomposite coatings in the cross-section (SEM): (a) 2D nanocomposite of the Ti/Zr(Nb) system; (b) 3D nanocomposite of Ti/AlSi system.

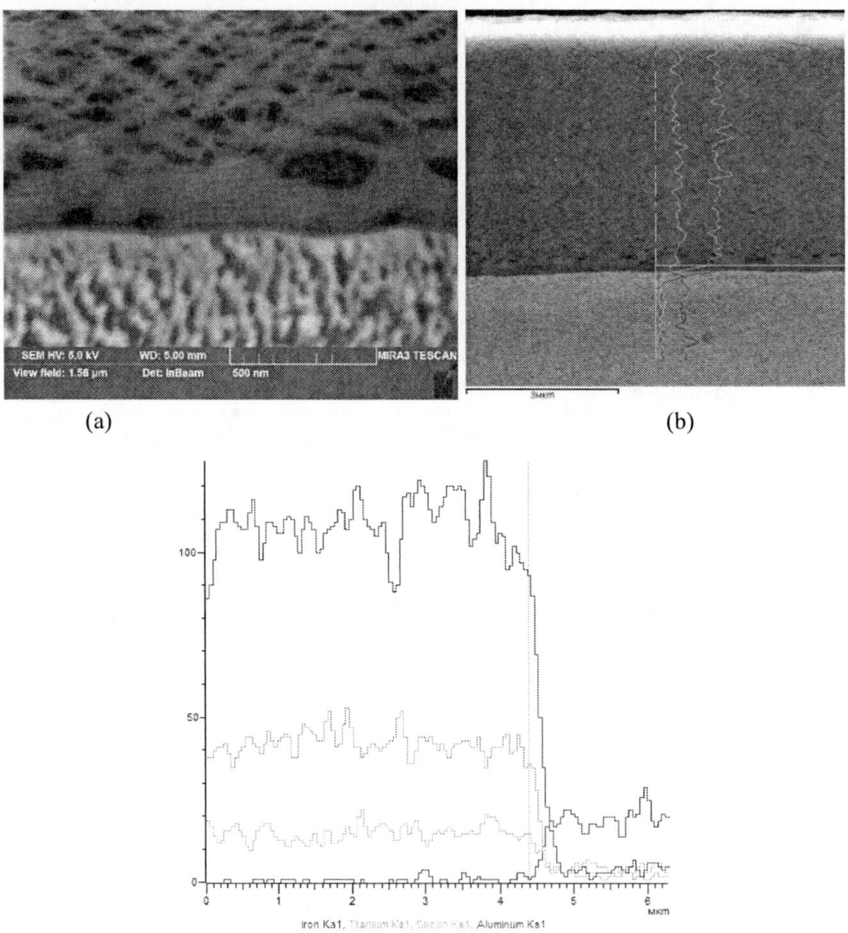

(a) (b)

(c)

Figure 2. Structure of the nanocomposite coating of the Ti/AlSi system (a) and its elemental composition (EDMA data): (b) alignment of the distribution curves of elements with the coating structure; (c) the distribution of elements along the depth of the coating and the base metal (in relative units); in all illustrations (a-c) the same sample is presented.

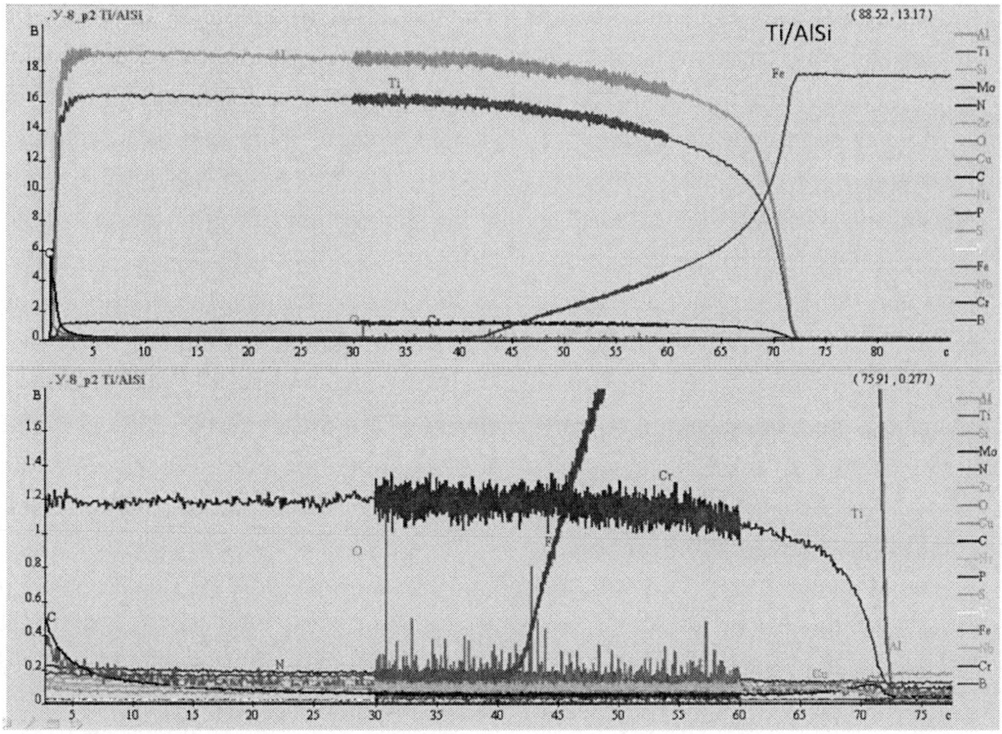

Figure 3. Element composition of the nanocomposite coating of the Ti/AlSi system (data of the spectral analyzer GD-PROFILER 2).

A formal evaluation of the morphology of particles in Ti/AlSi coatings shows that the scatter of particles by size in the cross-section of the coating is sufficiently large (Figure 1 b). The shape of the inclusions is nonuniform with the approximate ratio of the axes 1: (2.0 – 2.5), whereby the major axis is oriented predominantly parallel to the surface of the base metal. The results of computer statistical metallographic analysis, performed on microphotographs, are present in Table 1. This table, as well as Figure 2, gives data on the elemental composition of coatings.

The elemental composition of the coating is very important for understanding of the processes that take place in the formation of the Ti/AlSi coating structure. However, due to the specific shape of the probe and its relatively large size, characteristic for EDMA, for heterogeneous structures this method can give significant discrepancies. The glow discharge spectrometer GD-PROFILER 2 provides a higher level of data reliability and encompasses a wider range of elements. Its results are shown in Figure 3.

Summarizing the distribution of chemical elements in the coating obtained by different methods, it should be noted that the main element in the coating is Al, and the amount of Si is so insignificant that the investigated coatings of the Ti/AlSi system can in fact be considered as two-component Al-Ti. Apparently, the low Si content in the coating is due to the specificity of the magnetron excitation of the target atoms upon evaporation of the elements in the ion-plasma setting. It should also be emphasized that both methods used to analyze the elemental composition have methodological limitations in determining the composition of nanoobjects. In other words, from the data obtained it is impossible to accurately determine the chemical composition of the matrix and particles of the 3D-nanocomposite coating of the Ti/AlSi system. This question requires an analytical study.

The starting point for this research is that the parameters of ion-plasma technology used by us usually lead to the formation of multilayer coatings (2D-composite in Figure 1a). However, if the composition of the coating includes Al, the structure of the coating is transformed from the 2D-composite into a 3D-composite (see Figure 1b and 2a). Moreover, the structural transformation of the coating does not start from the surface of the base metal, but from a certain coating thickness of the order of 200 – 300 nm, within which the coating is formed according to the traditional layered 2D-principle. As a discussion of these experimental results, the following reconstruction of the structure formation of the coating can be proposed.

Elemental analysis data show that the coating is dominated by Al, which is a much more fusible element than Si and Ti. At the first stages of ion-plasma deposition, when the thickness of the coating is small, the formation of layers is due solely to the forces of interatomic interaction. As the thickness of the coating increases, it acquires the status of a thermodynamic (more precisely, a physico-chemical) system. In this case, the complex of control parameters of such a system (temperature, volume, pressure) becomes significant. The structure of the coating is determined not only by the interaction forces of individual ions, as in the first stage of formation, but also by the thermodynamic state of the coating, as a system. In this case, starting with a certain volume of the coating, the formation of its structure obeys the thermodynamic laws of phase equilibria, that is, it is determined by the phase diagram of the components. Comparing the above structures (Figure 1a, b), it is difficult to imagine any other physical mechanism for transforming of a 2D structure into a 3D structure, except for a nonvariant one. Moreover, the dominance of the low-melting component (Al) in the Al–Si–Ti system and the observed structural transformation of the coating suggest that the coating during the growth process most likely passes through the liquid-phase stage. This, in particular, explains the fact that the coating is not immediately formed in the form of 3D spatial modification. After all, for the flow, both primary and secondary crystallization, the thermodynamic system must have a certain volume, sufficient for the processes of nucleation and growth of phases in it.

According to the spectral analysis, the Si content in the coating is insignificant (Figure 3). Therefore, we assume that all the Si present in the coating is in the state of a solid solution with Al. In accordance with the phase diagram of Al–Si, the maximum solubility of Si in the Al lattice is 1.5 at.% at a eutectic temperature of 577°C. If it is instrumentally possible to lower the eutectic temperature to the room temperature, that the metastable content of silicon in the solid solution Al(Si) can reach up to 20 at.% in quasi-eutectics [7, 8]. Thus, we can assume that Si will not have an influence on the processes of crystallization and phase transformations in the coating. In this case, their flow should correspond to the phase diagram of Al–Ti shown in Figure 4.

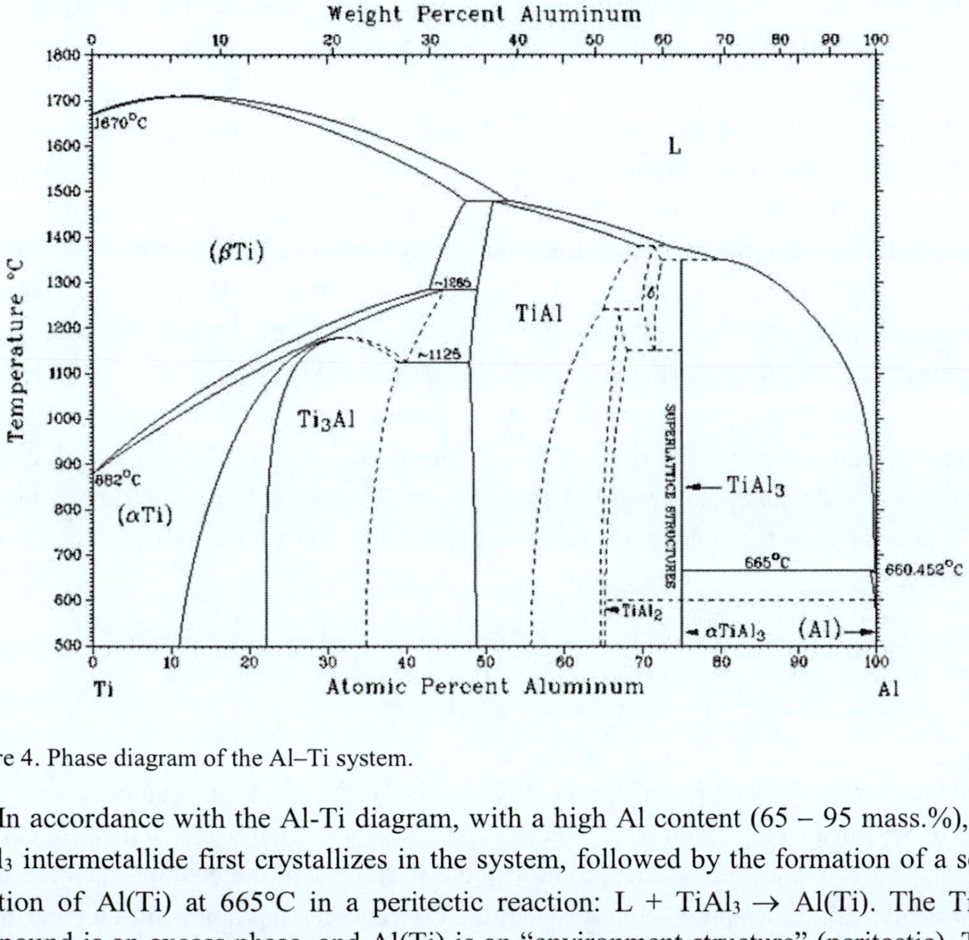

Figure 4. Phase diagram of the Al–Ti system.

In accordance with the Al-Ti diagram, with a high Al content (65 – 95 mass.%), the $TiAl_3$ intermetallide first crystallizes in the system, followed by the formation of a solid solution of Al(Ti) at 665°C in a peritectic reaction: L + $TiAl_3$ → Al(Ti). The $TiAl_3$ compound is an excess phase, and Al(Ti) is an "environment structure" (peritectic). This phase-formation mechanism fully corresponds to the 3D-structure of the coatings in Figure 1b and 2a, where dispersed particles appear to be $TiAl_3$ intermetallides, and the composite matrix is the Al(Ti) peritectic.

In the reconstruction of the phase formation, the question of transformation temperatures remains open, since the given equilibrium temperatures of the peritectic transformation in the Al–Ti system (665°C) or the eutectic transformation in the Al–Si system (577°C) are not achieved when coating is applied. During the technological cycle of coating deposition, the temperature of the surface of the base metal (sample) varies nonlinearly: the initial temperature increase is related to the electron heating of the sample for ion cleaning of its surface, during which the temperature falls. In the future, when the coating is applied, the temperature stabilizes and usually does not exceed 300°C.

Since the coating is considered as a thermodynamic system during the formation, it is necessary to take into account not only the temperature parameter *T*, but also the pressure *P*, which in the working chamber is 0.3 Pa, which is close to the boundary of the medium and high vacuum. The general known regularity, which can be observed on the phase diagrams of any inorganic systems in the *P* – *T* coordinates, is the lowering of the temperature of the phase equilibrium lines with decreasing pressure. In some systems, this dependence may be violated for the lines of phase equilibria of the polymorphic modifications of substances, that is, when all the phases are in the crystalline state. For the lines of equilibrium "vapor – liquid" or "vapor – crystal" in *P* – *T* diagrams of inorganic substances, there are no such exceptions. Therefore, within the framework of the discussion of the results obtained, there is every reason to believe that under very low pressure conditions in a system with low-melting components (Al), the temperatures of the phase transformations passing through the vapor and liquid phases can be substantially reduced. In particular, peritectic transformation in the Al – Ti system can take place in the temperature range of coating formation (250 – 300°C). In this case, the proposed reconstruction of the structure formation of the Ti/AlSi coating quite satisfactorily explains the transformation of the 2D-configuration of the composite, incorporated by the technology of ion-plasma deposition, into a 3D-configuration, formed in reality.

CONCLUSION

Thus, the main features of the coating of the Ti/AlSi system studied in the work should be considered a nontrivial 3D-nanocomposite structure and a low Si content. These features are closely related. With a high Si content in the coating, its structure is amorphized [2], with a low – the low-melting Al becomes the dominant element of the chemical composition. The latter circumstance has a decisive influence on the process of structure formation, which obeys the laws of phase equilibria in the Al–Ti system (in a region with a high Al content). Thus, under vacuum conditions, the temperature of peritectic transformation in the Al–Ti system can apparently be reduced to the temperature of heating of the base metal in the zone of deposition of the coating (250 – 300°C). Due to the large amount of low-melting Al, a liquid phase is provided, from which the nano-sized particles of the intermetallide $TiAl_3$ crystallize and the solid solution Al(Ti, Si) is crystallized from the peritectic reaction. As a result, the composite coating structure is formed in a 3D-format with an Al(Ti, Si) matrix and a filler (reinforcing element) in the form of highly dispersed $TiAl_3$ particles.

ACKNOWLEDGMENT

The reported study was funded by RFBR according to the research project no. 18-08-00546.

REFERENCES

[1] Varavka, V.; Kudryakov, O.; Ryzhenkov, A.; Kachalin, G.; Zilova, O. *Thermal Engineering*, 2014, *vol. 61(11)*, 797 – 803.

[2] Varavka, V.; Kudryakov, O.; Ryzhenkov, A. In: *Piezoelectrics and Nanomaterials: Fundamentals, Developments and Applications*, Ivan A. Parinov, Ed., Nova Science Publishers, New York, 2015, 105 – 132.

[3] Varavka, V.; Kudryakov, O.; Zabiyaka, I.; Morozkin, I. In: *Advanced Materials - Techniques, Physics, Mechanics and Applications*, Springer Proceedings in Physics, Ivan A. Parinov, Shun-Hsyung Chang, Muaffaq A. Jani (Eds.). Heidelberg, New York, Dordrecht, London: Springer Cham. 2017, *vol. 193*, 407 – 418.

[4] Vasiliev, A.; Swain, M.; Aizikovich, S.; Sadyrin, E. *Archive of Applied Mechanics*, 2016, *vol. 86(7)*, 1247 – 1254.

[5] Thornton, A. J. *Thin Solid Films*, 1977, *vol. 40*, 335 – 344.

[6] Anders, A. *Thin Solid Films*. LBNL Paper LBNL-2768E, 04-01-2010: http://www.escholarship.org/uc/item/3261x4bm.

[7] Murray, J. L.; McAlister, A.J. *Bull. Alloy Phase Diagrams*, 1984, *vol. 5(1)*, 74 – 84, 89, 90.

[8] http://www.himikatus.ru/art/phase-diagr1/Al-Si.php.

In: Proceedings of the 2018 International Conference … ISBN: 978-1-53615-862-5
Editors: Ivan A. Parinov et al.

Chapter 2

ORGANIC COATINGS FORMATION CONDITIONS AND PROTECTIVE PROPERTIES IN SULPHURIC ACID

S. P. Shpanko*[*], *E. N. Sidorenko*, *L. E. Kuznetsova*, *K. G. Abdulvakhidov* and *D .C. Obuhov
Southern Federal University, Rostov-on-Don, Russia

ABSTRACT

The protective effect of imidazole class organic compound and its compositions with chloride and rhodanide anions on the low-carbon steel corrosion in sulphuric acid solution was studied. It has been established that haloid ion concentration increase contributes to the steel corrosion inhibition coefficient growth. The rhodanide anions concentration increase, on the contrary, reduces the composite inhibitor protective effect. As the time of adsorption film formation increases, the protection degree increases for all studied systems. The mixed mechanism of mixed inhibitors protective effect with a certain anodic reaction inhibiting action predominance has been established. The negative values of electric capacitance and volt-ampere characteristics sections with the presence point of negative differential resistances to the hopping mechanism of electrical films conductivity have been studied.

Keywords: corrosion, inhibition, inhibition coefficient, volt-ampere (*V-I*) characteristics, conductivity, film

[*] Corresponding Author's E-mail: spsha@mail.ru.

1. Introduction

The modern production development states high demands to the structural materials reliability and durability. One of the fairly simple and affordable ways to protect metals from corrosion is the aggressive media modification by individual or composite corrosion inhibitors [1]. Adsorption type organic surfactants are usually used as inhibitors for acidic media. The inhibition method is characterized by high efficiency and reliability with the correct additives choice. It is known that inorganic anions affect the metals corrosion organic inhibitors protective action [2, 3]. Works in this direction continues. Issues related to the research subjects expansion and formed organic coatings physicochemical characteristics study are particularly promising in this area.

2. Research Methods

Organic imidazole class compound – dihydrochloride1-(3,4-dichlorophenyl)-2-{2-imino-3-[2-piperidine-1-yl)ethyl]-2,3-dihydro-1*H*-benzimidazole -1- yl} ethanol (surface-active substance, SAS) was investigated as low-alloy steel St3 corrosion inhibitor in 1M H_2SO_4 at concentration 10^{-3}mol/l and mixture with chloride and rhodanide anions. The anions were introduced into the electrolyte in the form of their potassium salts. The values of salts concentration C were varied within the limits of: $C_{KCl} = 0.01 – 0.12$ mol/l, $C_{KNCS} = 0.01 – 0.09$ mol/l.

The corrosion rate j was determined by samples mass decreasing in a time t from the surface area S by the formula:

$$j = (m_0 - m_i)/S \cdot t, \tag{1}$$

where m_0 and m_i are the sample mass loss in the background and in the inhibitor presence. The corrosion tests time ranged from 1 to 5 days. The protective action of the inhibitor was evaluated by the coefficient of deceleration of corrosion:

$$K = j_0/j_i, \tag{2}$$

where j_0 and j_i are the corrosion rates of the samples in the background and in the presence of the inhibitor.

Polarization curves were taken in a three-electrode cell in the galvanostatic regime. The working electrode is St3, the counter electrode is Pt, the reference electrode is silver chloride. The working solutions are sulphuric acid inhibited solutions.

3. Results and Discussion

The imidazole class organic compound and its mixtures with inorganic anions inhibitory effect on the low-carbon steel corrosion in acid solution was studied. Chloride and rhodanide anions have been studied as inorganic ions. The paper examines two variable factors influence on the inhibitors protective characteristics: steel and acid solution contact duration and inorganic salts concentration.

It has been established that the imidazole derivative is St3 corrosion inhibitor in acid solution. Its protective effect significantly increases in the first three days, then this increase slows down, and after 5 days even slightly decreases. The steel corrosion inhibition is also increased when inorganic anions are introduced into the corrosive medium. The anion nature is not affected on the protection effects time dependence: with steel and inhibited acid solution contact time increasing the mixed inhibitors protective effect increased regardless of the anion chemical nature.

An entirely different behavior is observed in the ions concentration effect study. An increase of haloid ion concentration contributes to the enhancement of the effect of protecting the organic surfactant, and an increase in the concentration of the rhodanide anion, on the contrary, reduces the degree of protection (Figure 1). It can be assumed that the results obtained are related to different mechanisms of interaction of anions with the metal surface. Chlorine anions, adsorbed on the iron surface, recharge it, promoting the inhibitor cationic form adsorption. The rhodanide ions effect depends on their concentration. In the small concentrations region (up to 5 mmol/l), their action is analogous to haloid ions.

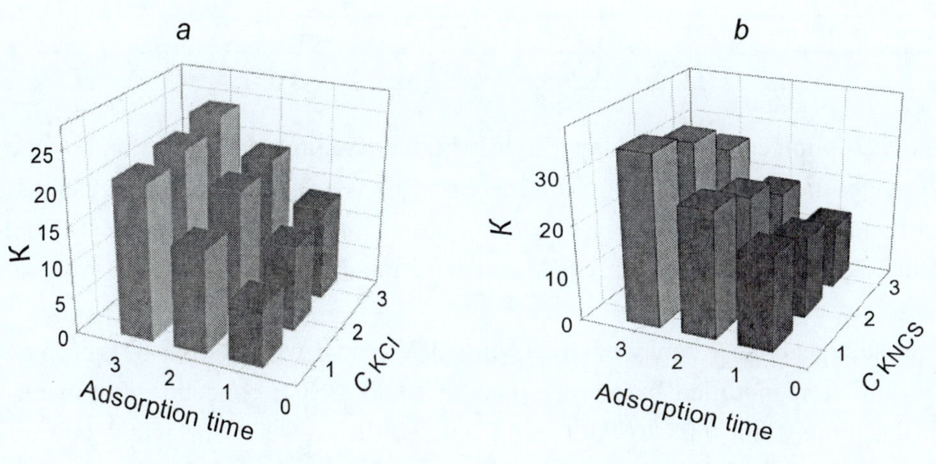

Figure 1. Steel corrosion inhibition coefficient dependence H_2SO_4 of K on the salts concentration (C_{SAS} =10^{-3} mol/l): (*a*) C_{KCl}, mol/l: 1 – 0.01; 2 – 0.03; 3 – 0.12; (*b*) C_{KNCS}, mol/l: 1 – 0.03; 2 – 0.06; 3 – 0.09; adsorption time, day: 1 – 1; 2 – 2; 3 – 4.

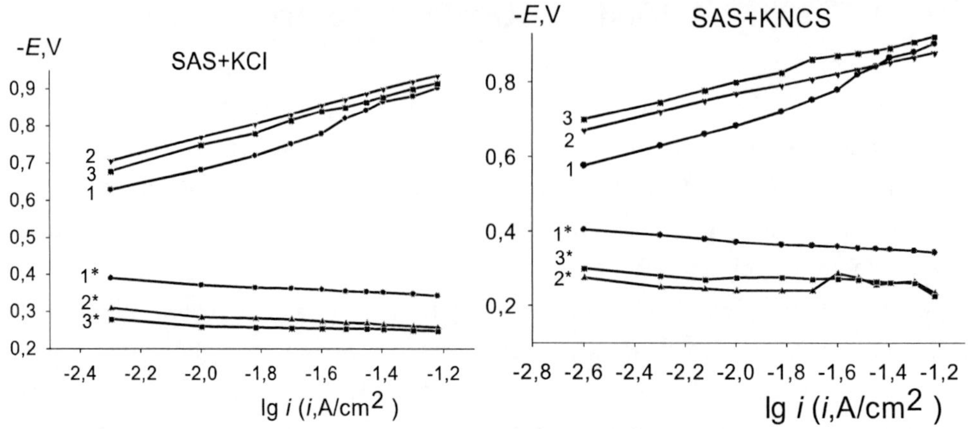

Figure 2. Cathodic (1, 2, 3) and anodic (1*, 2*, 3*) polarization curves at the steel electrode in the background (1 and 1*) and in the mixed inhibitors SAS + KCl and SAS + KNCS presence at C_{KX}, mol/l: 2, 2* – 0.03, 3; 3* – 0.09.

Table 1. Values of corrosion potentials E_{kor}, tafel constants $в_К$ and $в_A$, rates of cathodic i_K and anodic i_A reactions, reliant on the KCl and KNCS concentration in the mixture of inhibitors

C_{KX}, mol/l	$-E_{kor}$,V	$в_K$	$в_A$	i_K, mA/cm^2 (E_K= –0.75V)	i_A,mA/cm^2 (E_A= –0.29V)
Media (H_2SO_4)	0.437	0.155	0.041	21.0	-
C_{KCl}, mol/l					
0.03	0.411	0.200	0.050	8.7	7.9
0.09	0.387	0.203	0.051	11.9	2.,5
C_{KNCS}, mol/l					
0.03	0.426	0.144	0.050	10.0	2.5
0.09	0.421	0.142	0.050	5.0	5.0

Chemisorption complexions layer tightly bound with iron atoms, formed on the metal surface under the KNCS concentration increase. It has polymeric nature presumably [4]. The polymer film effect increases with C_{KNCS} increasing. As a result, the inhibitor molecules adsorption makes difficult, which is inevitably accompanied by protection degree decrease.

The electrochemical behavior of St3 in sulfuric acid was studied in the presence of organic inhibitor, modified by anions. The KCl and KNCS concentrations were varied within the limits from 0.03 to 0.09 mol/l against the constant surfactant concentration background. The experimental results are present in Figure 2 and in Table 1.

The composite SAS+KCl inhibitor introduction into sulphuric acid increases the corrosion potential E_{kor} to a greater extent, the higher the chlorine anions concentration. In a point of fact, both the polarization and the conjugate reactions polarizabilities (the constants $в_К$ and $в_A$) increase. Therefore, SAS + KCl is a mixed type inhibitor. However, the higher C_{KCl} concentration corresponds to the greater E_{kor} corrosion potential and anode reaction inhibition. Thus, this inhibitor inhibits the anodic reaction to a greater extent than the cathodic reaction.

The SAS + KNCS inhibitor also reduces the rate of both anodic and cathodic reactions. However, an increase in the tafel constant *в* with decreasing C_{KCNS} is observed only for the anodic process. Consequently, the inhibition of the anodic reaction of the total electrode process, and in this case, is decisive.

In the study of the average conductivity of disordered systems, their volt-ampere (VA) characteristics are measured. The study of the VA reversal characteristics of films, obtained during the adsorption of SAS, showed that forward and reverse curves traces $I = I(U)$ and $R = R(U)$ are practically the same (Figure 3a). It indicates the hysteresis effect absence from electric field exposure to the film. The $I = I(U)$ dependences are relatively equal to straight lines, along with small fluctuations not related to the experimental error. The film static resistance, on the one hand, small depends on the applied voltage (in the region of 0.5 Ω) and, on the other hand, it is very sensitive to any voltage changes. The $R = R(U)$ dependences are practically horizontal lines with a pulsating R value in the range of 0.01 – 0.02 Ω.

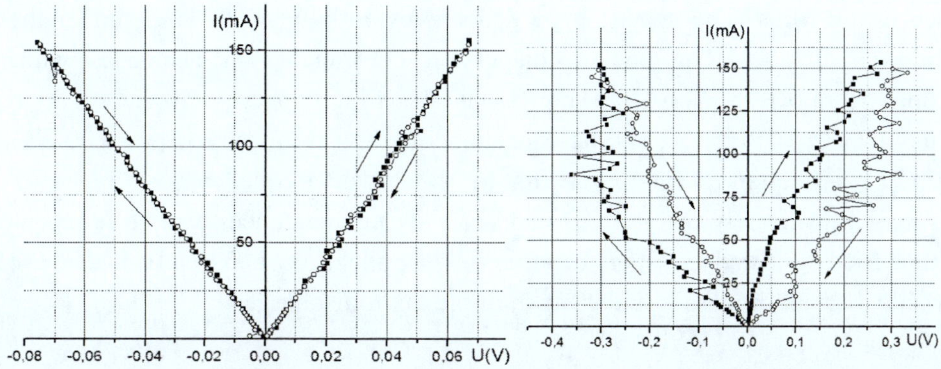

(a) (b)

Figure 3. Films reverse characteristics VA: (a) SAS and (b) SAS + KCl (C_{KCl} = 0.12 mol/l); adsorption time – 1 day.

The number of larger fluctuations on the $I = I(U)$ and $R = R(U)$ curves increases with films formation time, the chlorine anions concentration and the measurement cycles number increase. The VA characteristics are strongly broken lines (Figure 3b) with alternating sections, corresponding to positive and negative differential film resistances (PDR and NDR). The forward and reverse $I = I(U)$ curves traces, obtained with the voltage growth and decrease, do not coincide. The VA characteristics of films similar features are observed under the voltage polarity reversion. The films static resistance (R) depends heavily on the applied voltage and the R surge amplitude can reach 0.5 Ω and more. So, the external view of the VA characteristics of investigated films, regardless of the anion nature and its concentration, indicates a conductivity of hopping type, typical of organic materials. In them, the charge transfer occurs by means of random wanderings over localized centers, which are characterized by spatial and energy disorder [5].

The films electrical capacitances *C*, measured in the frequency range $2\times10^2 - 10^6$ Hz at room temperature are negative regardless of the adsorption time, the anion concentration and its nature, true for capacitances of other anticorrosive films, investigated earlier [3].

The materials with $C< 0$ are indicated by the radio electronics theorem from the semiconductor devices with NDR. This theorem uniquely relates the form of the VA characteristic and the type of feedback of the device under study [6]. It says: "Electronic devices with internal positive current feedback have a negative resistance section on static VA characteristic and reactivity of inductive character."

So, the investigated films can be considered as electronic microdevices (materials) with $C< 0$. First, electric microdischarges occur in the films under the influence of the applied voltage, and its VA characteristics contain sections with NDR. Secondly, films with micro-discharges are characterized by internal positive feedback between current and voltage.

The obtained *C*(*F*) spectra indicate the relaxation type permittivity dispersion presence in the films. The spectra type depends essentially on the C_{KCl} under the film formation (Figure 4). The spectra dispersion coefficient K_D decreases with C_{KCl} increasing. Here $K_D = C_{HF}/C_{LF}$. The C_{HF} is the film capacity, measured at a high frequency ($F_{HF} = 10^6$ Hz), C_{LF} is the capacitance at low frequency ($F_{LF} = 200$ Hz). The curves *C*(*F*) inflection points F_0 shifts to the lower frequencies region with C_{KCl} increasing. Unfortunately, $K_D = f(C_{KCl})$ and $F_0 = f(F)$ dependences cannot be repeated on films that tried repeated constant electric field application, apparently because of the irreversible thermal processes under the multiple microdischarges.

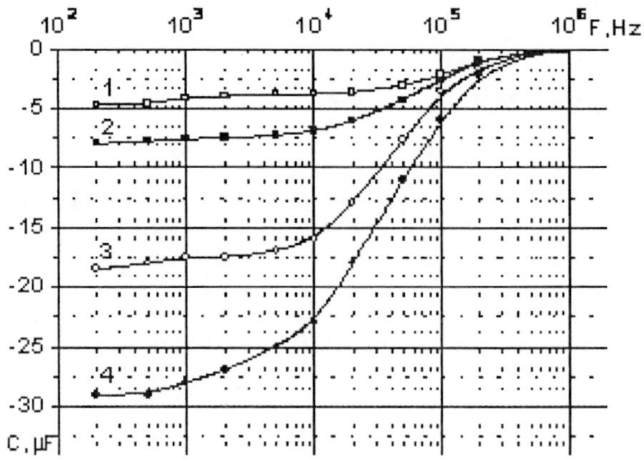

Figure 4. Films electrical capacity frequency dependences, obtained in composition electrolytes: SAS + KCl at C_{KCl}, mol/l: 1 – 0; 2 – 0.01; 3 – 0.03; 4 – 0.12.

CONCLUSION

The organic compound protective properties are determined by the inorganic anions nature and concentration and the protective layer formation time. The protection degree increases with *t* increasing for all the inhibitory systems studied, increasing from the SAS to its mixtures with anions.

The protective effect ambiguous dependence on the chloride and rhodanide anions concentration was explained on the ground of the Fe – X^{-}– SAS interaction forces nature in the adsorption layer.

The organic films electrical conductivity hopping character was established. As the adsorption time and the concentration of KCl increase in the compound inhibitor, the residual phenomena, caused by thermal processes in microdischarges, increase in the films. The films reactivity is inductive because of the positive internal current feedback.

REFERENCES

[1] Kuznetsov, YuI. *Corrosion: Materials, Protection*, 2018, No. 1, 1 – 17 (In Russian).

[2] Shpanko, SP; Sidorenko, EN; Semenchev, AF; Lyanguzov, NV; Anisimova, VA. *Protection of Metals and Physical Chemistry of Surfaces*, 2017, vol. 53(2), 335 – 342.

[3] Shpanko, SP; Sidorenko, EN; Kabirov, YuV; et al. In: *Proceedings of the 2016 International Conference on Physics and Mechanics of New Materials and Their Applications*, Ivan A. Parinov, Shun-Hsyung Chang, Muaffaq A. Jani (Eds.). Nova Science Publishers. New York., 2017, 345 – 352.

[4] Avdeev, YG; Tyurina, MV; Kuznetsov, YI; Pronin, YuE; Kazansky, LP. *Corrosion: Materials, Protection*., 2013, No 6, 17 – 23. (In Russian).

[5] Fateev, MP. *Fiz. Tverdogo Tela*, 2010, vol. 52(6), 1053 – 1059 (In Russian).

[6] Garyainov, SA; Abezgauz, ID. Semiconductor Devices with Negative Resistance. *Energy*, Moscow., 1970, pp. 1 – 320 (In Russian).

In: Proceedings of the 2018 International Conference … ISBN: 978-1-53615-862-5
Editors: Ivan A. Parinov et al.

Chapter 3

THE ROLE OF SURFACE ALLOYING OF POWDER PARTICLES ON THE TEMPERATURE OF SINTERING OF PIEZOCERAMICS

***A. A. Nesterov*[1,*] *and M. I. Tolstunov*[1]**
[1]Department of Chemistry, Southern Federal University,
Rostov-on-Don, Russia

ABSTRACT

The most important problems of modern materials science in the field of piezoelectric materials, based on lead-containing ferroelectric phases, are concerned with the environment. This is due to the high toxicity of all the known lead-containing compounds, regardless of whether they are in the form of vapors and dust in an atmosphere of industrial premises or a part of a solution. Technologies of these types of piezoceramic materials are high-temperature and lead to a sharp increase of the concentration of lead-containing compounds in the environment. In this context, the main way to solve the environmental problems of this production is to reduce temperatures of synthesis of powders of initial phases and temperatures of sintering of ceramics, in a combination with a decrease in the duration of these processes. Two variants of the method for reducing the temperature and sintering time of piezoceramics are considered in this chapter, which is provided by a deliberate change in the defectiveness of the surface layers of the particles of the original powders. This effect is due to the surface doping of these powder particles by the phases of the following composition: (i) MeFeO$_{3-x}$ (Me = Ca, Sr or Ba) and (ii) MeVO$_{3\pm x}$ (Me = Pb or Ba).

Keywords: lead-containing ferroelectric phases, surface activation of powders, low-temperature sintering technologies

* Corresponding Author E-mail: lanesan2448@ya.ru.

1. INTRODUCTION

At present, the tasks aimed at reducing the temperatures of the synthesis of lead-containing powders of ferroelectric phases in the $PbZrO_3$ – $PbTiO_3$ (PZT) system and $PbTiO_3$ (LT), as well as the temperatures of sintering of piezoceramic materials, produced from these powders, are still relevant. A need to solve these tasks is concerned with a number of shortcomings of the traditional technologies suitable to manufacture these materials, for instance, (i) the relatively low reproducibility of the dielectric and piezoelectric parameters of ceramic samples, and (ii) high concentrations of lead-containing compounds in industrial air. A progress on solving the issues of low-temperature synthesis of ultra-disperse powders of lead-containing ferroelectric phases is achieved in the past decade, but successful attempts to reduce the sintering temperatures of ceramic materials are still rare [1-7].

In our opinion, this is due to the lack of theoretically grounded algorithms in this research field, which are to be based on a deep understanding of the thermodynamics and kinetics of the processes of synthesis and sintering. In particular, this concerns algorithms of the mechanisms of sintering, which are implemented precisely within the framework of the lead-containing system, as well as their successive changes at the transition from one sintering stage to another (induction period, primary recrystallization, grain growth, and secondary recrystallization).

2. METHODS

2.1. Objective

The aim of the present chapter is to study the effect of two types of additives on the sintering temperature of two groups of materials. The ferroelectric LT and PZT phases are used as basic phases.

Additives of the first type are phases with the $MeFeO_{3-x}$ composition (Me = Ca, Sr, or Ba), and additives of the second type are described by the formula $MeVO_{3\pm x}$ (Pb or Ba). The formation of solid solutions in the $PbTiO_3$ – $MeFeO_{3-x}$ systems solves two problems of materials based on LT as follows:

(i) reduces the ratio of the parameters of the tetragonal unit cell c/a (LT phase), which reduces the probability of cracking of the ceramic based on LT;
(ii) the technologically controlled defectiveness of the anion sublattice leads to increasing the speed of sintering of the ceramic, which is manufactured from powders of lead-containing phases.

Using the phases of the $PbZrO_3 – PbTiO_3 – MeVO_{3±x}$ system for the manufacturing of ceramics is not only important to change the rates of the diffusion processes, but also changes a ratio of the main phases in the morphotropic region of the PZT system. The necessity to control the rate of diffusion processes in these systems during the sintering is accounted for by the fact that the diffusion of the Me^{2+} and O^{2-} ions is the limiting mechanism of this process (along with the transfer of O_2 through the gas phase with the migration of the Me^{2+} ion and electrons through the reaction zone). During the sintering process, two opposite tasks are solved as follows: (i) reducing the defectiveness of the ceramic grains for their stabilizing and reducing the rate of the secondary recrystallization, and (ii) increasing the surface defectiveness of particles of ferroelectric powders to provide low activation energy at the formation of intergranular boundaries. As follows from the composition of alloying additives, their formation in the studied systems is concerned with various types of chemical processes. The $MeFeO_{3-x}$ phases contain Fe^{4+} ions in compositions, and the formation of these ions is possible only in an oxidizing medium, whereas the formation of V^{4+} ions should proceed in a reducing medium.

2.2. Objects of Study and Technology of Their Manufacture

Taking into account the specifics of the mechanisms of the formation of alloying additives and their localization in systems, a mixture of $Me[(Fe(OH)_4(OH_2)_2]_2$ and $Me(NO_3)_2$ (molar ratio 1:1) was used to cover a surface of LT particles by $MeFeO_{3-x}$ phases. In the case of doping of the powders of the phases in the PZT system, a suspension consisting of an ultrafine $MeVO_{3.5}$ powder and ammonium oxalate was used at the molar ratio 2:1. Initial ultrafine LT powders were manufactured according to the technology, developed earlier by us, and this technology envisages the use of complex LT-based compounds as precursors [8, 9]. The precursor with the $Me[(Fe(OH)_4(OH_2)_2]$ composition, used to form the surface phases of the $MeFeO_{3-x}$ type, was synthesized during the interaction between $Fe_2O_3{\bullet}xH_2O$ and $Me(OH)_2$ (molar ratio of reactants 1:1). To form a defective surface layer on LT particles, in the first stage, the components of the dopant {i.e., $Me[(Fe(OH)_4(OH_2)_2]_2$ + $Me(NO_3)_2$} were mixed in a planetary mill at a molar ratio of 1:1 in the presence of water. The resulting suspension at a calculated molar ratio was added to the LT powder, and this mixture was again treated to obtain a dispersion in the planetary mill. After drying the resulting mixture at 120 – 140°C for 2 hours, cylindrical billets (d = 22 mm and h = 5 mm) were produced from the produced powders. Firing these powders was carried out in a stepwise manner:(i) at 25 – 150°C, the rate increase of the temperature (V_T) was 15°C/min; (ii) at 150 – 320°C, V_T = 3°C/min; (iii) at 320 – 1050°C, V_T = 25°C/min; (iv) isothermal sintering at 1050°C within 20 min; (v) cooling to 950°C for 10 min; and (vi) sintering at 950°C for 60 – 90 min.

Samples of the $PbZrO_3$ – $PbTiO_3$ – $MeVO_{3\pm x}$ system were produced in a similar manner, using a mixture of V_2O_5 and hydroxides of lead and barium to prepare the $MeVO_{3.5}$ suspension. The base phase of the PZT system had the composition given by the formula $Pb_{0.95}Sr_{0.05}Ti_{0.46}Zr_{0.50}W_{0.02}Cd_{0.02}O_3$, and the ultra-disperse powders of this system were manufactured using the low-temperature technology [7]. After blending and grinding (in ethanol) of pre-dosed mixtures of $MeVO_{3.5}$ and $Pb_{0.95}Sr_{0.05}Ti_{0.46}Zr_{0.50}W_{0.02}Cd_{0.02}O_3$ powders, billets with the same sizes (i.e., d = 22 mm and h = 5 mm) were produced by pressing. Firing of samples of this type was also carried out in a stepwise manner:

(i) at 25 – 350°C, V_T = 6°C/min;
(ii) at 350 – 110°C, V_T = 25°C/min;
(iii) isothermal firing at 1100°C within 15 min;
(iv) cooling to 920 – 950°C for 10 min; and
(v) sintering at 920 – 950°C for 60–90 min.

Optimal firing regimes for pressed samples, of both the species, manufactured at various parameters of sintering, were obtained, based on analysis of their density and electrophysical parameters (EPPs).

3. Experimental Data

In the first stage of the process of our manufacturing, optimal modes of sintering the samples were determined. The purpose of these studies was to define the effect of the duration of the final stage of sintering of ceramic samples, with a fixed concentration of the dopant, on their microstructure (Figure 1).

Our analysis of the data on the microstructure of the manufactured ceramic $PbTiO_3$ – $MeFeO_{3-x}$ samples shows that, irrespective of the type of the Me^{2+} cations and their molar fraction in the system from 0.02 to 0.10, heating of billets in the temperature range from 320 to 1050°C at the ratio of 25°C/min is sufficient to form numerous ceramic grains.

As is known from work [10], piezoceramics, based on the individual LT phase, have no practical significance. This is due to the fact that during the phase transition (high-temperature first-order paraelectric – ferroelectric phase transistion at 540°C), the volume of the perovskite unit cell changes significantly, and this leads to a destruction of a ceramic sample. The additives introduced into LT enable us to reduce the tetragonal distortion of the unit cells of the phases of solid solutions and to reduce the degree of cracking of the ceramic sample. A probability of the formation of cracks and their development in the samples increases with the poling of such piezoceramics due to the electrostriction of its individual grains.

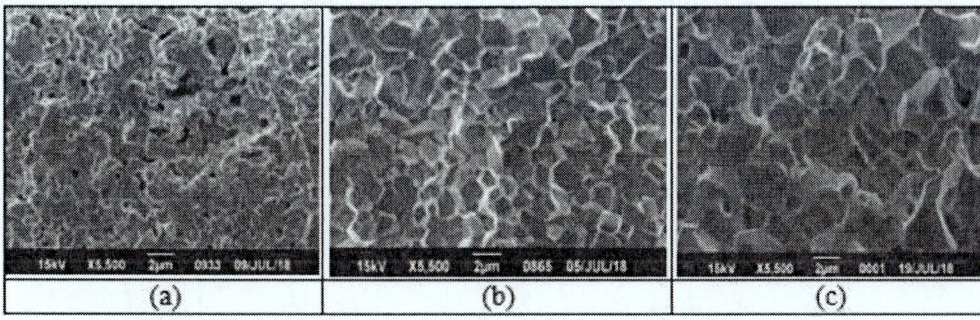

Figure 1. Microstructure of ceramic samples based on the ferroelectric Pb0.9Ca0.1Ti0.95Fe0.05O3-x phase, time of the final stage of sintering at 950°C: (a) 30 min; (b) 60 min; (c) 90 min (micrographs obtained by means of an electron microscope JSM-6390LA).

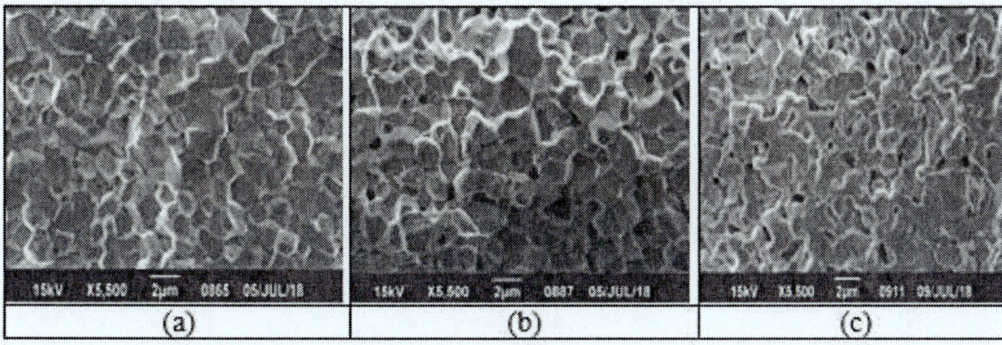

Figure 2. Changes in the microstructure of ceramic Pb0.9Ca0.08Sr0.02Ti0.92Fe0.08O3-x samples with a change in the strength of the poling field Ep: (a) a non-poled sample; (b) Ep = 4.5 kV/mm; (c) Ep = 6.5 kV/mm.

Our investigations showed that the destruction of ceramic samples of the $PbTiO_3$ – $\mathit{Me}FeO_{3-x}$ systems proceeds in two stages (Figure 2). The first of them occurs at the intensity of the poling field to 4.5kV/mm and is accompanied by the destruction of intergranular boundaries in the region of small grains (Figure 2). When the poling field strength reaches 6.5kV/mm, the large grains of ceramics and their druses are destroyed, i.e., cracking is observed (Figure 3).

In turn, the change in the microstructure of the samples reduces their mechanical quality factor Q_m, planar electromechanical coupling factor K_p, transverse piezoelectric modulus d_{31}, and promotes increasing the volume (or hydrostatic) piezo-parameters (d_v and g_v). Disadvantage of the known technologies, suitable for these types of materials, is the low reproducibility of the EPPs of the samples being manufactured. This fact is accounted for by the probabilistic nature of the formation of the microstructure of the individual ceramic samples, which is due to the heterogeneity of the intergranular boundaries and due to the different number and volume of grains in the local parts of the system. The method put forward by us, in addition to solving the environmental problems of the manufacturing of these materials, partially also solves the problems of the reproducibility of their EPPs.

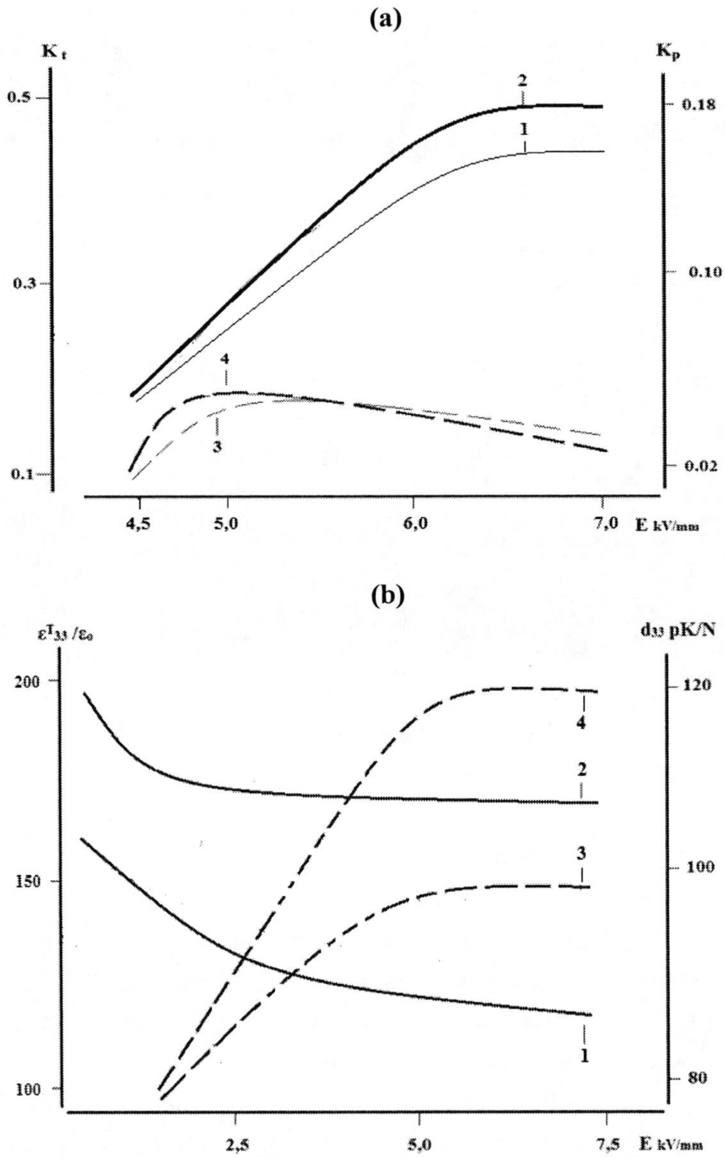

Figure 3. Changes in EPPs of ceramic $Pb_{0.88}Ca_{0.12}Ti_{0.93}Fe_{0.07}O_3$ samples (curves 1 and 3) and ceramic $Pb_{0.88}Ba_{0.12}Ti_{0.92}Fe_{0.08}O_3$ samples on increasing the poling electric field E: (a) electromechanical coupling factors (thickness K_t, curves 1 and 2, and planar K_p, curves 3 and 4); (b) relative dielectric permittivity $\varepsilon^T{}_{33}/\varepsilon_0$, curves 1 and 2 and d_{33}, curves 3 and 4.

This is due to the fact that the formation of ceramics in this case begins with the formation of a large number of primary nano-sized grains of approximately the same volume and defectiveness. In connection with these characteristics of primary grains, their growth rates are approximately equal like conditions for the formation of the intergranular boundaries therein. The EPP values of some piezoceramic materials made from ultradispersed powders of the $PbTiO_3 - MeFeO_{3-x}$ phases, where *Me* = Ca, Sr, or Ba, are shown in Figure 3 and Table 1.

It should be noted that the samples of this system, with a $MeFeO_{3-x}$ content of over 10 molar percent, have high dielectric loss tangent (tanδ) values, which makes their effective polarization impossible. For the remaining samples, the maximum of the piezoelectric parameters is achieved after poling in a siloxane liquid at the electric field E = 6 – 6.5kV/mm for 40 – 60 minutes at a temperature from the range 100 – 120°C.

Table 1. EPPs of poled (E_p = 6kV/mm) piezoceramic samples, based on the $Pb_{0.85}Ca_{0.15-x}Sr_xTi_{1-x}Fe_xO_3$ phases

EPPs / X	$\varepsilon^T_{33}/\varepsilon_0$	d_{33}, pC/N	d_{31}, pC//N	d_v, pC//N	g_v, 10^3 V·m/N	d_vg_v, 10^{12} Pa^{-1}
0	135	75	14	47	39	1.83
0.02	150	84	12	60	45	2.70
0.04	165	98	11	76	52	3.95
0.06	175	106	9	88	57	5.02
0.08	185	119	10	99	60	5.94
0.10	200	103	13	77	44	3.39

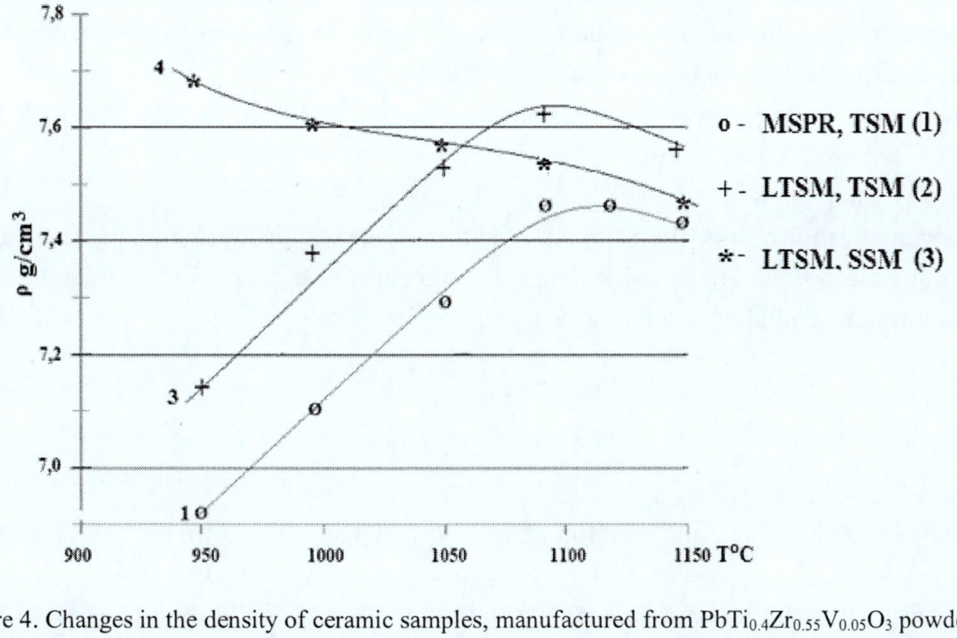

Figure 4. Changes in the density of ceramic samples, manufactured from $PbTi_{0.4}Zr_{0.55}V_{0.05}O_3$ powder, were obtained by means of different methods: MSPR, solid phase reactions; LTSM, low-temperature synthesis; the samples of ceramics were sintered of the methods: traditional – TSM; stepped sintering – SSM.

Table 2. EPPs of ceramic samples, based on ($0.98PbTi_{1-x-y}Zr_xV_yO_3 + 0.02PbCd_{0.5}W_{0.5}O_3$) phases

Values of x and y		Values of EPPs				
x	y	$\varepsilon^T_{33}/\varepsilon_0$	tan δ, %	d_{33}, pC/N	$-d_{31}$, pC/N	d_v, pC/N
0.55	0	670	2.1	220	90	40
0.55	0.05	610	0.6	255	105	45
0.55	0.07	630	1.5	235	98	39
0.55	0.09	690	3.9	200	81	38
0.53	0	560	2.3	180	74	32
0.53	0.05	510	0.7	190	75	40
0.53	0.07	540	1.2	195	80	35
0.53	0.09	580	4.1	185	75	35
0.58	0	620	1.9	210	93	24
0.58	0.05	710	0.8	280	121	38
0.58	0.07	790	0.7	310	135	40
0.58	0.09	760	2.9	270	116	38

The maximum values of the volume piezo-parameters (d_v and g_v) are stated for ceramic materials with the $Pb_{0.85}Ca_{0.15-x}Sr_xTi_{1-x}Fe_xO_3$ composition (Table 1). Analogous data were obtained for PZT piezoceramics, doped with $MeVO_{3\pm x}$ (a change of the microstructure of the samples, when changes: (i) the composition of the ferroelectric phase, and (ii) sintering regimes of the ceramics (Figure 4).

As is seen from Figure 4, for example, at a content of V^{4+} in ($0.98PbTi_{1-x-y}Zr_xV_yO_3 + 0.02PbCd_{0.5}W_{0.5}O_3$) over 4 mol. %, the optimal sintering temperature is reduced by 150 – 200°C (by using the SSM).

For applications, only the samples of the type, in which the molar fraction of V^{4+} does not exceed 0.09, are of value, since a further increase of this molar fraction in the system promotes rapid increasing tanδ in the samples (Table 2).

CONCLUSION

It has been stated that the addition of $MeFeO_{3-x}$ and $MeVO_{3\pm x}$ (Me = Pb, Ca, Sr or Ba) additives to ultrafine powders of lead-containing ferroelectric phases leads to lowering the sintering temperature and reduces the sintering time of ceramic samples based on the aforementioned phases.

The EPPs of these samples are larger than the similar parameters measured on the non-doped samples.

ACKNOWLEDGMENT

In the present chapter, we publish the results on the research project No. 11.5814.2017/8.9 in the internal grant (Southern Federal University) in the field of the scientific activity.

REFERENCES

[1] Hayashi, T., Inoue, T., Akiyama, Y., *J. Eur. Ceram. Soc.*, 1999, *vol.* 19, 999 - 1002.

[2] Chen, I. W., Wang, X. H., *Nature*, 2000, *vol. 404,* 168 - 171.

[3] Shiratsuyu, K., Hayashi, K. *Jpn. J. Appl. Phys.*, 2000, *vol. 39*, 5609 - 5612.

[4] Panadda, S., Pongkiti, W., Naratip, V., et al. *Integrated Ferroelectrics,* 2013, *vol. 149(1),* 39 - 48.

[5] Tandon, R. P., Singh, V., Narayanaswami, N., Hans, V. K., *Ferroelectrics,* 1997, *vol. 196*, 437 - 440.

[6] Gu, Y., Huang, J., Sun, D., et al. *International Journal of Modern Physics B*, 2009, *vol. 23(6-7)*, 1021 - 1027.

[7] Kaneko, S., Dong, D., Marakumi, K., *J. Am. Soc.,* 1998, *Vol. 81*, 1013 - 1018.

[8] Nesterov, A. A., Panich, A. E., Dolya, V. K., et al. *Piezoelectric Materials and Devices*, I. A. Parinov (Ed.), Nova Sci. Publ., New York, 2011, 145 - 183.

[9] Nesterov, A. A., Topolov, V. Yu., Panich, E. A., *Ferroelectrics Letters Section,* 2016, *vol. 43,* 1 - 7.

[10] Kitamura, N., Takeuchi, M. *Techn. Rep. CS News*, 1985, *vol. 44(2),* 35 - 39.

In: Proceedings of the 2018 International Conference … ISBN: 978-1-53615-862-5
Editors: Ivan A. Parinov et al.

Chapter 4

Influence Composition Analysis on the Process of Making Sugar Cane Husks and Coal Briquette as Alternative Fuel

***Muhyin*[1] *and Dwi Yuli Rakhmawati*[2,*]**
[1]Department of Mechanical Engineering, University of 17 Agustus 1945 Surabaya, Surabaya, Indonesia
[2]Department of Industrial Engineering, University of 17 Agustus 1945 Surabaya, Surabaya, Indonesia

Abstract

Sugar cane husks and coal briquette is one of the alternative fuels. The need of energy sources in modern times is still increasing in line with the level of technological progress, and the growing population of the world is increasing rapidly. Even though the energy supply is very limited, gradually the petroleum supply will be brushed aside because it is not balanced with the consumption and usage used by humans. Therefore, it is necessary to do an analysis of the composition effect on the process of making bagasse and coal waste briquettes to heating values and flame temperatures as alternative fuels. The process of making briquettes starts with mixing bagasse, coal and tapioca starch as an adhesive. The variations composition of those materials are 65%: 25%: 10%, 55%: 35%: 10%, 45%: 45%: 10%. The pressure used in the process of making this briquette is 88.99 kg/cm^2, where then the formed briquettes are heated with a constant temperature 1000° C for one hour. Particles 40, 60, 80 mesh with a composition of 65%: 25%: 10%, have the highest flame temperature i.e., 335° C with a heating value of4952.46 Kcal/kg, 5085.90 Kcal/kg, 4764.88 Kcal/kg, respectively.

Keywords: briquette, coal, combustion, dregs, tapioca starch

---

* Corresponding Author's E-mail: dwi_yuli@untag-sby.ac.id.

1. Introduction

The need for energy sources continues to increase in line with the level of technological progress and the growing population of the world. As we know, the energy supply is very limited, gradually the oil supply will be depleted because it is not balanced with the consumption and use used by humans. The idea arose to use energy sources derived from alternative fuels from non-petroleum sources including bagasse. So we are trying to make alternative bagasse and cylindrical coal briquettes alternative fuels that are easy to carry and use. Bagasse and coal briquettes can be used as fuel for stoves and stoves, so they can reduce the use of petroleum fuels.

2. Experiment

Coal is a combustible organic mineral, formed from the remnants of the ancient plants that settles changing shape due to the physical and chemical processes that take place over millions of years. Therefore, the coal is included in the category of fossil fuels [1]. It also defined as solid hydrocarbon fuel formed from plants growing in oxygen-free environments and exposed to the influence of the heat and pressure that lasts a long time [2]. The chemical composition of coal used for this research in May is from Tambang Batubara Bukit Asam company at Gresik, East Java province. Where the coal has the chemical content of as in the table below:

Table 2.1. The chemical composition of coal used

Parameter	Unit	Element value
Total Moisture	%	28
Inherent Moisture	%	15,0
Ash Content	%	8,0
Volatile Matter	%	40
Fixed Carbon	%	37,0
Total Sulfur	%	0,8
Gross Calorific Value	Kcal/kg	5800

Source: PT. Tambang Batubara Bukit Asam (Persero) Gresik- Jawa Timur.

Sugar cane husks is solid fiber material coming out of the final after grinding processes to milking nira. Dregs of sugar cane around 35%-40% of the weight of the sugar cane grinding which is produced by sugar industry. Given so much waste, sugar cane husks will have value added for the sugar factory after given a further treatment. According to Husin [3] the chemical composition of sugar cane husks are as follows:

Table 2.2. Sugar cane husk chemical composition

Composition	Level (%)
Ash	3,82
Lignin	22,09
Cellulose	37,65
Quintessence	1,81
Pentosan	27,97
SiO_2	3,01

In this case, the bagasse is a material that is easily destroyed during its use. This causes difficulties in its use as fuel, one way to overcome this difficulty is to make bagasse pulp briquettes with coal and adhesives.

Adhesive is a material that has the ability to unite similar or not similar objects through surface bonds. Addition of adhesive aims to increase the strength of the briquette product [4]. The use and selection of adhesives is carried out based on several things, which are good absorption of water, relatively cheap and easy to obtain. The selection of adhesives affects the quality of the briquettes [4]. A simple and easily available natural adhesive is starch. The starch composition of the raw material for briquettes is as much as 5% of the weight of the briquette material itself and starch adhesives are not easy to rot due to the influence of fermentation [5].

The drying process can be interpreted as a process to separate water from a material. Drying can also be used for the preservation process, because when the water content is reduced to below 10%, microorganisms cannot grow and multiply [6]. The way to dry briquettes can also determine the fuel calorific value [6]. The influence of drying method on heat value fuel can be seen in table 2.5.

Table 2.3. Influence of drying method on heat value fuel

Drying method	Heat value fuel (Cal/gr)
- Control (heating using oven)	2435,24
- Sun direct heating	2425,95
- Undirect sun heating	2425,85

Source: Widyaningsih, 1997.

The materials used in this experiment are a mix of coal, sugar cane husks and organic adhesive composition:

- coal 65% : husks 25% : tapioca starch 10%
- coal 55% : husks 35% : tapioca starch 10%

- coal 45% : husks 45% : tapioca starch 10%

The first step in making briquettes is to make coal powder and bagasse charcoal with sizes: 40 mesh, 60 mesh and 80 mesh. The number of samples from each size is 5 pieces. On second step, all samples were given a pressure of 88.99 kg/cm² and burned at a temperature of 100° C for 30 minutes. Furthermore, we do testing the compressive strength, density, calorific value and combustion temperature.

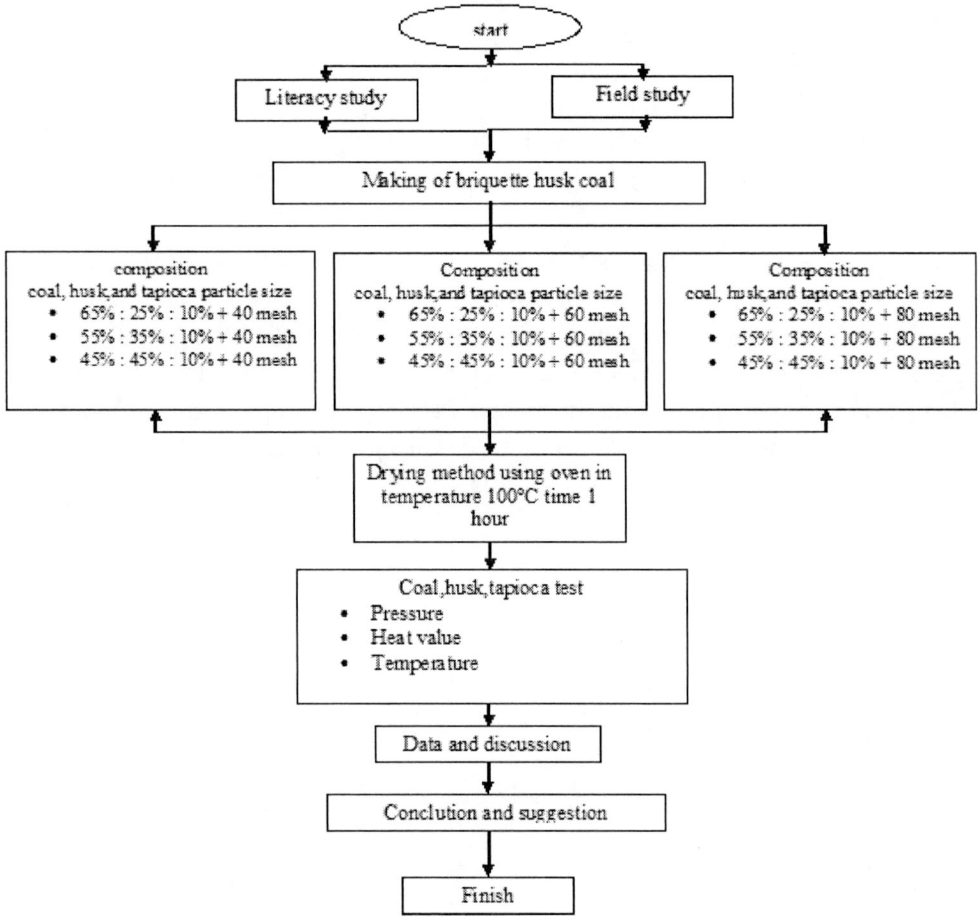

Figure 2.1 Scheme of Experimental Design.

3. Results and Discussion

Graphs from the results of testing and data analysis can be seen as follows.

Figure 3.1 shows that the value of briquette compressive strength decreased from size 40 mesh to 60 mesh and increased in size to 80 mesh. The highest strong press value occurs in briquettes with a composition of 65%:25%:10% of 0.693 kg/cm², this is because the size of 80 mesh is the ideal size for the bonding process between particles.

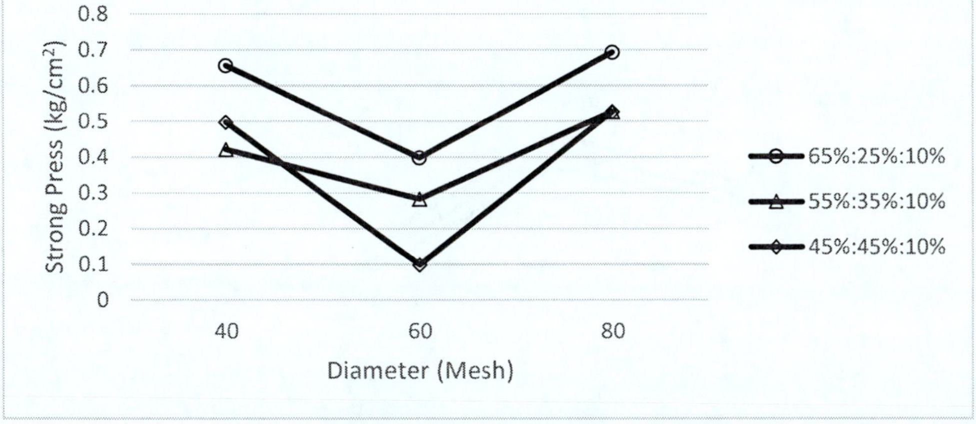

Figure 3.1 Influence of mesh size and briquettes composition on strong press.

Figure 3.2 shows the effect of mesh size and composition of briquettes on density values. The highest density value occurs in 80 mesh briquettes with a composition of 65%: 25%: 10% which is 1.184 gr/cm³. This happens because the size of 80 mesh has a larger surface area so that the bond between particles is more homogeneous and the particle size is very soft which will cause the bonding of particles to be strong and the greater percentage of coal composition results in higher density values compared to other compositions.

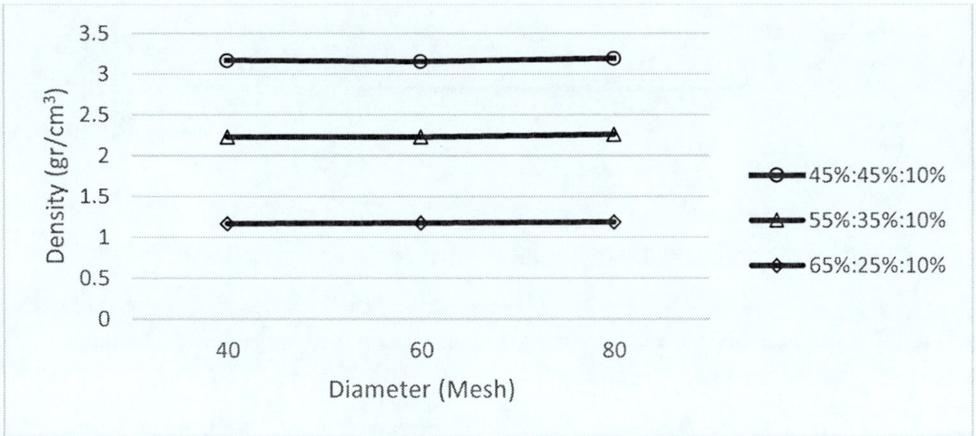

Figure 3.2. Influence of mesh size and briquettes composition on density.

Figure 3.3 shows that the mesh size and composition of the briquette affect the calorific value. The maximum calorific value produced by briquettes is 60 mesh and the composition is 65%: 25%: 10% for 5085.90 Kcal / kg. This happens because the calorific value is influenced by a mesh diameter that is not too large and not too small with a greater percentage of coal composition resulting in a higher heating value compared to other compositions.

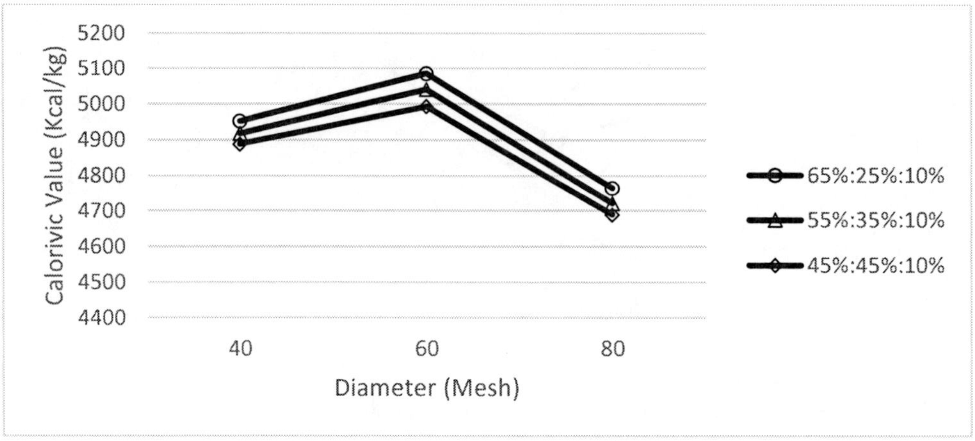

Figure 3.3. Influence of mesh size and briquettes composition on calorivic value.

Figure 3.4 shows that briquettes with coal particle content have more influence on the briquette temperature while mesh diameter does not affect the briquette flame temperature.

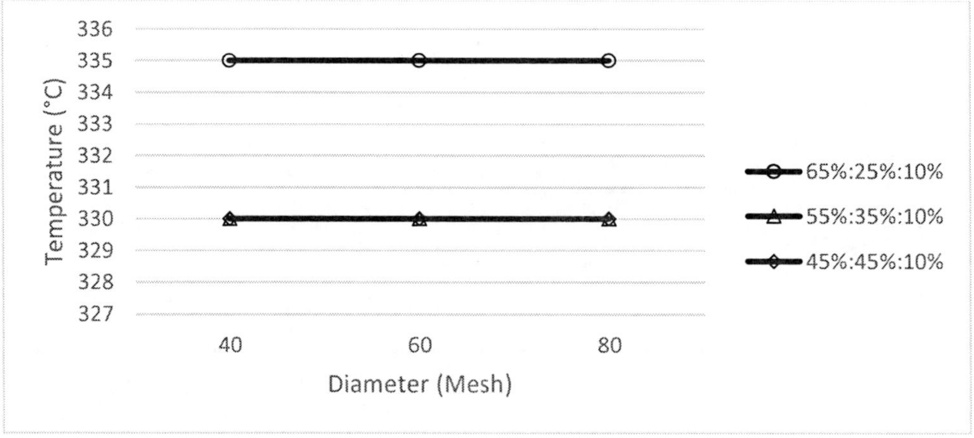

Figure 3.4. Influence of mesh size and briquettes composition on the temperature of the flame.

CONCLUSION

The conclusion of this research consist in that the greater number of powder coal briquettes in furnace, the higher flame temperature. Briquettes with 80 mesh and composition of 65%: 25%: 10% has the highest value of strong press and density value. Particle 40, 60, 80 mesh with the composition of 65%: 25%: 10%, pressure 88.99 kg/cm^2, temperature flame i.e., 335°C have the heat value 4952.46 Kcal/kg, 5085.90 Kcal/kg, 4764.88 Kcal/kg, respectively. Strong press and density have connection with mesh diameter. The smaller the mesh diameter and the greater the composition of the coal, the greater the compressive strength and density. The amount of calorific value and combustion temperature are influenced by the density, water content and mass of the briquettes themselves when testing the calorimeter bomb in the combustion process.

REFERENCES

[1] Raharjo. (2006). *Teknologi UCG Pengelolaan Batubara Ramah Lingkungan* [*UCG Technology Eco-friendly Coal Management*]. On Minithesis of Fitriani, D in University of Sriwijaya.

[2] Kepmen, LH. (2003). *Teknologi UCG Pengelolaan Batubara Ramah Lingkungan* [*UCG Technology Eco-friendly Coal Management*]. On Minithesis of Fitriani, D in University of Sriwijaya.

[3] Husin, A. A. (2007). *Pemanfaatan Limbah Untuk Bahan Bangunan* [*Utilization of Waste for Building Materials*]. http:// www.kimpraswil.go.id/ balitbang/ puskim/ Homepage%20Modul%202003/modulc1/MAKALAH%20C1_3.pdf. Accessed on February 1st 2007.

[4] Budhi, A. S. (2003). *Pembuatan Briket Arang dari Faeces Sapi dan Tempurung Kelapa Sebagai Alternatif Sumber Energi* [*Making Charcoal Briquettes from Cow and Coconut Shell Faeces As Alternative Energy Sources*]. Minithesis in Institute Technology of Sepuluh Nopember, Indonesia.

[5] Tejolaksono, M. N. (1998). *Pemanfaatan Limbah Isi Rumen dan Manure Sebagai Sumber Energi Biomassa Briket Arang* [*Utilization of Rumen and Manure Waste as Energy Sources of Biomass Charcoal Briquettes*]. Minithesis in University of Airlangga, Indonesia.

[6] Widyaningsih, A. (1997). *Cara Pengeringan, Pembuatan Briket, dan Uji Kalor Limbah Padat Organik* [*Ways of Drying, Briquette Making, and Organic Solid Waste Heat Test*]. Minithesis in Institute Technology of Sepuluh Nopember, Indonesia.

In: Proceedings of the 2018 International Conference … ISBN: 978-1-53615-862-5
Editors: Ivan A. Parinov et al.

Chapter 5

PHYSICAL AND MECHANICAL PROPERTIES OF CONCRETE BASED ON TECHNOGENIC RAW MATERIALS FOR CREATING MONOLITHIC STRUCTURES

***N. I. Buravchuk*[1,*]*, O. V. Guryanova*[1]*, M. A. Jani*[2] *and E. P. Putri*[2,3]

[1]I. I. Vorovich Mathematics, Mechanics and Computer Sciences Institute, Southern Federal University, Rostov-on-Don, Russia
[2]University of 17 Agustus 1945 Surabaya, Indonesia
[3]Khon Kaen University, Khon Kaen, Thailand

ABSTRACT

It is shown that the replacement of mineral raw materials by man-made waste is accompanied by an improvement in the physical and mechanical properties of concrete while saving cement. Ash and slag waste and burnt mine rocks are used in the composition of concrete as fine-grain additives, fine-grain and large aggregates. Concretes with industrial raw materials are tested in the construction of load-bearing structures, namely mine support on monolithic and sprayed-concrete technology. The presence of fine additives increases the number and area of contacts during the formation of the structure of hardening material. The result is a dense, strength, well-cemented concrete of high frost resistance and water resistance. The concrete is suitable for hydro-technical building.

---

[*] Corresponding Author's E-mail: burav@math.rsu.ru.

Keywords: technogenic raw materials, fly ash, ash and slag waste, burnt mine rocks, monolithic concreting, sprayed-concrete, physico-mechanical properties

1. Introduction

One of the directions of the concept of sustainable development, taking into account the interest of not only the present, but also future generations, is the use of waste industry. Such wastes accumulate in the dumps and are technogenic raw materials. Resource conservation becomes comprehensive and becomes part of the economic development of each country. Building companies are especially acute faced with the problem of environmental management, which allows them to meet production needs, not forgetting about the protection and reproduction of the environment. The main trend in the development of the production of building materials in the XXI century is associated with the use of man-made waste as sources of raw materials. This is due on the one hand to the depletion of reserves of natural raw materials, on the other hand connected with the intensive accumulation of industrial waste. Of all the formed and accumulated wastes, the most large-tonnage waste arises from the coal mining and fuel industry, namely rocks of mine dumps and ash and slag waste. Technogenic waste is distinguished by a large variety of properties and differently affects the properties of products based on them. This leads to the need to develop special methods for their preparation for each type of waste. As rightly pointed out by researchers [1 – 3], an efficient use of technogenic waste requires a fundamentally new approach to their disposal, which requires additional costs to create an information and analytical data bank on waste, technologies and their sanitary and hygienic assessment, to develop air-conditioning and storage methods.

Unfortunately, the problem of recycling industrial waste is not received enough attention. With significant amounts of man-made wasters, the level of their utilization is low. The main consumer of industrial waste may be the building industry. The volume of technogenic raw materials in the manufacture of building materials can be significant. The production of building materials is the most material- and energy-intensive branch of human activity. In this direction of production activity, humankind uses natural resources, which are maximal ready for use, since they require minimal labor input. Extraction of natural resources from interconnected natural states, where their presence ensures equilibrium and environmental sustainability, introduces some imbalance in the system of the self-organization processes of geosystem. This imbalance can be reduced by changing the natural resource base of the building industry by replacing or/and replenishing it with a new type of raw material, namely technogenic one. The results of scientific research and existing practical experience [4 – 6] indicate that many industrial wastes are valuable raw materials. For example, the accumulation of the wastes of mining and combustion of coal in the quantity and quality of the contained mineral raw materials, suitable for effective use in the field of material production, are technogenic deposits. The use of these wastes can expand the raw materials base of the building and non-metallic materials industry.

2. Problem Formulation

The influence of technogenic raw materials from ash and slag wastes and burned mine rocks on the physico-mechanical properties of heavy concrete for the manufacture of monolithic constructions was studied. With this aim, we carried out analytical and laboratory studies of the composition and physico-mechanical properties of technogenic raw materials. Technological and industrial tests were carried out, for which a full assessment of the quality of raw materials and concrete was given in accordance with the requirements of regulatory and technical documentation.

In this paper, the objects of research are ash and slag wastes and burnt mine rocks, accumulated in dumps, located in the territory of the Rostov region (Russia). Ash and slag wastes are represented by ash of dry selection from electrostatic precipitators (fly ash) and ash – slag mixture from dumps. Preliminary waste preparation is not required. Burnt mine rocks from dumps are crushed and grinding. Fillers of various fractions, screenings of crushing and fine fraction (ground burnt mine rocks) are suitable for use.

3. Volume Fraction Optimization

3.1. Properties of Burned Mine Rocks and Ash – Slag Wastes

Analysis of the experience of using technogenic wastes in concrete technology [7] confirms that burnt mine rocks and ash – slag wastes may become a promising raw materials from the viewpoint of the main technological processes and properties of the resulting binders and concrete. Ash and slag wastes and burnt mine rocks are related to a class of technogenic deposits, the development of which is a strategic way to resource saving and improvement of the natural environment in the area of operation of thermal power plants and coal-industrial areas.

Fly ashes and ash – slag wastes from coal combustion and burnt mine rocks are chemically mineralogical, particle size and phase mineralogical composition in many aspects identical to the natural mineral raw materials [8, 9]. The chemical composition of the averaged samples of ash – slag waste is given in Table 1; Table 2 presents burned mine rocks of some mine dumps, located in the Rostov region.

Table 1. Chemical composition of ash – slag wastes, %

Probe	SiO_2	Al_2O_3	Fe_2O_3	CaO	MgO	SO_3	TiO_2	$\sum(K_2O, Na_2O)$	p.p.p.*
Fly ashes	52.45	21.32	7.45	4.96	2.30	0.69	0.98	3.65	5.82
Ash – slag wastes from the dump	51.52	23.15	6.53	5.75	2.07	0.78	0.94	2.36	6.44

* p.p.p. are the losses at calcining.

Table 2. Chemical composition of burnt mine rocks, %

Mine Title	SiO_2	Al_2O_3	Fe_2O_3	CaO	MgO	SO_3	TiO_2	$\sum(K_2O, Na_2O)$	p.p.p.*
Komsomolskaya Pravda	55.24	21.78	6.25	1.61	1.82	2.17	1.15	4.70	5.26
Petrovsky	55.82	22.84	6.42	1.13	1.44	1.73	1.11	3.85	4.92
Vorovsky	54.77	22.18	7.21	1.16	1.47	2.37	1.18	3.98	5.32
Maiskaya	56.58	21.32	5.5	1.27	2.54	2.78	1.00	4.07	4.32
No. 26	54.33	21.98	6.67	1.33	1.86	2.73	1.12	4.08	5.22
Burgustinskaya	54.53	23.08	5.85	1.47	1.52	2.58	1.22	3.97	5.40

* p.p.p. are the losses at calcining.

The chemical composition is represented mainly by oxides of silicon, aluminum and iron, and the studied technogenic raw materials are acidic, with a high content of coloring oxides. The content of acid oxides is more than 70%. Such technogenic raw materials can be characterized as siliceous and aluminous.

In the mineral part of solid fuels, as in the mine rocks, clay minerals are predominate. The most typical type of clay minerals in coal seams is kaolinite ($Al_2O_3 \cdot 2SiO_2 \cdot 2H_2O$). The base of the activity of burnt mine rocks and ash and slag waste is the change of clayey matter during the physicochemical processes that occur during the firing of mine rocks in the dumps and the combustion of solid fuel in boilers. In the result of thermal exposure, dehydration and disruption of molecular bonds of aqueous aluminum-silicates and other minerals and the appearance of active components occur. Due to the nature of their formation and the thermodynamic state of ash and burnt mine rocks they have an increased stock of internal energy and corresponding reactionary activity. This new state is the cause of the ability of mine rocks and ashes to absorb lime from a solution. The activity of the ashes is explained by the presence of a glassy phase, and presence of active clay in the case of burnt mine rocks. These features of ashes and burnt mine rocks are the base for their use in concrete technology. Their use should have a positive impact on both the economy of natural resources and the improvement of the physical and mechanical properties of concrete.

3.2. Experimental-Industrial Building of Concrete Constructions by Using Technogenic Raw Materials

For the building of monolithic constructions, movable concrete presents interest. Two types of such concrete have been investigated: with traditional aggregates and fine-grained without sand and rubble. In such a concrete, fly ash replaces part of the cement, and ash – slag wastes are used as fine-grained aggregate. In the presence of fly ash, high mobility and good workability of the concrete mixture are achieved. Fine-grained aggregate of ash and slag waste ensures the stability and uniformity of the concrete mixture. All the pointed features of the ash concrete possess to it the properties of cast concrete and greatly facilitate the transportation of the concrete mixture through the concrete pipeline and laying the mixture in the formwork.

In the laboratory conditions were developed ash concrete. In the compositions of concrete used large and small aggregate. The decrease in cement consumption compared with the control was 15 – 35%. The technological process for monolithic concrete was worked out on a model set-up simulating conditions similar to those in production. The removal of the formwork on the model set-up was performed via 6 hours after the start of concreting. In parallel, testing the control cubes was carried out. Compressive strength after 6 hours was 1.20 – 1.30 MPa for ash-bearing concrete with rubble, and 0.8 – 0.9 MPa for fine-grained ash concrete.

The test of fine-grained ash concrete was carried out by monolithic concreting of the support of the ventilation shaft of mine. By concreting the test site, fly ash and ash – slag wastes of hydraulic dumps were used. Part of the site was concreted by ash with large aggregate and ash – slag wastes instead of sand. Cement fraction is reduced by 30% (143 kg per 1 m^3). Another part of the test site was made of fine-grained ash concrete. The main components of fine-grained ash concrete: cement, fly ash, ash – slag wastes. The economy of cement was 30%, and economy of sand and rubble was 100%. The fineness modulus of ash – slag wastes was 2.02. The technology of preparing ash concrete for monolithic concreting is similar to the technology of preparing ordinary concrete. The mobility of the concrete mixture according to the rule of cone is 15 – 17 cm, the designed grade of concrete is 200 on compressive strength (concrete class B15), and grade on concrete water tightness is not less than W2.

Through one day after concreting, the formwork was separated, and no deviations from the conventional concreting technology were observed. The formwork was completely removed after 7 days in accordance with the existing regulations. Testing of ash concrete was performed through 28 days after concreting by a non-destructive testing. Statistical processing of control measurements at different marks of the entry showed that this fine-grained ash concrete has an average compressive strength of 21.3 MPa, which corresponds to the class B15 of concrete on strength. The simultaneous control tests of ordinary concrete also showed characteristics of concrete with compressive strength of class B15. It should be noted that ash concrete hardens more slowly, but has a longer time for increasing its strength in comparison with ordinary concrete. Therefore, the strength of the ash-concrete, established for a period of 28 days, does not attain a final value.

Through one year after concreting the mine shaft, ash-concrete was re-tested. A statistical processing of control measurements at different marks of entry was carried out. The strength of fine-grained ash concrete increased 2.2 times, and ash concrete with rubble 2.4 times. The increase in strength of the control composition of concrete did not exceed 7%.

A species of the monolithic concrete construction is a sprayed-concrete, processed without formwork. For such a concrete, the ratio between coarse and fine aggregates and the particle size distribution of the mixture with the lowest rebound value are important. The sprayed-concrete technology was used to carry out repair work at the Mayskaya mine (Rostov Region) by using experimental compositions. A clinkerless binder consisting of finely ground burning mine rock or fly ash and low-modulus liquid glass was used. The specific surface of the rock was 300 – 350 m^2/kg, the modulus of liquid glass was 0.8 – 2.0, the density was 1.30 – 1.45 g/cm^3. To improve workability and reduce the rebound of materials during concreting, natural sand was additionally introduced. The ratios of dry components were the following: ground burnt mine rock (ash) : natural sand : burnt mine rock : crushed stone from burnt mine rocks = 1 : 1 : 1 : 2 – 4. The optimal size of the coarse aggregate fraction was 5 – 15 mm, mixture of dry components with liquid glass in the amount of 8 – 10%, consumption of milled additives was 200 – 250 kg/m^3, concrete strength was 28.7 – 34.5 MPa, the specific surface area of the rock was 300 – 350 m^2/kg, liquid glass module was 0.8 – 2.0, density was 1.30 – 1.45 g/cm^3. To improve workability and reduce the rebound of materials during concreting, natural sand was additionally introduced. Ratios of dry components the following: fine-milled burnt mine rocks (ash) : natural sand : sand from burnt mine rock : crushed stone from burnt mine rocks = 1 : 1 : 1 : 2 – 4. The optimal size of the large filler fraction was 5 – 15 mm. Mixture of dry components with liquid glass in an amount of 8 – 10%, consumption of fine-milled additives was 200 – 250 kg/m^3, concrete strength was 28.7 – 34.5 MPa.

In the process of building the support, a moist mixture of binder and aggregates is applied on the concrete surface with compressed air under pressure. Due to the speed of the jet, the concrete mixture is laid and compacted, strongly strengthens the surface and solidifies. The presence of fine-milled additives with a plasticizing effect increases the adhesive ability of sprayed-concrete. Particles of ash and ground burnt mine rock as micro-fillers are involved in the formation of a dense structure of the laid concrete. The created structure is a single "support – rock" system, the bearing capacity of which exceeds this indicator, as in comparison with monolithic concrete, laid in the usual way behind formwork. The construction of support by spraying-concreting allows one to control the thickness of the support. This is very important when the mine shaft crosses a non-uniform rocks in strength. It is advisable to use such a method of building support at great depths (over 800 m), where it is technically and economically not feasible to create powerful structures with increased carrying capacity. The use of ash and slag waste and burnt mine rocks in the technology of sprayed-concrete reduces dustiness, reduces the consumption of cement and material rebound, enhances adhesion to smooth surfaces, and reduces the consumption of natural mineral raw materials.

3.3. Physical and Mechanical Properties of Concretes, Containing Technogenic Raw Materials

Table 3 shows the main characteristics of heavy, fine-grained and sprayed-concrete containing finely ground additives of ash and burnt mine rock. The aggregate contains ash – slag wastes, crushed sand from burnt mine rock, natural sand, natural crushed stone and crushed stone from burned mine rocks. Such concretes are characterized by an increase in the volume of mixed cementitious binding. For rigid mixtures, this value is 320 – 380 litre/m^3, for mobile mixtures it is 420 – 470 litre/m^3. This is reflected in the rheological properties of concrete mixtures: increased cohesion, plasticity, water retention, improved uniformity and workability. The surface quality of goods is improved by reducing the number of large pores and cavities. The plasticizing effect of fine fraction burned mine rocks is lower than that of ash. This is due to the shape, porosity and surface roughness of the rock particles. Spherical particles predominate in the ashes, thereby reducing internal friction in the concrete mixture. A characteristic feature of such concretes is the slow increase in strength at the beginning of hardening.

Table 3. Physical and mechanical properties of concrete based on technogenic raw materials

Type of Concrete	**Strength, MPa**				**Grade of Concrete**		**Economy of Cement, %/ kg**
	Hardening Time, days						
	28	180	28	180	Frost Resistance	Water tightness	
	Compression	Bend	Compression	Bend			
Ash concrete							
Heavy Concrete	38.0	6.35	54.0	7.79	F300	W10	30/70
Fine-grain Concrete	36.3	6.29	50.2	7.63	F250	W8	30/102
Concrete from burned mine rocks							
Coarse-aggregate Concrete	34.5	5.79	45.7	7.66	F200	W8	20/69
Sprayed-Concrete	31.2	5.28	41.3	6.87	F200	W8	20/77

After reaching the design strength, concrete hardening continues for a long time. Through one year, the strength of concrete can exceed the design value by more than 2 times at economy of cement more than 20%. The high-strength concretes having low water permeability are suitable for hydrotechnical constructions.

3.4. Structure Formation in Concrete with Technogenic Raw Materials

The processes of structure formation in the ash concrete are described by the physicochemical mechanics of dispersed structures [10]. Concrete is a capillary-porous system. At the stage of hardening, concrete is a microheterogeneous dispersion system. The main role in shaping the structure of concrete is played by the strength of contacts between particles. The formation of the structural frame takes place in several stages. The initial interaction between dispersed particles is due to long-range van der Waals forces and the formation of coagulation structures. For the further process of structure formation, the maximum contacting and the transformation of coagulation structures into condensation-crystallization structures, ensuring the mechanical strength of concrete, is characteristic. The use of fine-grain micro-fillers (ash or burnt mine rock) and aggregates from technogenic raw materials allows one to obtain high-qualitative fine-grain dense structure of concrete. An additional amount of hydrosilicates, which condense and strengthen the structure of the cement stone, arises due to the pozzolanic reaction [11], which is characteristic of finely dispersed ash and burnt mine rocks. The smaller the grain of cementing compounds, the higher the strength and durability of concrete [12]. The more perfect the concrete structure states the lesser probable accumulation of local destruction zones in concrete with the formation of cracks under various variable loads or climatic influences, for example, during alternate freezing and thawing, influence of aggressive media, etc.

CONCLUSION

Dumps of ash and slag wastes and burnt mine rocks attract attention as promising sources of raw materials for various industries. The prospect of using technogenic deposits is obvious, since their use allows us to simultaneously solve a number of economic, social and environmental problems. Involvement in the processing of technogenic raw materials provides:

(i) reducing the cost of exploring new deposits;
(ii) expanding the resource base, primarily for the construction industry;

(iii) reducing the cost of production and improving the conditions of extraction of raw materials, as man-made deposits located on the surface of the Earth;
(iv) release of land occupied by dumps, their re-cultivation and the elimination of sources of environmental pollution, reduction of man-made loads in industrial areas;
(v) production of building constructions that are not inferior in quality to similar goods from natural raw materials, but have lower cost.

Creation of new materials, working in extreme climatic conditions and alternating loads, is one of the most important problems of our time and equally relevant for any civil sector of the national economy.

ACKNOWLEDGEMENT

This research was carried out within the framework of financing of the Ministry of Education and Science of the Russian Federation for the project No. 11.6767.2017/BCh.

REFERENCES

[1] Solomatov, V.I. *Construction Materials, Equipment, Technologies of the 21st Century*. 2000, *No. 1*, 28-29 (In Russian).
[2] Lesovik V.S. *Proc. Int. Conf. Problems of Building Materials Science and New Technologies*. BelGTASM, Belgorod, 1997, *Part 5*, 45 – 49 (In Russian).
[3] Trubetskoy, K.N.; Umanets, V.N. *Gorny Zhurnal*, 1992, *No. 1*, 12 – 16 (In Russian).
[4] Buravchuk, N.I., Rutkov, K.I. *Recycling and Use of Waste Mining and Combustion of Coal*. Publishing House SKNTS VS Press, Rostov-on-Don, 1997, pp. 1 – 224 (In Russian).
[5] Buravchuk, N.I. *Resource Saving in the Technology of Building Materials*. Southern Federal University Press, Rostov-on-Don, 2009, pp. 1 – 224 (In Russian).
[6] Buravchuk, N.I.; Budnitsky, V.M.; Brazhnikov, V.F.; Melentyev, S.A. Resource Conservation in the Technology of Binders and Concretes. Publishing House SKNTS VS Press, Rostov-on-Don, 1999, pp. 1 – 176 (In Russian).
[7] Dvorkin, L.I.; Dvorkin, O.L. *Construction Materials from Industrial Wastes*. Phoenix, Rostov-on-Don, 2007, pp. 1 – 368 (In Russian).
[8] Abdrakhimov, V.Z.; Abdrakhimova, E.S. *News of Universities. Building*. 2006. *No. 1*, 36 – 39 (In Russian).

[9] Ukhaneva, M.I.; Khobotova, E.B. *Visnik of the Kharkiv National University. Chemistry*. 2010. *No. 895. Iss. 18 (41)*, 260 – 268 (In Ukrainian).

[10] Rebinder, P.A. *Surface Phenomena in Dispersed Systems. Physico-chemical Mechanics. Selected Works*. Science, Moscow. 1979, pp. 1 – 384 (In Russian).

[11] Kokubu, M.; Yamada, D. In: *Fifth International Congress of Cement Chemistry*. Stroiizdat, Moscow. 1973, 405 – 416.

[12] Bazhenov, Yu. M. *Concrete Technology*. High School, Moscow. 1987, pp. 1 – 415 (In Russian).

In: Proceedings of the 2018 International Conference … ISBN: 978-1-53615-862-5
Editors: Ivan A. Parinov et al.

Chapter 6

DESIGN OF CENTRIFUGAL CASTING MACHINE HORIZONTAL TO CASTING ALUMINUM IN LABORATORY SCALE

Gatut Prijo Utomo[1,*],
Pramoda Agung Sumadijono[1]
and Imah Luluk Kusminah[1]
[1]Department of Mechanical Engineering, University of 17 Agustus 1945 Surabaya, Jawa Timur, Indonesia

ABSTRACT

Centrifugal Casting Horizontal is a casting technique by utilizing a continuous spin motion resulting in cylindrical casting material. The continuous rotation makes the material pushed or pressed to the outer radius of the radius resulting in a cylindrical cast material with a better surface than other techniques of casting. Aluminum is often used in the cast industry because has many advantages such as resistance to corrosion rate, is a good electrical conductor and others. Therefore, it is widely used for alloys with other metals or as core in cast. Designing the Horizontal Aluminum Casting Centrifugal Engine is done from the design process until the drawing of the work with the general typical shape on the frame has dimensions of 1000mm, width 500 mm, height 500mm and electric motor used to rotate the liner mold of 1 HP with maximum rotation speed 1500rpm.

Keywords: aluminum, casting, centrifugal

* Corresponding Author's E-mail: gatut_pu@untag-sby.ac.id.

1. Introduction

Numerous problems in the foundry are found mostly in the gravity casting. Gravity casting is the basic method of metal casting, because this technique has many shortcomings if not performed perfectly [1]. Therefore, there are many mistakes found in the process of casting because the flow of foundry flue caused by drainage hot liquid material and it does not quickly and precisely become to fully liquid [2]. Air cavity defects occur because there are trapped air in the material that cools very quickly because the material temperature has not a high melting point. The entering speed of molten metal is too slow that leads to trapped air cavity, cracked defects in casting caused by imprecise mold and imperfect material cooling. The improvement of casting method in the world is developing, attaining periodic perfection found same method of centrifugal casting technique. Centrifugal casting can cover the deficiencies of gravity casting and can be called a refinement because centrifugal casting uses rapid rotation techniques so that the hot material liquid poured is pushed and pressed from the rotation axis to the outer side of the radius so that the material becomes more perfect similar to a linear cylinder. Surface castings are better than gravity casting, in a whole [3].

Centrifugal casting's mechanism relies on rotation process as its basic technique. It is different from static casting. The mold is immovable and only liquid material is poured. In centrifugal casting, pouring is done when the mold is rotating, although in certain types, especially in perpendicular centrifugal casting, the casting begins when the mold is stationary. Then the mold is rotated to a certain speed so that the metal cooling occurs at the time of the mold spinning. In the horizontal casting method, the casting is done at the mold rotation and following pouring the material into the rotating mold [4]. The round used can be varied as it is needed. At the time of casting, the cooling phase of the material from high temperature to low temperature is very fast so before it is poured the linear mold, it is heated first to avoid the foundry defect caused by the temperature difference between the high temperature of the material and room temperature of mold. If the mold has not been preheated before the pouring, then there will be mistakes in casting and castings become defective and imperfect. At centrifugal casting, the castings can have high hardness on the outer surface compared with the surface at static casting; therefore, centrifugal molds have more advantages [5].

2. Experiment

2.1. Centrifugal Casting

Frames in centrifugal casting are usually made by using welding. However, now many prefer frame of welding rods because they are less pressing manufacturing costs.

By just welding the construction from iron bars, we can assemble various types of casting centrifugal molds that we want to connect with welding techniques and all they can be performed simply and certainly safe. The advantages of centrifugal casting are:

(i) reduce weight by 30%;
(ii) create a thin frames;
(iii) can use various kinds of metals;
(iv) easy design changes because the machine is made in the form of a circuit that can be changed so as to minimize the costs incurred;
(v) if an error occurs in the design then this error can be corrected easily;
(vi) it can be given additional circuits in some points if needed.

The frame serves to support components used for casting such as drive motors, anchors, shafts, rollers, pulleys. The frame used is made as rule from low-carbon steel in the form of "L".

2.2. Shaft

The shaft is a rotating cylindrical rod with usual cross-section or rod also attached to components such as rollers, pulleys, and V-belts that work synchronously driven by a driving motor to drive the mold. Shaft function in a machine consists in that it is a receiver to power by utilizing rotational power. Each element of a rotating machine, such as chakra strap, belt on a pulley, motor belt, including cast mold will rotate according to the rotating speed on the given shaft. As the examples, we consider:

(i) *Axial Strength.* The shaft will accept twisting moment, bending moment or combination of twisting and bending loads. In the shaft planning, it is necessary to pay attention to several factors, namely fatigue, collision and the influence of stress concentration when using a powered shaft or a pin on the shaft. The designed shaft must be safe enough to withstand these loads.
(ii) *Axial Stiffness.* Even though a shaft has a capacity that is safe enough to withstand loading but the bending or deflection that is too large will result in inaccuracy (in machine equipment), vibration and sound (noise). Therefore, in addition to paying attention to the strength of the shaft, the shaft stiffness must also be considered and adjusted with the engine, which transmit loading to the shaft.
(iii) *Critical Rotation.* If the engine speed is increased, it will cause vibration of the engine. The boundary between engine rotations that have a normal number of revolutions with engine rotation that causes high vibration is called the critical

rotation. This can occur in turbines, fuel motors, electric motors, etc. In addition, the emergence of high vibration can cause damage to the shaft and other parts. So in the design of the shaft it is necessary to consider the working rotation of the shaft to be lower than the critical rotation.

(iv) *Shaft Material.* Shafts commonly used for high speed and heavy loads are generally made of alloy steel with a hardening process (case hardening) so that they are resisted to wear. Some of them are nickel chrome steel.

2.3. Transmission

Transmission is a tool for passing power from one axis to another by using corresponding device, for example the tool chain, belt, gear etc. In this centrifugal casting of aluminum tool, we use a transmission belt that is connected using pulleys.

The belt is a flexible machine element that can be used easily to transmit torque and rotating motion from one component to another. Belt is used to move power between two parallel axes. Separate shafts at a certain minimum distance depend on the type of belt to work efficiently. The belt has the following characteristics:

(i) drunk can be used for long axis distances;
(ii) because of the slip and slow belt movement, the ratio of the angles between the two shafts are not constant and equal to the ratio of the pulley diameter;
(iii) if a V-belt is used, several variations in angular velocity comparisons can be obtained using a small pulley with a spring-loaded side; pulley diameter is a function of belt tension and can be changed by changing the axis distance;
(iv) when the belt is in operation, a slight adjustment of the distance between the axles is usually required;
(v) using a multilevel pulley an economical speed modifier can be obtained.

Processes followed from the beginning are designing the centrifugal casting machine, calculation, manufacturing, and RPM testing, finished by aluminum casting using centrifugal casting machine.

3. Results and Discussion

Designing the aluminum cast with centrifugal casting engine is conducted from the design process until the drawing of the work with the general typical shape on the frame has dimensions of 1000mm, width 500mm, height 500mm and electric motor, used to rotate the liner mold of 1 HP and creating maximum rotation speed 1500rpm. The impact testing results are presented in Figures 1-4.

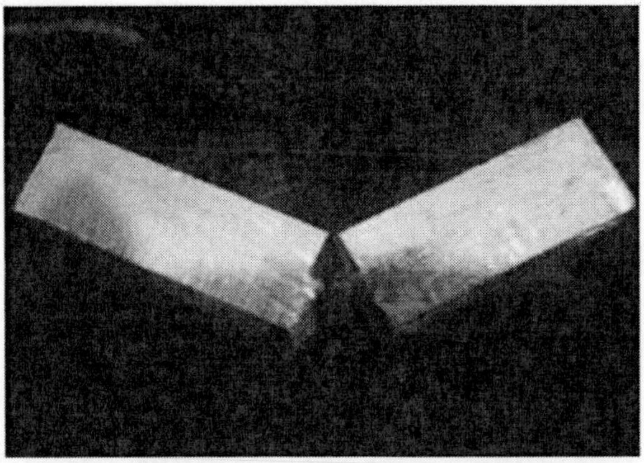

Figure 1. Centrifugal casting product (left); impact testing result (right) with 500 rpm speed.

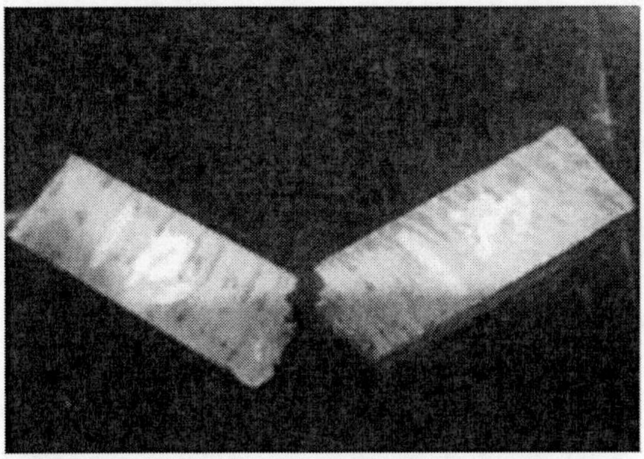

Figure 2. Centrifugal casting product (left); impact testing result (right) with 1000rpm speed.

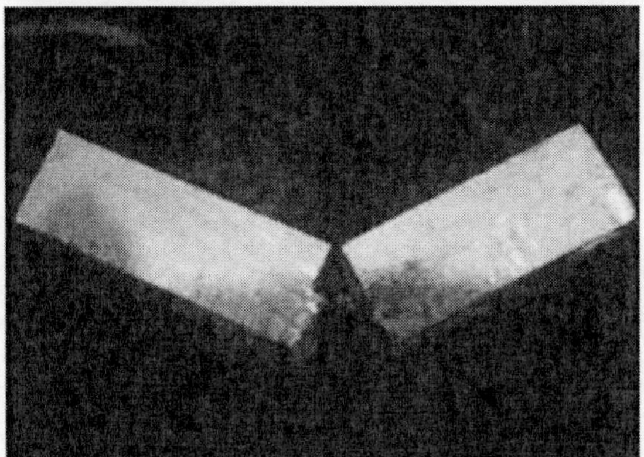

Figure 3. Centrifugal casting product (left); impact testing result (right) with 1500rpm speed.

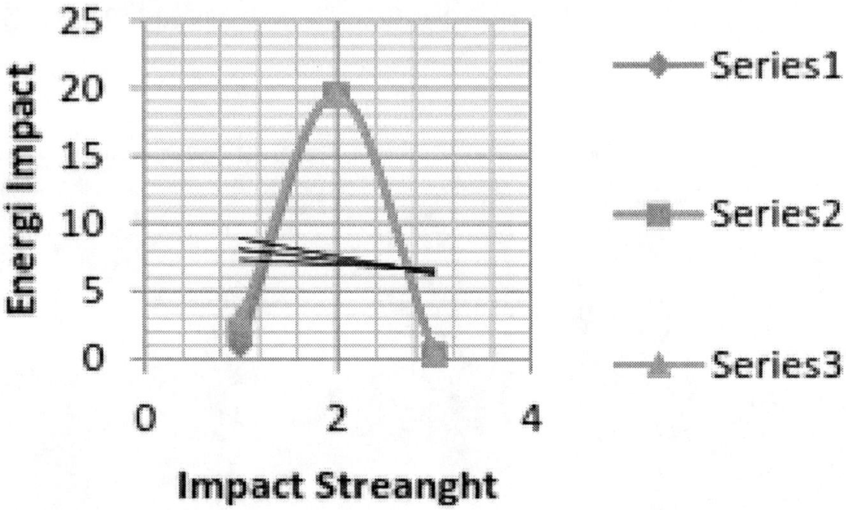

Figure 4. Impact strength *vs* impact value.

CONCLUSION

1. Designing a horizontal centrifugal casting machine begins with stating the following parameters: a length of 600mm, a width of 500mm, a height of 450mm, a power of 1 HP, a maximum rotation speed 1500rpm.
   The casting machine is expected to produce cast cylindrical sample, which is better than due to the die-casting because it has unique properties in the casting technique.
2. The preparation of frame, two axial shafts, molds, pulleys and motors, covering of mold, liner mold, base plate, and lock for the safety precede the manufacturing process of centrifugal casting horizontal.
3. The end of aluminum casting produces good performance, when the process takes place without speed drop or decrease of engine speed. So this machine has enough motor power to do the casting process.
4. This cast molding machine can be used in long life as well as require care is not difficult so it can last long enough.

ACKNOWLEDGMENTS

The authors thank The Ministry of Research, Technology, and Higher Education of Republic of Indonesia for funding this research.

References

[1] Baikhaky, Aby Al., *Design of Plastic Bottle Counter Machine, Final Project,* DIII Department of Mechanical Engineering, Universitas Diponegoro, 2015 (In Indonesian).

[2] Husen, A., *Modification of Straw Counter Machine, Final Project,* DIII Department of Mechanical Engineering, Universitas Diponegoro. 2012 (In Indonesian).

[3] Khurmi, R. S.; Gupta, J. K. *Machine Design*, Eurasia Publising House, New Dehli. 1982.

[4] Sularso, *Basic Planning and Selection of Machine Elements*, Pradnya Paramita, Jakarta. 1997 (In Indonesian).

[5] Sunday, A. M.; Oyewole Adedipe, *Design and Fabrication of a Centrifugal Casting Machine*, 2011, *vol. 3(11)*, 8204.

In: Proceedings of the 2018 International Conference … ISBN: 978-1-53615-862-5
Editors: Ivan A. Parinov et al.

Chapter 7

EFFECT OF PRESSURE LOADS AND DURATION ON SQUEEZE CASTING TO THE HARDNESS AND TOUGHNESS OF BOLT AND NUT FROM COMPOSITE AL 6061 – BOTTOM ASH

Edi Santoso* and Muhyin
Department of Mechanical Engineering,
University of 17 Agustus 1945 Surabaya, Indonesia

ABSTRACT

Metal composite or better known as metal matric composite (MMC) is a combination of two or more materials, in which the metal as a matrix and ceramic as a reinforcement are used to obtain the desired characteristics. To produce solid components and fine microstructure in its manufacture, the squeeze casting process has an approximate capability. In the manufacture of aluminum matrix composites, to obtain better mechanical properties such as hardness and strength, it is required to perform T6 heat treatment process. The purpose of this research is to know the effect of pouring temperature variation and the duration of emphasis on squeeze casting process on shape change, dimension, hardness distribution and microstructure of aluminum 6061 – aluminum oxide composite after T6 treatment. The research method was conducted by casting aluminum 6061, then melt it at 660°C. The molten metal next is added with aluminum oxide from electrolysis plating and magnesium. The temperature was increased to 700°C. After pouring, wait 60 seconds and then it is pressed with a pressure load varying as 15 kgf, 20 kgf and 25 kgf and pressing duration varying as 60 seconds, 120 seconds and 180 seconds. After that T6 heat treatment was conducted. Experiments included Rockwell B hardness testing and impact toughness. The results of the Rockwell B hardness testing show that the heavier load weighs and the longer the duration of the

* Corresponding Author's E-mail: edisantoso@untag-sby.ac.id.

emphasis will make the atomic structure more compact and resulting in declining porosity of the castings characterized by the narrowing of the grain size of the microstructure that causes the violence to rise. The results of the test on impact toughness indicate that the greater the load will affect the hardness increase and the toughness decrease.

Keywords: "aluminum 6061 – aluminum oxide" composite, hardness, hardness distribution, T6 heat treatment, toughness, squeeze casting

1. Introduction

Metal composite or better known as metal matric composite (MMC) is a combination of two or more materials, in which the metal as a matrix and ceramic as an amplifier are used to obtain the desired characteristics [1, 2]. To produce solid components and fine microstructure in its manufacture, the squeeze casting process has an approximate capability. In the manufacture of aluminum matrix composites, to obtain better mechanical properties such as hardness and strength, it is required to perform T6 heat treatment process. The purpose of this research is to know the influence of pouring temperature variation and the duration of emphasis on squeeze casting process, shape change, dimension, hardness distribution and microstructure of "aluminum 6061 – aluminum oxide" composite after T6 treatment [3]. The research method is conducted by casting aluminum 6061, melt it at 660°C, and then added aluminum oxide from electrolysis plating and magnesium. Then the temperature was increased until temperature 700°C. After pouring, wait 60 seconds and then pressed with a pressure load with varying as 15 kgf, 20 kgf and 25 kgf and pressing duration varying as 60 seconds, 120 seconds and 180 seconds. Then T6 heat treatment was conducted. Experiments included Rockwell B hardness testing and impact toughness [4, 5]. The results of the Rockwell B hardness testing show that the heavier load weighs and the longer the duration of the emphasis will make the atomic structure more compact and resulting in declining porosity of the castings characterized by the narrowing of the grain size of the microstructure that causes the violence to rise. The results of the test on impact toughness indicate that the greater the load will affect the hardness increase and the toughness decrease.

2. Experiment

2.1. Research Steps

a) Preparation of tools & materials;

b) the process of electrolysis plating;
c) squeeze casting process;
d) Rockwell hardness test before T6 heat treatment;
e) impact toughness test before T6 heat treatment;
f) T6 heat treatment process;
g) Rockwell hardness test after T6 heat treatment;
h) impact toughness test after T6 heat treatment;
i) the process of machining and the manufacture of bolts and nuts.

2.2. Hardness Testing

Hardness testing with the Rockwell method aims to determine the hardness of a material as a material resistance to indenters by using steel balls or diamond cones that are applied to the surface of the test material. Rockwell hardness testing is one of the hardness tests that are used often, because Rockwell's hardness testing is simple, fast, does not require a microscope to measure traces, and is relatively non-destructive [6]. Rockwell hardness testing is done by pressing the specimen surface (test object) with an indenter. The indenter pressure into the test object is carried out by applying the initial load (minor load), then added to the main load (major load), then the main load is released while the minor load is still maintained. The hardness testing with Rockwell method is according to DIN 50103 standard steps:

- installation of the indenter in accordance with the type of test, namely steel or cone indenter;
- selection of the load used for the loading process;
- statement of the specimen tested on hardness;
- testing the object pressed by the indenter with a minor load;
- testing the object pressed with a major load;
- major load is taken so that to remain a minor load.

2.3. Impact Testing

Impact test is testing using fast loading; it measures material resistance to impact loading. This is what distinguishes the impact testing with tensile and hardness testing, where loading is put slowly. The impact testing is an attempt to simulate material operating conditions that are often encountered in transportation equipment or construction, where the load does not always occur slowly but comes suddenly, for example deformation on the car bumper at the time of the accident collision.

In the impact test occurs a large energy absorption process when the load hits the specimen. The energy absorbed by this material can be measured by using the principle of potential energy difference [7, 8]. The base of the test is the absorption of potential energy from the load pendulum that swings from a certain height and hits the specimen so that the test object is deformed. In testing this impact, the amount of energy absorbed by materials for fracture is a measure of the impact resistance or toughness of the material.

Strength properties of a component can be detected from tensile testing and impact testing, but under different load conditions. The load on impact testing as described above is abrupt, while the tensile testing is slow [9]. From the results of tensile testing, we can estimate for the results of impact testing. However from impact testing can be known the properties of the toughness of metal and impact influences for different temperatures, ranging from very low temperatures (–30°C) to high temperatures. Whereas in the tensile experiment, the working temperature is room temperature.

There are two kinds of impact test methods, namely the Charpy and Izod methods. The fundamental difference of the method is the specimen location. The Charpy test is more accurate because the specimen holder also absorbs the energy, so that measured energy is not absorbable material [10].

Research steps include:

a) preparation of specimens, according to the desired dimensions (standard) with making a notch on each specimen right in the middle;
b) the pendulum is placed in the initial position for testing;
c) the pointer needle is set to position 0;
d) specimen is taken and placed at proper place;
e) the pendulum is lifted to an angle of 110° from the upright line;
f) the load locking pin is pressed, so the pendulum slides over the specimen;
g) the brakes are pressed when the pendulum is about to swing for the second time;
h) observation and recording the magnitude of the angle and the amount of energy shown by the pointer.

3. Results and Discussion

3.1. Hardness Testing

We obtained the results of the hardness testing of Rockwell B hardness with the number of specimens of 18 pieces. Testing was performed in one area, namely in the upper part at 5 points. The result of hardness testing are shown in Figure 1 and 2.

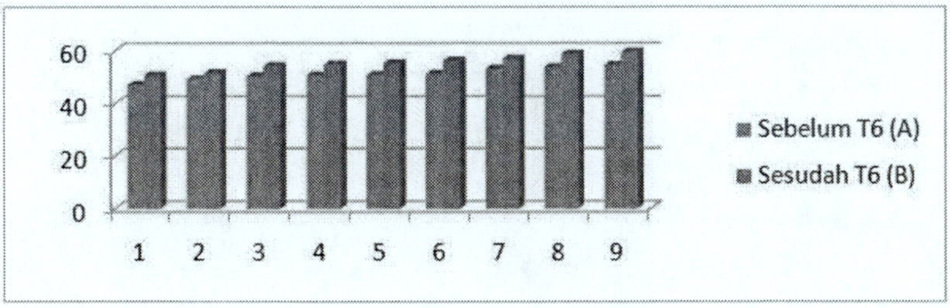

Figure 1. Rockwell B hardness results before and after T6 heat treatment (blue – before T6, red – after T6).

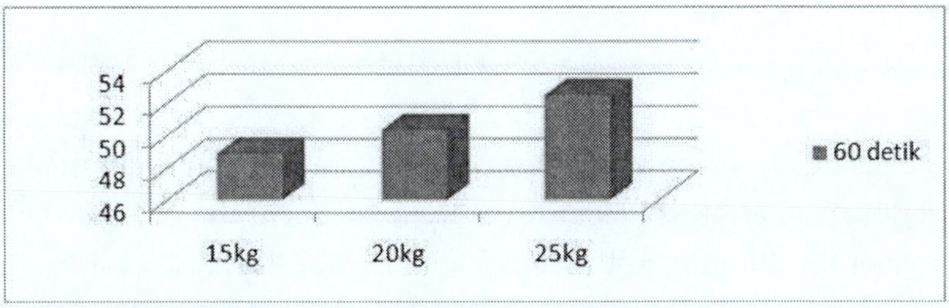

Figure 2. Rockwell B hardness results in the same pressing duration.

From the hardness testing result shown in Figs. 1 and 2, it can be seen that the values of Rockwell B hardness (HRB) on specimens after T6 heat treatment was evenly higher than for the specimens that were not subjected to the heat treatment.

3.2. Impact Testing

From the results of impact stress testing in Figs. 3 and 4, it can be seen that the values of impact energy for the specimens after T6 heat treatment are higher than for the specimens that were not subject heat treatment. It is directly proportional to the impact loading shown in the figures and is higher in 1.8 times for the specimens after T6 heat treatment compared to those without T6 heat treatment.

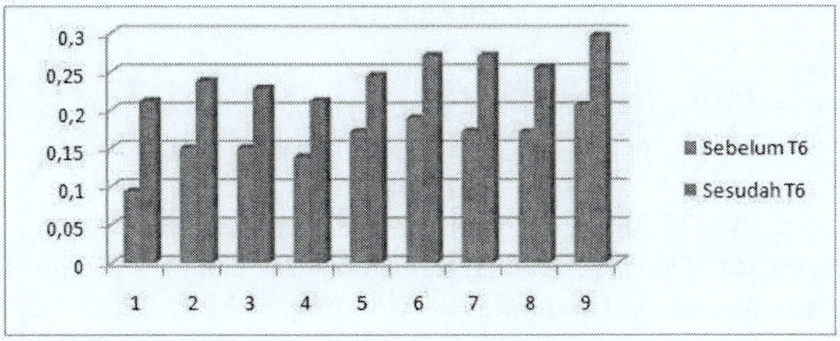

Figure 3. Impact energy results before and after T6 heat treatment (blue – before T6, red – after T6).

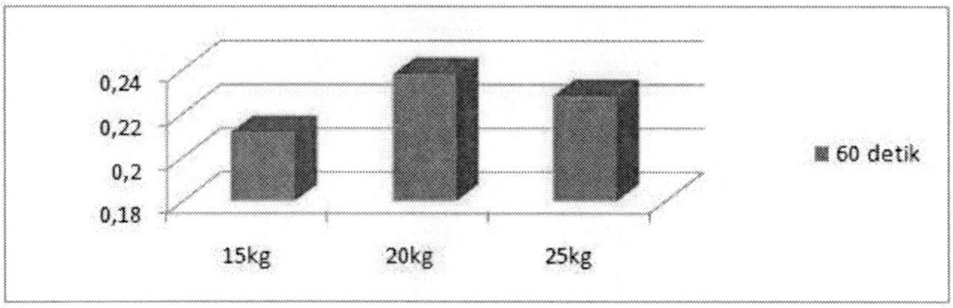

Figure 4. Impact energy results in the same pressing duration.

CONCLUSION

From the data of the compressive loading tests and the duration of the stress, it is followed effect on the hardness value of the test specimen: the heavier the compressive load and the longer the duration of the stress, the value of the hardness rises.

The highest hardness value was obtained for the specimen with code 9B under compressive load of 25 kgf and the duration of loading equal to 180 seconds.

The lowest hardness value was obtained for the specimen with code 1A under compressive load of 15 kgf and the duration of loading equal to 60 seconds.

Thus, the heavier the compressive load and the longer the duration of the stress will make the atomic structure more dense and decrease the porosity of the cast accompanied by the smaller size of the grain in the microstructure providing the hardness increase.

ACKNOWLEDGMENTS

The authors thank The Ministry of Research, Technology, and Higher Education of Republic of Indonesia for funding this research.

REFERENCES

[1] Radhika, N.; Balaji, T. V.; Palaniappan, S. *Studies on Mechanical Properties and Tribological Behavior of LM25/A$_2$O$_3$ Composite*. Department of Mechanical Engineering, Amrita School of Engineering, Amrita Vishwa Vidyapeetham, Coimbatore 641 112, India. 2015.

[2] Kumar, Ankesh; Kanhaiya, Kumar; Suman, Saurav dan Siva Sankur Raju, R. *Study of Physical, Mechanical and Machinability Properties of Aluminium Metal Matrix Composite Reinforced with Coconut Shell Ash Particulates*. Gunupur India. Student, Department of Mechanical Engineering, GIET. 2016.

[3] Attar, Sallahauddin; Madeva, Nagaral; Reddappa, H. N.; Auradi, V. *A Review on Particulate Reinforced Aluminum Metal Matrix Composites*. R&D Centre, Department of Mechanical Engineering, Bangalore Institute of Technology, Bangalore, Karnataka, India. 2015.

[4] Kittali, Praveen; Satheesh, J.; Anli Kumar, G.; Madhusudhan, T. *A Review on Effects of Reinforcements on Mechanical and Tribological Behavior of Aluminum Based Metal Matrix Composites.* P. G. Scholar, Mechanical Department, SJB Institute of Technology, VTU, Belagavi, Karnataka, India, Mechanical Department, SJB Institute of Technology, VTU, Belagavi, Karnataka, India. 2016.

[5] Patel, G. C.; Robins Mathew, Manjunath; Prasad Krishna; Mahesh B. Parappagoudar. *Investigation of Squeeze Cast Process Parameters Effects on Secondary Dendrite Arm Spacing Using Statistical Regression and Artificial Neural Network Models*. Department of Mechanical Engineering, National Institute of Technology Karnataka, Surathkal, India. 2014.

[6] Dhanashekar, M.; Kumar, Senthil. *Procedia Engineering*. 2014. *vol. 97*, 412 – 420.

[7] Veerabhadrappa, Alguar; Balaraj, V.; Nagaraj, Lori. *Effect of T6 Type Heat Treatment on the Mechanical Characterization of Al-6061 – Al_2O_3 Particulate Composites*. Department of Industrial and Production Engineering, Rao Bahadur Y. Mahabalesh, Warappa Engineering College, Bellary, India. 2015.

[8] Fazlur, Rahman K.; Benal, M. M. *Effect of Heat Treatment on the Coefficient of Thermal Expansion of Aluminum 7075alloy-Sicp (5wt %) Composites*. Depertement of Mechanical Engineering, Anjung Institute of Technology and Management, Bhatkal, Karnataka. India. 2012.

[9] Rahman, Md. Habibur; Rashed, H. M. M. A. *Effect of Magnesium on Wear Characteristics of Silicon Carbide and Alumina Reinforced Aluminum - Metal Matrix Composites.* Department of Materials and Metallurgical Engineering, Bangladesh University of Engineering and Technology, Dhaka, Bangladesh. 2014.

[10] Mahalingegegowda, H. B.; Mahesh, B. S. *Mechanical and Wear Behavior of Al-6061 – Al_2O_3 composites and AL-6061 – Al_2O_3 – Gr Hybrid Composites*. Department of Manufacturing Science Engineering. Ghousia College of Engineering, Ramanagara, Karnataka, India. 2014.

In: Proceedings of the 2018 International Conference … ISBN: 978-1-53615-862-5
Editors: Ivan A. Parinov et al.

Chapter 8

DIMENSIONAL CHANGES OF ALUMINIUM 6061 – BOTTOM ASH COAL COMPOSITE CASTED BY SQUEEZE CASTING AND T6 HEAT TREATMENT

I. Made Kastiawan[1,*], Ninik Martini[1] and Ahmad Fauzan Adziimaa[1]

[1]Department of Mechanical Engineering, Universitas 17 Agustus 1945 Surabaya, Jawa Timur, Indonesia

ABSTRACT

Squeeze casting is a casting by pouring liquid metal in a mold. This study aims to determine the effect of fluid volume variations with tool temperatures on squeeze casting and also changes in shape, dimensions and surface roughness in dependence on T6 heat treatment of coal bottom ash base aluminium alloy- composites. In this study, the temperature of the liquid when is poured is 700°C with the volume are 450ml, 500ml, and 550ml during15 seconds, and it is given the pressing under 100kgf for 90 seconds with the temperatures of dies equal to 300°C, 400°C and 500°C. The weight percentage of composites includes 96.5% of aluminiumalloy, 2.5% of coal bottom ash and 1% of magnesium. After the T6 heat treatment process, the results of the study stated changes in shape. The largest sizes occur on a specimen with an area of 1190.5 mm³ and volume variation of 550ml liquid by a tool temperature of 500°C.This sample also has the highest surface roughness that equal to 1.90μm.

Keywords: surface roughness, aluminium alloy composite, T6 heat treatment, shape and dimension changes

---

* Corresponding Author's E-mail: madekastiawan@untag-sby.ac.id.

1. Introduction

One of the ways to increase mechanical properties of metal-based material is to use a heat treatment. Besides increasing mechanical properties, heat treatment also distorts material after the quenching process [1]. This research shows how the distortion arises in Al – bottom ash coal composite after T6 heat treatment, taking into account variation of specimen thickness and cooling media. The composite is fabricated by casting and has a thickness of 2, 3 or 4mm, and cooled by water, saltwater or oil of SAE 40. The specimens testing is performed before and after T6 heat treatment. T6 heat treatment is carried out by solution heat treatment at 540°C, holding time for 4 hours, and aging at 185°C with holding time for 2 hours. The result depends on the cooling medium. For 2mm thickness specimens, the highest distortion is achieved when using the oil SAE 40 medium; in the 3 mm thickness, the highest distortion is found on the water medium, and for 4 mm thickness, the highest distortion occurs on saltwater medium [2].

Heat treatment is a process of heating and cooling of controlled materials directed to changing physical and mechanical properties for particular purposes [3]. The features of metals with primary mechanical properties depend strongly on their microstructure and chemical composition. The heating or cooling of metal materials and alloys with a certain speed, it could be changed microstructure, that allows one to improve the mechanical properties. While heat treatment can improve the mechanical properties of a material, one of the disadvantages resulting from the heat treatment process is the distortion of hot material [4]. The distortion is especially true after the quenching process in the hardening process. Quenching can distort due to non-uniform cooling of the material; this uniformity is affected by the material thickness [5].

The objective of this research is to know how great the change of dimensions that occurs in aluminium ash composite material in dependence on thickness variation and quenching media (cooling), after previously performed T6 heat treatment process.

2. Experiment

2.1. CMM Testing

To guarantee the casting machining material (CMM) engine construction is made very rigid, one of the ways is to use granite as a reference sample. Test estimations of changes in shape and dimensions of CMM were performed before and after T6 heat treatment, which aimed to determine the difference in colour and dimensions before and after T6 heat treatment. Before the CMM testing, we give the code for each variation used.

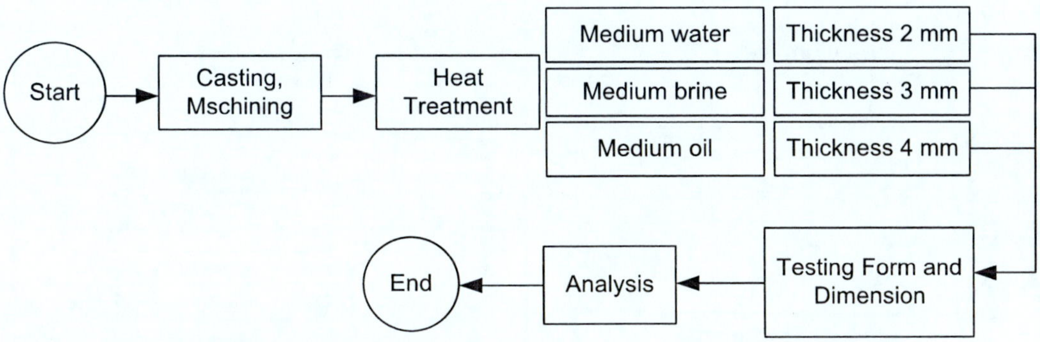

Figure 1. Simple flowchart of experimental works.

After the measurement, we study how many changes in shape and dimensions occurr in the test specimen. The flowchart of experimental works is present in Figure 1.

2.2. Surface Roughness Testing

Surface roughness defines the sizes of surface as the result of the machining process.The surface roughness is an essential property about the quality of the product. The surface roughness of the object. varies depending on the structure and the machining process.

The obtained parameters in surface roughness testing, are:

(i) total roughness (R_t) is the distance between the reference line and the basic line;
(ii) roughness estimate (R_p) is the distance between the reference line and the measured line;
(iii) arithmetic average of roughness (R_a) is the arithmetic value of the midline and the measured line.

3. Results and Discussion

Figure 2 presents graphical results of the dimensional changes in different cases of medium and thickness. The calculation of the specimen volume changes before and after T6, in the dependence on the thickness variation for the same cooling medium was performed. In Figure 2, the greatest shape change of specimen occurs in the thickness of 2mm for the oil cooling medium. for 3mm thickness, the greatest change takes place for the water cooling medium, and for 4mm thickness, the most significant change is achieved at the brine cooling medium.

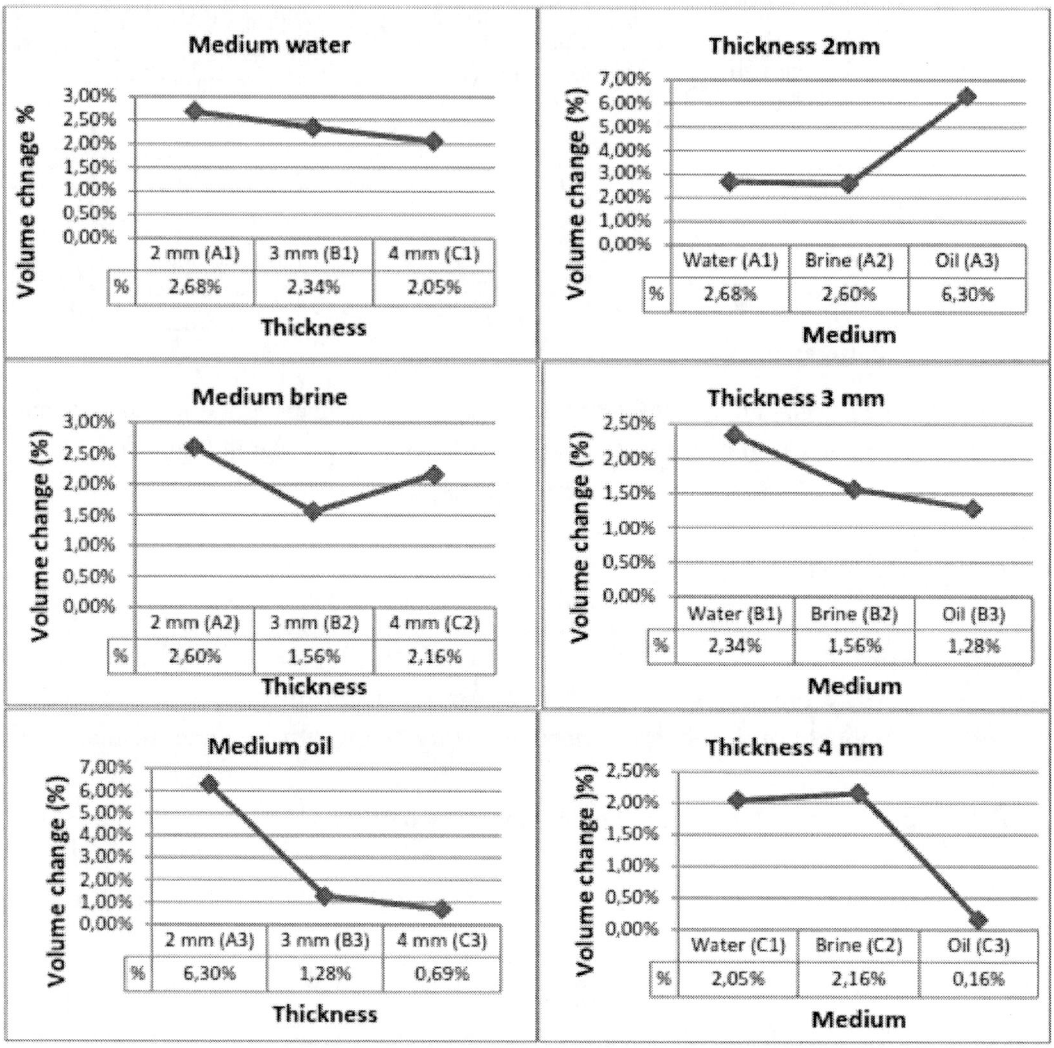

Figure 2. Graphical representation of results of the dimensional changes in different cases of medium and thickness.

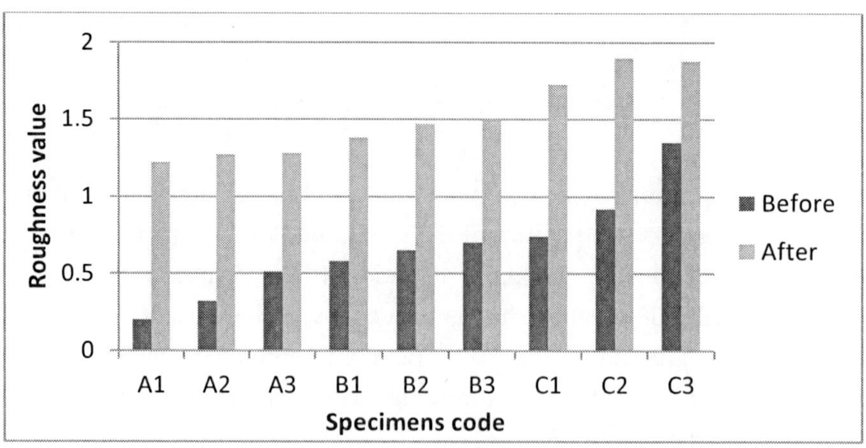

Figure 3. Values of surface roughness before and after T6.

From the above surface roughness testing data, Figure 3, it can be seen that the surface roughness before and after the T6 heat treatment process have significant differences. In other words, the surface of the specimen after the T6 heat treatment process is coarser. The T6 heat treatment process that warms up the sample to high temperatures and quenches immediately with high to lower air pressure can result in a chemical reaction on the surface. Samples become more opaque than before T6 heat treatment process.

CONCLUSION

The greatest shape changes and variably occur in specimens with a thickness of 2mm for the oil media; for 3mm thickness, the most significant shape change happens in water media; and for 4mm, the greatest shape change occurs in the brine medium.

There are changes in the shape and dimensions of the test specimen after the T6 heat treatment process. This result is due to the expansion and shrinkage caused by the heat and rapid cooling during the T6 heat treatment process, so there is an increase in the value of surface roughness on the test specimen. This condition is due to the chemical reaction on the specimen surface. At the highest tool temperature of 500°C, there is the most significant dimension change, while the smallest shape changes exists at tool temperature of 300°C. The test results show that at the lowest tool temperature 300°C the most subtle surface roughness arises with a value of 0.23μm. And, in the highest tool temperature of 500°C, the coarsest surface roughness is a value of more than 1.39μm. The higher temperature defines the higher roughness value. The short cooling rate and metal flow will produce small groups of grains and vice versa.

ACKNOWLEDGMENTS

The authors thank The Ministry of Research, Technology, and Higher Education of Republic of Indonesia for the funding of this research.

REFERENCES

[1] Aspiyansyah. Effect of Squeeze Casting Parameter Process (Melt Temperature, Die Temperature and Al – 3.22%Si) on Microstructure, Hardness and Tensile Strength in Thin Wall Casting. 2016.

[2] Fang Li. *China Foundry,* 2015. vol. 12(5).

[3] Respati Bondan; Sri Mulyo. In: *A Review on Particulate Reinforced Aluminum Metal Matrix Composites*, Attar, Sallahauddin; Madeva Nagaral; Reddappa, H. N.; Auradi, V. (Eds.). R&D Centre, Department of Mechanical Engineering, Bangalore Institute of Technology, Bangalore, Karnataka, India. 2016.

[4] Kittali, Praveen; Satheesh, J.; Anli Kumar, G.; Madhusudhan, T. *A Review on Effects of Reinforcements on Mechanical and Tribological Behavior of Aluminum Based Metal Matrix Composites*. P. G. Scholar, Mechanical Department, SJB Institute of Technology, VTU, Belagavi, Karnataka, India, 2016.

[5] Patel, G. C.; Manjunath; Robins Mathew; Prasad Krishna; Mahesh B. Parappagoudar. *Investigation of Squeeze Cast Process Parameters Effects on Secondary Dendrite Arm Spacing Using Statistical Regression and Artificial Neural Network Models.* Department of Mechanical Engineering, National Institute of Technology Karnataka, Surathkal, India. 2014.

In: Proceedings of the 2018 International Conference … ISBN: 978-1-53615-862-5
Editors: Ivan A. Parinov et al.

Chapter 9

EFFECT OF COMPOSITION AND TEMPERATURE OF CASTING COMPOSITE AL-BOTTOM ASH ON THE TRANSFORMATION OF GRAIN SIZE FOR THE FOOTSTEP PRODUCT

Zainun Ahmad*[*] *and Ismail
Department of Mechanical Engineering, University of 17 Agustus 1945
Surabaya, Indonesia

ABSTRACT

The use of aluminum materials in the engineering process has increased remarkably since the advent of aerospace technology. As time goes by, a lot of modern machines work very easy and efficient, especially motor vehicles. Footstep component plays an important role in motor vehicles. This research to study the influence of ash composite aluminum casting temperature on footstep casting grain. The casting method is used in the process of preparing the specimen, using 3 cases of the compositions. The first variation is aluminum - 95%, magnesium - 0.8%, coal base ash - 4.2%, at temperatures of 600°C, 700°C, 800°C; The second is aluminum - 90%, magnesium - 0.8%, coal base ash - 9.2%, at temperatures of 600 °C, 700°C, 800°C. The third variation is aluminum - 85%, magnesium - 0.8%, coal base ash - 14.2%, at temperatures of 600°C, 700°C, 800°C. From the result it is shown that, the largest average diameter of the grains equal to 141.88 μm (for 50 grains) was in the composition with a casting temperature of 800°C. The smallest diameter of the grains equal to 26.7 μm (for 1400 grains) was in the composition with a casting temperature of 600°C.

Keywords: aluminum, bottom-ash, casting, footstep, grain size

---

[*] Corresponding Author's E-mail: zainun@untag-sby.ac.id.

1. INTRODUCTION

Nowadays technological developments are increasingly rapid, a lot of thinking and equipment are the equipment and machines, created by experts to facilitate human activities. Along with the times, the development of lot of machines and modern expertise by means of work or use become very easy and efficient, especially in motorized vehicles [4]. With the emergence of high-tech Japanese-branded motorized vehicles, there are more choices or alternatives for motorcycle users [3].

The use of aluminum materials in the engineering has experienced a tremendous increase since the development of aerospace technology. By adding alloying elements, aluminum material can be produced which has good mechanical properties and engineering properties, where such conditions are indispensable in the application. The application of aluminum materials in certain working conditions requires process and material engineering to obtain optimum performance.

The application of aluminum casting with casting mold is one of the breakthroughs in process engineering to obtain qualitative product and good efficiency of technical process. One of the mechanical components manufactured by die casting process was the author initiative to examine footstep motor vehicle as the main component of motorcycle. Footstep component plays an important role in motor vehicles, apart from to be a support in a motor vehicle. Footstep also serves to provide comfort on the vehicle. Moreover, the footstep also serves to facilitate motorists in shifting gear and improving the best material characteristics at the high operating loads [5]. Referring to these conditions, the footstep casting process must eliminate temperatures that might occur with the casting results in its mechanical properties. The purpose of this research is to find out the influence of temperature and composition of coal ash mixture of 4.2%, 9.2%, and 14.2% on casting aluminum ash composite of coal base on grain changes at footstep casting [6].

2. EXPERIMENT

Raw material for aluminum AA 356 was added with magnesium and basic ash of coal. We considered 3 compositions: (i) aluminum – 95%, magnesium – 0.8%, coal bottom ash – 4.2%; (ii) aluminum – 90%, magnesium – 0.8%, coal base ash – 9.2%, and (iii) aluminum – 85%, magnesium – 0.8%, coal base ash – 14.2%. Each composition was casted with temperature variations of 600°C, 700°C and 800°C.

The method used in this test is a large calculation method of Planimetric items (Jefferies), this method uses a circle which generally has an area of 5,000 mm^2. Magnifications are chosen so that there are at least 75 items in the circle. Then calculate the total number of all the items in the circle plus half of the number of items that intersect the circle. The item size is calculated by multiplying the number of items by the Jefferies multiplier (*f*) from Table 1.

Table 1. Multiply Jefferies Table [2]

Magnification	***f***
1	0.002
25	0.125
50	0.5
75	1.125
100	2.0
200	8.0
300	18.0
500	50.0
1000	200.0

Grain Size No.	N_A Grains/Unit Area		$\bar{A}$ Average Grain Area		$\bar{d}$ Average Diameter		$\bar{\ell}$ Mean Intercept		$\bar{N}_L$
G	No./in.^{2} at 100X	No./mm^2 at 1X	mm^2	μm^2	mm	μm	mm	μm	No./mm
00	0.25	3.88	0.2581	258064	0.5080	508.0	0.4525	452.5	2.21
0	0.50	7.75	0.1290	129032	0.3592	359.2	0.3200	320.0	3.12
0.5	0.71	10.96	0.0912	91239	0.3021	302.1	0.2691	269.1	3.72
1.0	1.00	15.50	0.0645	64516	0.2540	254.0	0.2263	226.3	4.42
1.5	1.41	21.92	0.0456	45620	0.2136	213.6	0.1903	190.3	5.26
2.0	2.00	31.00	0.0323	32258	0.1796	179.6	0.1600	160.0	6.25
2.5	2.83	43.84	0.0228	22810	0.1510	151.0	0.1345	134.5	7.43
3.0	4.00	62.00	0.0161	16129	0.1270	127.0	0.1131	113.1	8.84
3.5	5.66	87.68	0.0114	11405	0.1068	106.8	0.0951	95.1	10.51
4.0	8.00	124.00	0.00806	8065	0.0898	89.8	0.0800	80.0	12.50
4.5	11.31	175.36	0.00570	5703	0.0755	75.5	0.0673	67.3	14.87
5.0	16.00	248.00	0.00403	4032	0.0635	63.5	0.0566	56.6	17.68
5.5	22.63	350.73	0.00285	2851	0.0534	53.4	0.0476	47.6	21.02
6.0	32.00	496.00	0.00202	2016	0.0449	44.9	0.0400	40.0	25.00
6.5	45.25	701.45	0.00143	1426	0.0378	37.8	0.0336	33.6	29.73
7.0	64.00	992.00	0.00101	1008	0.0318	31.8	0.0283	28.3	35.36
7.5	90.51	1402.9	0.00071	713	0.0267	26.7	0.0238	23.8	42.04
8.0	128.00	1984.0	0.00050	504	0.0225	22.5	0.0200	20.0	50.00
8.5	181.02	2805.8	0.00036	356	0.0189	18.9	0.0168	16.8	59.46
9.0	256.00	3968.0	0.00025	252	0.0159	15.9	0.0141	14.1	70.71
9.5	362.04	5611.6	0.00018	178	0.0133	13.3	0.0119	11.9	84.09
10.0	512.00	7936.0	0.00013	126	0.0112	11.2	0.0100	10.0	100.0
10.5	724.08	11223.2	0.000089	89.1	0.0094	9.4	0.0084	8.4	118.9
11.0	1024.00	15872.0	0.000063	63.0	0.0079	7.9	0.0071	7.1	141.4
11.5	1448.15	22446.4	0.000045	44.6	0.0067	6.7	0.0060	5.9	168.2
12.0	2048.00	31744.1	0.000032	31.5	0.0056	5.6	0.0050	5.0	200.0
12.5	2896.31	44892.9	0.000022	22.3	0.0047	4.7	0.0042	4.2	237.8
13.0	4096.00	63488.1	0.000016	15.8	0.0040	4.0	0.0035	3.5	282.8
13.5	5792.62	89785.8	0.000011	11.1	0.0033	3.3	0.0030	3.0	336.4
14.0	8192.00	126976.3	0.000008	7.9	0.0028	2.8	0.0025	2.5	400.0

Figure 1. Grain size measurement [1].

Grain size calculation is calculated as

$$G = [3.322 \log (N_A) - 2.95] \text{ and } N_A = f(N_1 + N_2/2) \quad (1)$$

where G is the size of the item, referred to in the ASTM E-112 table (see Figure 1) to find grain diameter value; N_A is the number of items; N_1 is the number of items in a circle; N_2 is the number of items that intersect the circle line; f is the multiplier factor in Jefferies Table 1.

The results of grain size calculations are present in Figure 1.

3. Results and Discussion

Based on data above, the highest number of granules (N_A) are found in the C composition with 600°C pouring temperature with 1,400 items and the least number there is in the composition A with a pouring temperature of 800°C with 200 grains.

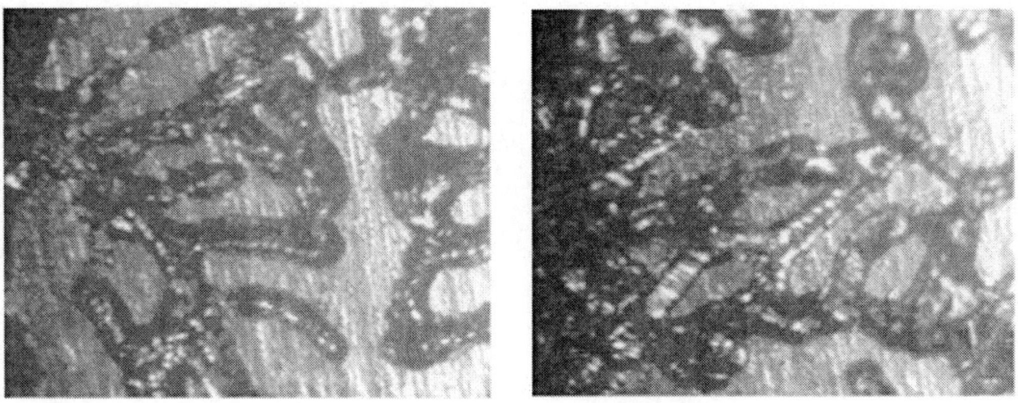

Figure 2. Microstructure of specimen: Al – 95%, Mg – 0.8%, bottom ash – 4.2%; pouring temperature – 800°C.

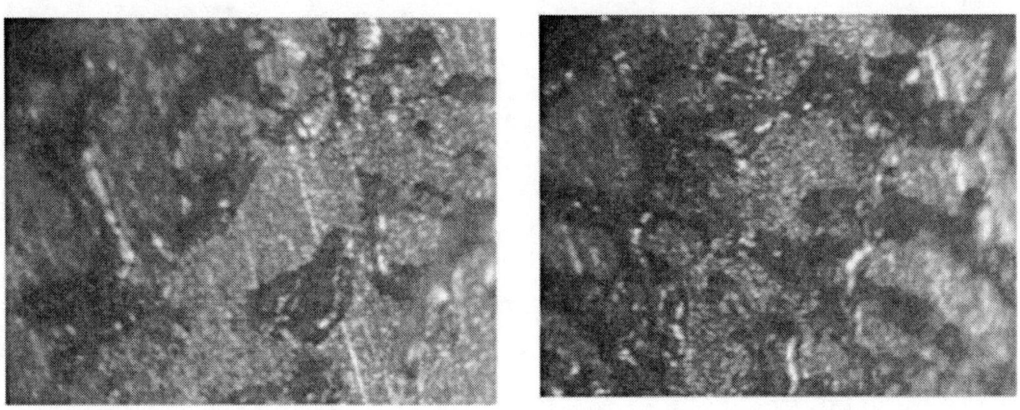

Figure 3. Microstructure of specimen: Al – 95%, Mg – 0.8%, bottom ash – 4.2%; pouring temperature – 700°C.

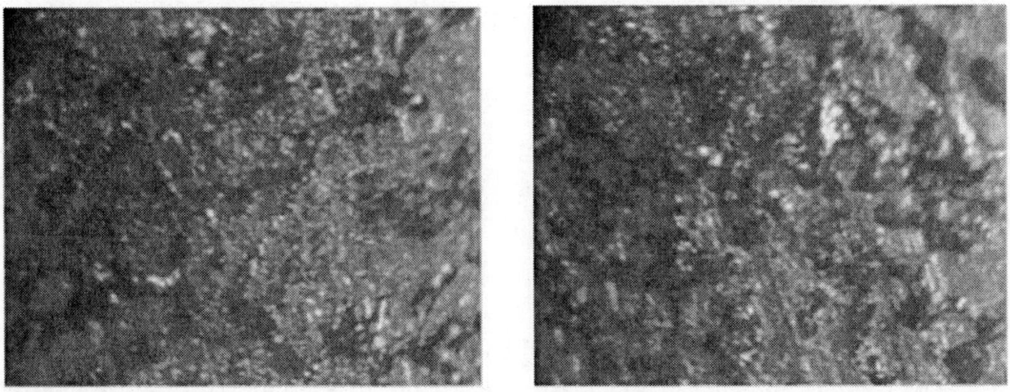

Figure 4. Microstructure of specimen: Al – 95%, Mg – 0.8%, bottom ash – 4.2%; pouring temperature – 600°C.

The highest dant number of granules (N_A) is found in the C composition with 600°C pouring temperature with 700 grains and the least number there is in the composition A with a pouring temperature of 800°C with 50 grains.

Based on the value of the average diameter, the grain composition A at a pouring temperature of 800°C has the greatest grain diameter of 70.94 μm. While the C composition has the smallest average diameter of 26.7 μm at pouring temperature of 600°C.

Based on the value of the average diameter, the grain composition A at a casting temperature of 800°C has the greatest grain diameter of 141.88 μm. While the C composition has the smallest average diameter of 37.8 μm at casting temperature of 600°C.

The highest number of grains (N_A) are found in the C composition with 600°C pouring temperature with 1,400 items and the least number there is in the composition A with a pouring temperature of 800°C with 200 grains.

The highest number of grains (N_A) is found in the C composition with pouring temperature of 600°C with 700 grains and the least number there is in the composition A with a pouring temperature of 800°C with 50 grains.

Based on the value of the average diameter, the grain composition A at a pouring temperature of 800°C has the greatest diameter of 70.94 μm. While the C composition has the smallest average diameter of 26.7 μm at the pouring temperature of 600°C.

Based on the value of the average diameter, the grain composition A at a casting temperature of 800°C has the highest diameter of 141.88 μm. While the C composition has the smallest average diameter of 37.8 μm at the casting temperature of 600°C.

Conclusion

The greatest average diameter of grains equal to 70.94 μm (for 200 grains) was shown for the composition: Al – 95%, Mg – 0.8%, coal base ash – 4.2% with pouring temperature of 800°C. The smallest average diameter of grains equal to 26.7 μm (for 1400 grains) was shown for the composition: Al – 85%, Mg – 0.8%, coal base ash – 14.2% with a pouring temperature of 600°C.

The highest average diameter of grains equal to 141.88 μm (for 50 grains) was shown in composition: Al – 95%, Mg – 0.8%, coal base ash – 4.2% with pouring temperature of 800°C. The smallest average diameter of grains equal to 37.8 μm (for 700 grains) was shown in the composition: Al – 85%, Mg – 0.8%, basic ash of coal – 14.2% with a pouring temperature of 600°C.

Thus, the composition (Al – 85%, Mg – 0.8%, ash coal base – 14.2% with a temperature of pouring of 600°C) has the smallest average diameter of grains and hence it is the hardest the material.

Acknowledgments

The authors thank The Ministry of Research, Technology, and Higher Education of Republic of Indonesia for the funding this research.

References

[1] Annual Book of ASTM Standards Designation: E112-12. Standard Test Methods for Determining Average Grain Size. *ASTM*, 2013.

[2] Annual Book of ASTM Standards Designation: E407-07. Standard Practice for Microetching Metals and Alloys. *ASTM*, 2015.

[3] David, W. *Journal of Engineering*, 2010, vol. 12(3), (In Indonesian).

[4] Diyanto, RI; Sulardjaka. *Journal of Engineering*, Mesin-Universitas Diponegoro., vol. 2 (5), 2013, (In Indonesian).

[5] Djatmiko, E; Budiarto. *Analysis of the properties and microstructure of* $Al_{78}Si_{22}$ *alloy products is a method of squeezing casting*. BATAN, Jakarta., 2011, 435-442, (In Indonesian).

[6] Fadhilah, Irfan. *Mechanical Engineering Journal*, vol. 5(4), 2015, (In Indonesian).

In: Proceedings of the 2018 International Conference … ISBN: 978-1-53615-862-5
Editors: Ivan A. Parinov et al.

Chapter 10

Analysis of Motorcycle Foot Step in a Grain Deformation and Microstructure of Aluminum-Bottom Ash Coal Composite with T6 Heat Treatment

Ichlas Wahid[1,*] and Sugeng Priyandokohadi[1]
[1]Department of Mechanical Engineering, University of 17 Agustus 1945 Surabaya, Jawa Timur, Indonesia

Abstract

The metal matrix composite is one of the alternative material for cast iron used for machine components, because it produces strong properties, has lightweight and corrosion resistance. This study aims to fabricate the aluminum-ash base coal composite; then we analyze its microstructure at given T6 heat treatment. The fabricated composite is used as a footstep or disc cradle in a motorcycle. The composite fabrication process of coal-base aluminum is using electrolysis plating and squeeze casting method. Then the specimens were applied with T6 heat treatment, heating at 540°C for 4 hours, followed by rapid cooling using some cooling medium, i.e., SAE 10 – 40 oil, water, and brine. Then was performed artificial aging treatment with time for 1 hour, 3 hours, and 5 hours. Tests conducted are microstructure testing. The specimens were then cut to the preparation of metallographic specimens by polishing and etching. The etching solution given was NaOH, HNO_3 and HF with a volume ratio of 5%, 75%, 20%. The metallographic process is performed by using an optical microscope. From the results of the metallographic testing, we classified the parts formed, the number of grains, and the average diameter of the grains. Specimens with water cooling medium with artificial aging time of 1 hour had the lowest value, i.e., number of grains G = 3.9 with mean

* Corresponding Author's E-mail: ichlaswahid@untag-sby.ac.id.

diameter of 55 μm. The longer artificial aging time, the greater the value of *G* and the mean diameter, as well as the salt water cooling and SAE oil.

Keywords: aluminum-bottom ash, artificial aging, composite, grain deformation, microstructure

1. INTRODUCTION

In Indonesian universities, in recent years, there are very many projects for the manufacture of automotive. Into framework of this project, the university also conducts research of the components that will be used in one motorcycle, to be able to produce their own spare parts and compete to produce the latest types and models [1]. Metal matrix composite material is one of the solutions that can be developed, because the metal matrix composite will produce strong, lightweight and corrosion-resistant properties [2, 3]. This type of composite is a widely developed metal matrix composite (MMC), which is aluminum matrix composite (AMC). Currently AMC is used in automotive industry for piston, disc brake, gear etc. [4].

The advantages of AMC consist in its lightweight, it has high hardness, high value of specific modulus and good wear resistance properties [5]. One of the most important components of a motorcycle is the foot step. The amplifier component must have a higher elastic modulus than the matrix component. There should be a strong surface bonding between the amplifier and matrix components.

According to the research that has been done the selected, treatment process includes the stages: the solution treatment at 540°C for 4 hours, quenching and aging process [6]. The variables used are temperature variation from 100°C to 200°C with aging time for 1, 3 and 5 hours with following testing. Resulting conclusions are: the effect of aging temperature from 100°C to 200°C on KML materials can improve mechanical properties especially in the value of tensile strength and hardness. While the effect of aging time from 1 hour to 24 hours may decrease its tensile strength even though its hardness value increases [7].

2. EXPERIMENT

Preparing a Casting Furnace:

(i) the furnace starts to heat up to a certain temperature after that aluminum material is inserted into the already hot furnace;

(ii) when aluminum has melted at a temperature of 600 °C (measured by thermocouple) the silica sand is mixed to separate the dirt/crust from the aluminum material;

(iii) after the aluminum material is clean from the dirt/crust, it is mixed coal ash bottom and Mg alternately and stirred;

(iv) liquid aluminum is poured into a preheated mold.

T6 Heat Treatment:

(i) solution heat treatment at 540 °C;

(ii) quenching with media salt water, pure water, and SAE Oil 40;

(iii) artificial aging process with temperature of 180 °C and variation of time 1, 3 and 5 hours;

(iv) then cooling to room temperature.

Metallography Testing:

(i) metallography testing is first performed by cutting the sample in accordance with the size of the specimen, then mounting the specimen so easy to hold;

(ii) the surface of the part tested is then smoothed using abrasive paper from the roughest grid to the finest grid, so in the end we can get a shiny specimen, clean the surface, and conduct polishing by using alumina;

(iii) before testing, etching solutions were made to dissolve grain boundaries on the surface of the specimen so that it could be seen using a microscope;

(iv) Furthermore, we perform observation of microstructure by using metallurgical microscope.

The microscope is set using the standard method, and takes the parts that are representative and the data can be processed.

3. Results and Discussion

The metallography results are present in Figures 1 - 4.

(i) We calculated grain size G for given values $n_1 = 26$, $n_2 = 4$ and $f = 8$ (see Figure 1):

$$N_a = f(n_1 + n_2/2) = 8(26 + 4/2) = 224;$$

$$G = 3.322 \log N_a - 2.95 = (3.322 \log 224) - 2.95 = 4.85.$$

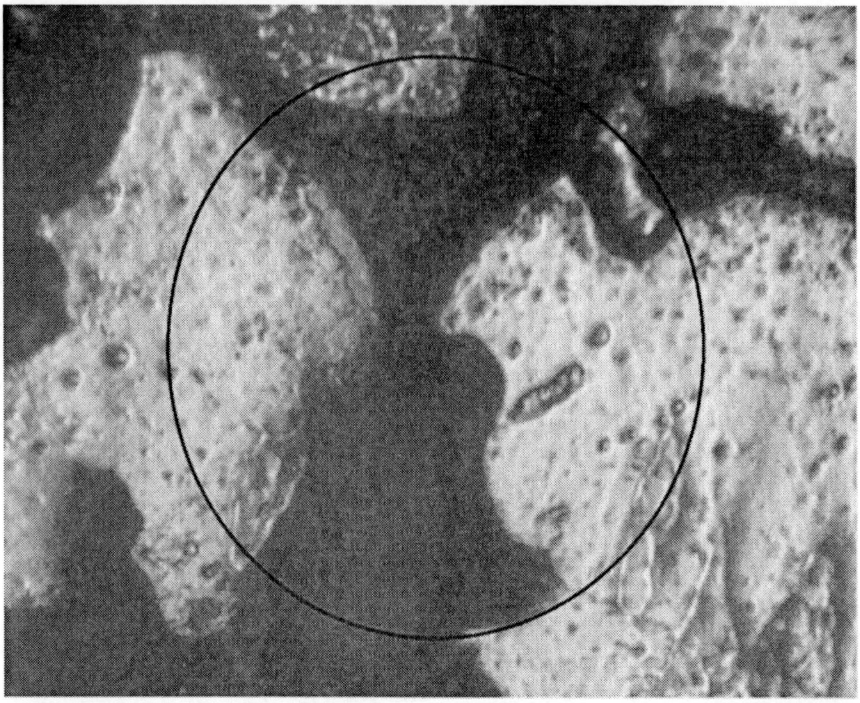

Figure 1. Microstructure of Al – bottom ash coal with pure water quenching and 1 hour time resistant (200 × magnification).

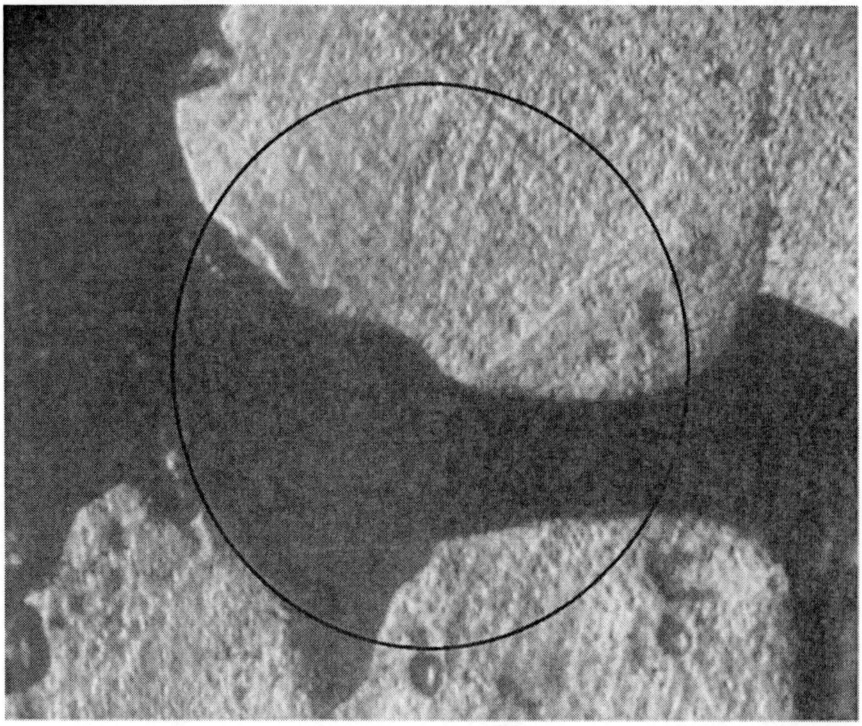

Figure 2. Microstructure of Al – bottom ash coal with brine (salt water) quenching and 1 hour time resistant (200× magnification).

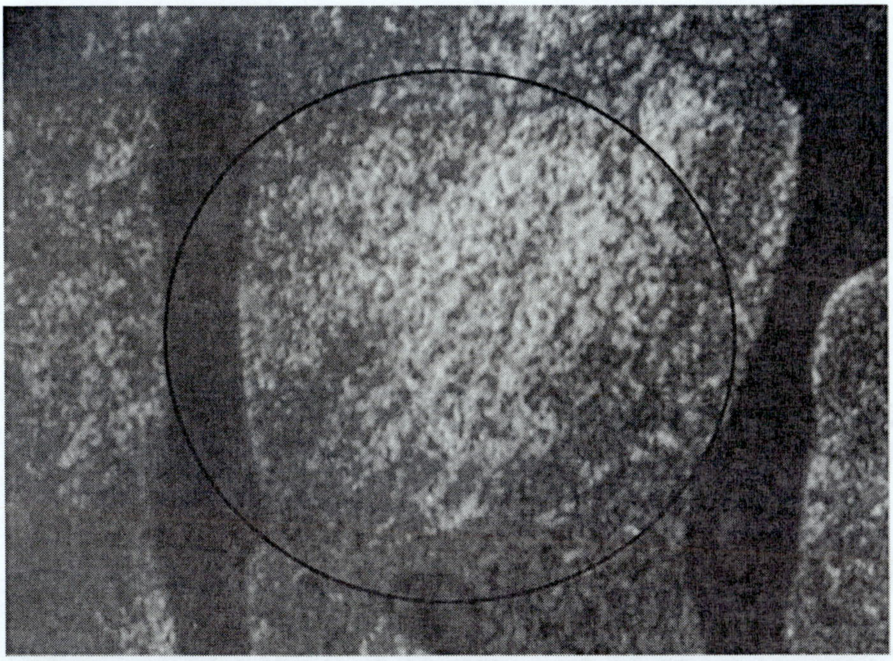

Figure 3. Microstructure of Al – bottom ash coal with SAE 10-40 quenching and 1 hour time resistant (200 × magnification).

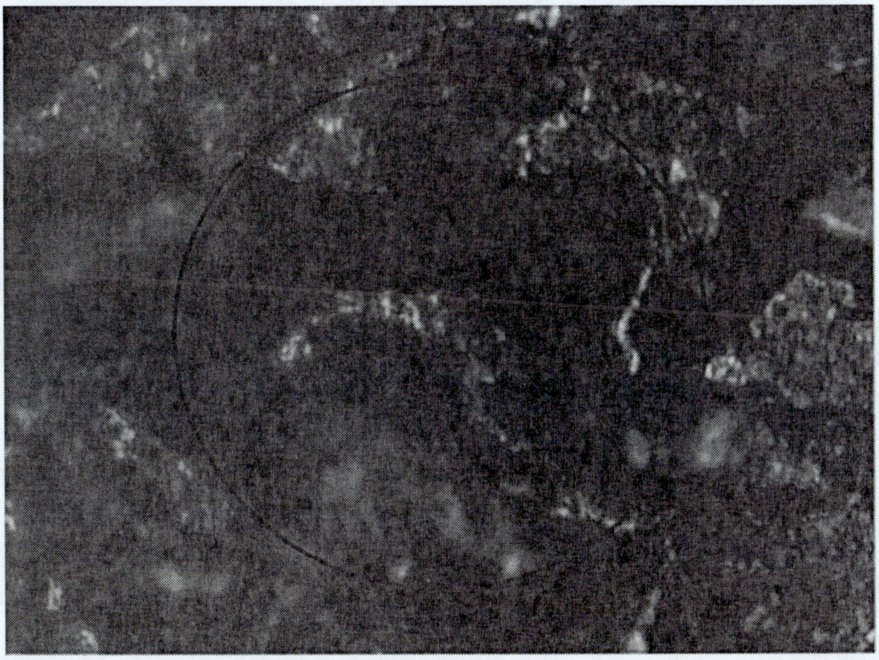

Figure 4. Microstructure of Al – bottom ash coal without T6 heat treatment (200 × magnification).

From ASTME-112 data, for $G = 4.85$, the average diameter of the grain is 65μm.

(ii) we calculated grain size G for given values $n_1 = 9$, $n_2 = 2$ and $f = 8$ (see Figure 2):

$N_a = f(n_1 + n_2/2) = 8(9 + 2/2) = 80;$

$G = 3.322 \log Na - 2.95 = (3.322 \log 80) - 2.95 = 3.37.$

From ASTME-112 data, for $G = 3.37$, the average diameter of the grain is 0.105μm.

(iii) we calculated grain size G for given values $n_1 = 82$, $n_2 = 14$ and $f = 8$ (see Figure 3):

$N_a = f(n_1 + n_2/2) = 8(82 + 14/2) = 712;$

$G = 3.322 \log N_a - 2.95 = (3.322 \log 712) - 2.95 = 6.52.$

From ASTME-112 data, for $G = 6.52$, the average diameter of the grain is 30μm.

(iv) we calculated grain size G for given values $n_1 = 11$, $n_2 = 7$ and $f = 8$ (see Figure 4):

$N_a = f(n_1 + n_2/2) = 8(11 + 7/2) = 116;$

$G = 3.322 \log N_a - 2.95 = (3.322 \log 116) - 2.95 = 3.90.$

From ASTME-112 data, for $G = 3.90$, the average diameter of the grain is 90μm.

Figures 1 - 4 show the microstructure tests of footstep components. The effect of giving aging time for 1, 3 and 5 hours and cooling medium of brine fluxing, pure water and SAE 10-40 oils to the number of grains and the mean diameter on aluminum-ash coal base composites demonstrates the following results.

Casting using water cooling medium with a holding time of 5 hours, the number of grains is equal to $G = 5.47$, the average diameter of the grains is equal to 55μm.

Specimen with the brine-cooling medium with the holding time of 5 hours, the number of grains is equal to $G = 5.02$, the average diameter of the grains is equal to 65μm.

The last specimen using SAE 10-40 hydraulic oil cooling medium with a holding time of 5 hours, the number of grains is equal to $G = 7.58$, the average diameter of the grains is equal to 27μm.

As a reference, the result of specimen without T6 heat treatment shows that the number of grains is equal to $G = 3.90$, the average diameter of the grains is equal to 90μm.

CONCLUSION

In the grain structure of foot step component, for the process without T6 heat treatment, we obtained number of grains $G = 3.90$ with average diameter equal to 90μm.

(i) In a water cooling medium with a holding time of 5 hours, the number of grains is $G = 5.47$ and average diameter of grains is equal to 55μm;
(ii) In salt water cooling medium with a holding time of 5hours, the number of grains is $G = 5.02$, and the average diameter of grains is equal to 65μm;
(iii) In SAE 10 - 40 hydraulic oil cooling medium with a holding time of 5 hours, the number of grains is $G = 7.58$, and the average diameter of grains is equal to 27μm.

The longer the aging time the greater the number of grains and the smaller the average diameter from the ASTM data, while the aging time with a short time the number of grains is also smaller but the average diameter from the ASTM data is also bigger.

ACKNOWLEDGMENTS

The authors thank The Ministry of Research, Technology, and Higher Education of Republic of Indonesia for the funding of this research.

REFERENCES

[1] Abdillah Fuad. *Heat Treatment of Al-Si Alloys in Piston-Based Piston Prototypes* (In Indonesian).
[2] Bekas. *Thesis,* Mechanical Engineering Department, Faculty of Engineering, Universitas Diponegoro, Semarang, 2010 (In Indonesian).

[3] *Aluminum Alloys: Aluminum Usage in The Transportation Industry*, http://www.azom.com/.

[4] Duskiardi; Soejono, Tjitro. *Effects of Pressure and Temperature Die Squeeze Casting Process on Hardness and Micro Structures in Local Commercial Piston Materials,* vol. 4(1), 2002 (In Indonesian).

[5] Haque, M. M.; Sharif, A. *Journal of Materials Processing Technology,* 2001, vol. 118, 69 - 73.

[6] JIS Handbook, *Non-ferrous Metal and Metallurgy*, Japan Standard Association.

[7] Zeren, M. *Material and Design,* 2007, vol. 28, 2511 - 2517.

In: Proceedings of the 2018 International Conference … ISBN: 978-1-53615-862-5
Editors: Ivan A. Parinov et al.

Chapter 11

EFFECT OF COOLING MEDIA AND TEMPERATURE IN T6 HEAT TREATMENT FOR THE PROPELLER PROTOTYPE FROM COMPOSITE AL-BOTTOM ASH ON THE TRANSFORMATION OF FORM AND DIMENSIONAL CHANGE

Djoko Sasono * ***and Mochammad Mufti***
Department of Mechanical Engineering, University of 17 Agustus 1945,
Surabaya, Indonesia

ABSTRACT

The development of material technology has created a new type of material built in stack form of several layers. This material is called composite material. Composite materials consist of more than one material and are designed to obtain the best combination of characteristics of each constituent component. The heat treatment process is one stage of processes series in metalworking before it can be used as products. In the heat treatment process, the constraint always is faced the emergence of distortion changes in shape and size (dimensions). In this research, we used T6 heat treatment with quenching media water, salt water and SAE 40 with studies at temperature quenching, room temperature, 50 °C and 100 °C. To found out the shape changes that occur, it is performed Coordinate Measuring Machine (CMM) testing and dial indicators are defined to know the dimensional changes that occur. Based on the performed CMM tests, the largest size change occurred in SAE 40 followed by salt water and then water. In the test using the dial indicator, it is seen the biggest change in the value of Z occurred in water followed by salt water and then SAE 40.

---

* Corresponding Author's E-mail: djokosm@untag-sby.ac.id.

Keywords: Al-Mg-Si, change dimension, composite, MMC, thermal stresses

1. INTRODUCTION

In the heat treatment process, the obstacle that is always faced is the emergence of distortion of changes in shape and size (dimensions). In this study, material that has corrosion resistance and small weight, accompanied with good properties is aluminum. Aluminum is a non-ferrous metal that is widely used in the manufacture of several aircraft components, several car parts, household applications and so on [1]. The mechanical properties of aluminum will increase with the addition of other alloys, such as Mg, SI, and others. However, the addition of these alloy elements is not enough to add strength and violence. In order to be able to compare the desired material, the material selector must be accordingly therefore in this study to find out the variation of the cooling medium for changes in dimensions and shape in the T6 heat treatment of coal base aluminum ash composites [2].

The development of material technology has created a new type of material with a stack shape of layers. This material is called composite material. Composite material consists of more than one material and is designed to get the best combination of characteristics from each constituent component. In a whole, composites can be defined as macroscopic mixtures of fibers and matrices. Fiber is a material component that is much stronger than matrix and serves to provide tensile strength. While the matrix serves to protect the fiber from environmental effects and damage due to load [3]. Carbon fiber composites have light and strong properties. This type of composite is widely used for aircraft structures, sports equipment, and continues to increase used as a substitute for damaged bones. Compared to conventional materials, composite is advantage having adjustable strength (tailorability), fatigue resistance, which is good corrosion resistance, and a high relative strength (strength to weight ratio). The main benefit of using composites is to get a combination of strength properties, high stiffness and low weight. By choosing the right combination of fiber and matrix materials, we can create a composite material with optimal properties, satisfying the requirements for a certain structure and specific purpose [4].

Aluminum can be easily combined with other metals. Important aluminum alloys include:

(i) *Al-Cu alloy*

Al-Cu types undergo to heat treatment. Through hardening/deposition, the mechanical properties of these alloys can match the properties of soft steel but their corrosion resistance is more than other types of alloy [5]. Copper is one of the important alloying elements, used with Al because it forms a solid solution at suitable heat treatment, which can increase its strength by forming precipitate.

(ii) *Al-Cu-Mg alloy*

This alloy contains 4% Cu and 0.5% Mg and is an alloy that possesses high strength. It is usually called duralumin. It is commonly used in aircraft constructions and other goods, which require a large ratio of strength to weight.

(iii) *Al-Si alloy*

Al-Si alloy is a very good alloy in liquid phase, has a good surface, without heat brittleness, shows a good castability and corrosion resistance, and is very light. Al-Si alloy is a good conductor of electricity and heat, because of its properties. It is widely used as material for the welding of Al alloy metals, both in cast alloys and wrought alloys.

(iv) *Al-Mg alloy*

AL-Mg alloy is an alloy with a good corrosion resistance. About 4% or 10% Al-Mg alloys have good corrosion resistance and mechanical properties. They have a tensile strength exceeds 30 kgf/mm^2, which increases more than 12% after heat treatment. This alloy is commonly used for parts of chemical industry equipment, ships, aircrafts and so on.

(v) Al-Mg-Si *alloy*

Aluminum-magnesium-silicon alloy can be used as covering and has the sufficient ability to cut, weld and corrosion resistance [2]. As in the case of magnesium silica (Mg_2Si), most aluminum alloys contain Si, so the addition of magnesium is needed to obtain the hardening effect of Mg_2Si. However, the nature of this alloy will lead to its brittleness, so to reduce this, the additives are restricted by the range of 0.03% – 0.01% [5].

2. Experiment

2.1. Heat Treatment

The development of material technology is growing rapidly towards strong, lightweight, inexpensive, environmentally friendly and safe construction. From a summary of Avner [6], it is stated that heating and cooling a solid metal or alloy are performed in such a way that to obtain desired conditions or properties. Heating for the sole purpose of hot working is excluded from the meaning of this definition. Heat treatment is a process of heating and cooling the metal in a solid state to change its mechanical properties. Through proper heat treatment, stress can be removed, grain size can be enlarged or reduced. Moreover, it is found to be reinforced or can produce a hard surface around the ductile core. To allow for proper heat treatment, the composition of the alloy metal must be known because changes in chemical composition can result to changes in physical properties.

Heat treatment is a heating process of a metal that aims to obtain optimum mechanical properties. In addition, heat treatment especially aluminum aims to obtain a uniform cast metal structure, improve engine-ability, dimensional stability, and eliminate residual stress due to contraction during melting.

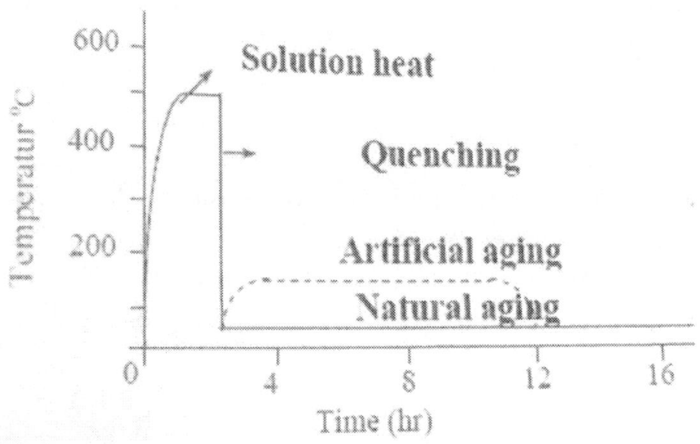

Figure 1. Schematics of T6 Heat Treatment [7].

The most suitable heat treatment as the finishing process for this coal base ash composite material is T6 heat treatment. There are three stages in T6 heat treatment process (see Figure 1), namely:

(i) solution heat treatment;
(ii) quenching;
(iii) aging (holding time)

2.2. Thermal Stress

When the body temperature drops, the outer layer decreases faster. The effects of thermal stress can be studied in low carbon steel, where we can ignore the possibility of martensite formation. Austenitic steel is also suitable for study. Frehser and Lowitzer conducted a series of studies on the effects of cooling variations on slabs with sizes 200 × 200 × 20 mm3 made of mild steel. Slab α in Figure 2 demonstrates a drastic decrease in temperature, showing a change in dimensions. Figure 2 shows the greater the temperature drop after cooling, the greater the deformation.

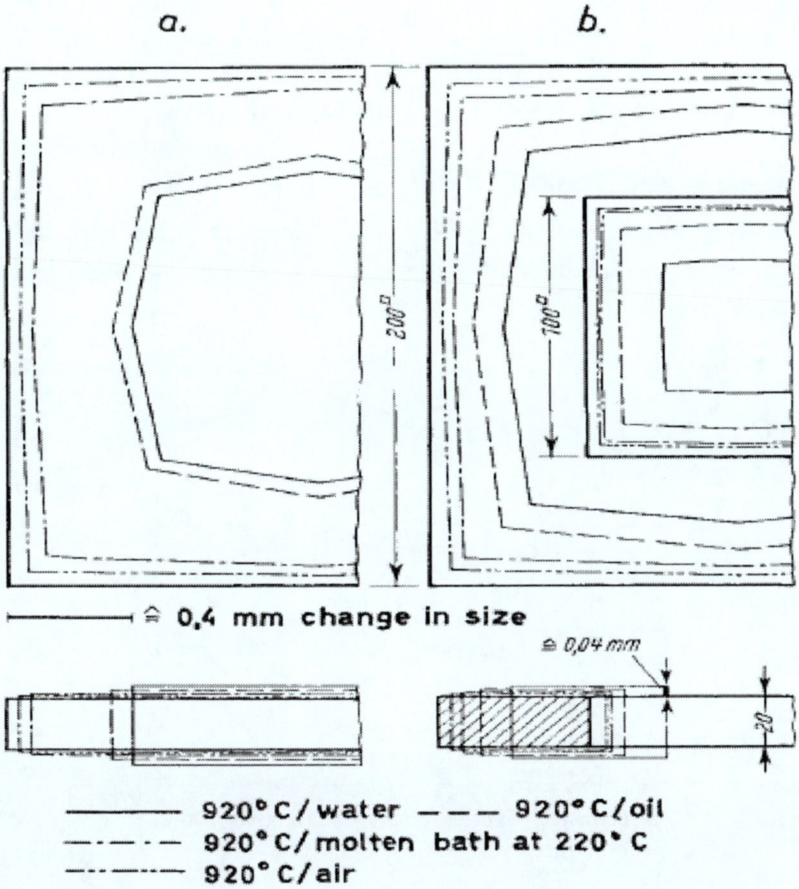

Figure 2. Metal deforming because of thermal stress.

2.3. Electroless Plating

Electroless is a plating process that does not use electric current in the coating process. Coating occurs due to the oxidation reaction and reduction on the surface of the goods, so that a metal layer from the metal forms. Because of it needlessness in electric current in electron exchange, the coating process occurring goes slower, so to accelerate the coating, the temperature in the process must be high, may be more than 90 °C.

Electroless plating is similar to a chemical reaction and without using a power source [6]. Applied for example to nickel sample, Electroless plating can be used for non-conductive metals such as plastics and ceramics. This process is more expensive than electro-plating, but it results in a more uniform coating.

Electroless plating is metal deposition of a solution using a reducing agent (RA) in a substrate solution or dissolution with free electrons. Electroless plating is usually used for mechanical and electronic components, especially to increase wear resistance and in some cases decrease corrosion (with special treatment). This process is very independent of specimen geometry [8].

3. Results and Discussion

From Figure 3a the average of the biggest change occurred in water cooling media with a quenching temperature of 25 °C with an average value of 9.207, the smallest change occurred in the medium of salt water and oil with an average value of 3.95.

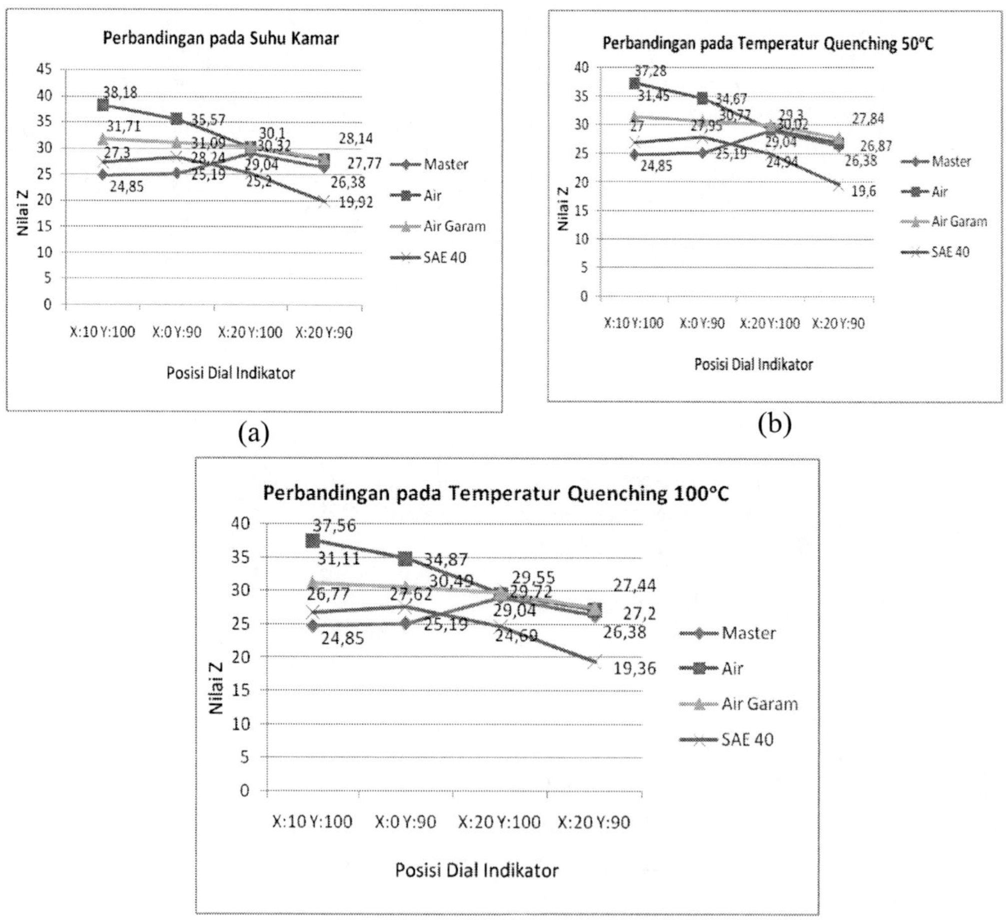

Figure 3. Comparison of dial indicator position in the temperature of: (a) 25 °C, (b) 50 °C, and (c) 100 °C.

From Figure 3b the average of the biggest change occurs in water cooling media with an average value of 5.66, the smallest change occurs in the medium of salt water with an average value of 3.655. From Figure 3c the average of the biggest change occurred in water cooling media with an average value of 5.93, the smallest change occurred in the medium of salt water with an average value of 3.382.

At the coordinates X10 Y100; X0 Y90; X20 Y100 and X20 Y90 all cooling media undergo changes in shape and dimensions. The biggest change occurred at the water cooling medium with an average value of 9.207 with a quenching temperature of 25 °C, while the smallest change occurred for SAE40 cooling medium with a quenching temperature of 100 °C and had the average value of 3.930.

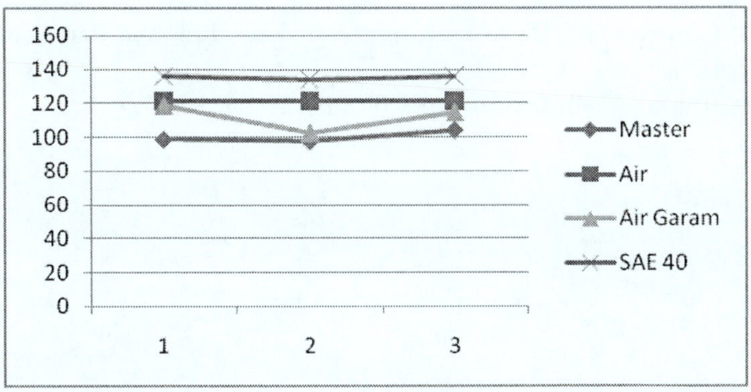

Figure 4. Hardness Testing Result.

CONCLUSION

1. The higher the temperature of the quenching medium the smaller the shape changes occur. At a temperature of the cooling medium 100 °C the average changes were equal to 5.93 for water, 3.65 for brine and 9.30 for oil.

2. From Micro-Vickers test, the differences of hardness values between master and specimen that have been underwent to T6 heat treatment are seen (Figure 4). The hardness value of 135.7 HV, which was the most optimal for propeller, was obtained by using oil cooling media because oil have viscosity higher than water and brine.

ACKNOWLEDGMENTS

The authors thank The Ministry of Research, Technology, and Higher Education of Republic of Indonesia for the funding of this research.

REFERENCES

[1] *ASTM E-92. Standard Test Methods for Vickers Hardness of Metallic Materials*. 1992.
[2] Azki Hakim, *Composite Material Technology*. Forum Sains, Indonesia. 2007 (In Indonesian).
[3] Eka Puji Hemawan, *Analysis of the Effect of Electroless Plating of Coal Base Ash as a Metal Matrix Composite Strengthener on Reaction Products Formed on the Surface of Coal Base Ash*. Indonesia. 2015 (In Indonesian).
[4] Thelning, K.E. *Steel and its Heat Treatment*. Butterworth-Heinemann. 2013.
[5] Nayiroh, Nurun. *Composite Material Technology*. Indonesia. 2013 (In Indonesian).
[6] Schwartz, M.M. *Composite Material Processing Fabrication and Applications*. Prentice Hall, USA. 1997.
[7] Surdia, T.; Chijiiwa, K. *Metal Casting Technique*. Pradnya Paramita, Jakarta, 1976 (In Indonesian).
[8] S. Deni, et al. *Makara Sain Journal,* 2008 (In Indonesian).

PART 2. PHYSICS

In: Proceedings of the 2018 International Conference … ISBN: 978-1-53615-862-5
Editors: Ivan A. Parinov et al.

Chapter 12

LAYER ANALYSIS OF PHOTOINDUCED LIGHT SCATTERING EXCITED IN $LiNbO_3$ SINGLE CRYSTALS

Diana V. Manukovskaya*
Tananaev Institute of Chemistry – Subdivision of the Federal Research Centre
Kola Science Centre of the Russian Academy of Sciences,
Apatity, Murmansk Region, Russia

ABSTRACT

Photoinduced light scattering (PILS) and round scattering images were studied for lithium niobate (LN) congruent single crystals doped by zinc in a wide range of concentrations $LiNbO_3$:Zn (0.018 – 2.05wt.% Zn in a crystal) and stoichiometric (Li/Nb = 1) single crystals, grown by two methods: (i) from a melt with extra Li, and (ii) from a congruent melt with an alkaline flux K_2O. A statistical analysis of the images of color components (RGB) revealed that each PILS and round scattering image contained a maximum on each color component. These maxima corresponded to compact areas on PILS and round scattering images.

Keywords: lithium niobate, photoinduced light scattering, image processing

1. INTRODUCTION

Photorefractive materials based on the single crystal lithium niobate (LN, $LiNbO_3$) have received much attention due to a possibility to vary its optical properties and

* Corresponding Author's E-mail: deenka@yandex.ru, manukovskaya@chemy.kolasc.net.ru.

structure. For example, doping of congruent LN single crystals by Zn in small concentrations leads to an increase in its cation sublattice ordering, a decrease in the amount of Nb_{Li} defects [1-4]. Structuring of the melt also changes the properties of grown crystals; the addition of K_2O flux allows us to obtain stoichiometric crystals from a congruent melt [5]. LN crystals of different composition have many possible uses as light modulators, optical shutters and nonlinear laser media. However, LN crystals possess a photorefractive effect. When irradiated perpendicular to the *Z* crystallographic axe, refractive index changes in the laser beam tracks and some areas of a crystal around it due to the appearance of laser-induced defects with changed refractive index. Photons scatter on these defects (it is a Rayleigh scattering) and form a pattern of photo induced light scattering (PILS) on the exit face of the crystal. The change of the refractive index leads to considerable loss in the laser beam power and blurring of information. This limits possible application and causes a need to research photo induced defects to be able to govern optical properties of LN based materials without harm for its structure.

Numerous works have been devoted to the study of PILS and the possibility to decrease photorefractive effect. For example, paper [6] claimed three types of speckles in a PILS pattern. Due to this fact, authors of [6] have distinguished three layers of a LN PILS pattern and explained that each layer has a different type of laser-induced defects: the brightest central spot was supposed to be light scattered on the crystal area, where refractive index changes completely. The second layer around the first one was light scattered on separate static laser-induced defects, and the furthest from the laser beam layer was light scattered on fluctuating laser-induced defects. However, many papers considered PILS only as a sign of existence of photorefractive effect in the crystal. Such PILS parameters like time and the angle of PILS opening are used as indicators of the photorefractive effect value (see, for example, works [1-3, 7-9] and other papers of Sidorov N.V., Palatnikov M.N. with co-authors). In addition, some authors started to evaluate photoelectric fields due to the angle of PILS opening recently (for example, see [10]), yet the information about laser-induced defects, carried by the PILS pattern itself, has received little attention.

The main difficulty in studying laser-induced defects via PILS patterns is that a researcher can investigate only light, scattered by the defects, but not the defects themselves.

Luckily, fractal dimension was found to reflect ordering of both PILS patterns and laser-induced defects [11-13]. These researches caused another problem. A fractal dimension can be found only for black-and-white images, and PILS is a complicated pattern with many illumination levels. The splitting of PILS patterns images revealed that it consists of six illumination levels. This fact was explained in works [15, 16] by statistical analysis of PILS patterns pictured in shades of gray. However, the question of why authors of paper [6] have distinguished 3 layers and a computer program has separated 6 layers [11-16] stays unclear.

To solve this problem, we carried out color analysis of the original PILS patterns pictures. In order to exclude possible errors that can appear due to the similarity of PILS patterns in the same type of LN crystals, we have analyzed the PILS of a number of doped $LiNbO_3$:Zn (0.018, 0.3, 0.53, 0.62, 0.88, 2.05wt.%) crystals of congruent composition, and stoichiometric crystals grown by different methods. In stoichiometric LN crystals, PILS was excited by two laser powers – 35 and 160mW. Note that in the crystal $LiNbO_3$:Zn (2.05wt.%) only a round scattering appears and PILS indicatrix does not open. The results on PILS are compared with the results on the round scattering.

2. Method of the Experiment

This work employs a statistical analysis of RGB components of PILS patterns pictures. In this section, crystal growth methods, a PILS patterns registration, an RGB components separation and their statistical analysis are described.

The crystals used in this study are materials widely applied in acousto-optics and laser physics. Stoichiometric LN and $LiNbO_3$:Zn single crystals possess weak coercive fields and low photorefractive effect. Therefore, they are perspective materials for the nonlinear laser media with periodically polarized domain structures for the holographic information recording and converting coherent and broad-band radiation [17]. The statistical counting method was chosen for the PILS patterns image processing because it is a simple and effective way to find laws of color distribution on images. Though it depends on many factors (such as image resolution, illumination level, and exposure time), it allows us a rapid and unambiguous definition of the color distribution laws.

All crystals researched were grown by Czochralski's method on air. $LiNbO_3$:Zn crystals were grown from a congruent melt; Zn was added during the usual method of direct doping. Stoichiometric crystals were grown by different methods: $LiNbO_{3stoich}$ was obtained from a stoichiometric melt, $LiNbO_{3stoich}K_2O$ was obtained from a congruent melt with an addition of 6wt.% of alkaline flux K_2O. Samples for PILS registration were parallelepipeds, axes of which were directed along crystallographic axes (*Z* is a polar axis of a LN crystal). PILS was excited by a MLL-100 YAG laser (λ = 530nm, P_{max} = 160mW) radiation. In order to obtain PILS patterns of the best quality, laser radiation power was decreased to a needed value. In experiments with $LiNbO_3$:Zn crystals, laser radiation 110mW was used. In experiments with $LiNbO_{3stoich}$ and $LiNbO_{3stoich}K_2O$, two laser power was used – the maximal possible being 160mW and a much smaller one at 35mW. The laser beam was directed along the *Y*-axis so that the electric-field strength vector $\boldsymbol{E}$ of the laser radiation was parallel to the crystal polar *Z*-axis. This provides the brightest PILS [18]. The PILS patterns pictures, used for the analysis, were shot at 60 s of irradiation. After this time, all diffusion and drift processes are usually finished and the picture becomes stable. Laser radiation, scattered by the crystal, fell on a semitransparent screen, located behind the crystal, and was recorded by a digital camera SONY NEX-F3. Camera parameters were the following: shutter speed 1/60sec, ISO speed 64, focal length 6mm, flash off. The resulting pictures were obtained in the format .jpeg. Light, other than from PILS, was absent from the room where the experiment was held. The speckle structure on the semitransparent screen in this experiment geometry was a superposition of the PILS speckle structure itself and the speckle structure of coherent radiation, scattered by other optical elements of the experiment, such as the edge of the laser output mirror. However, the latter could be neglected, because the PILS intensity was considerably greater than the intensity of a scattered coherent radiation. Each picture of the obtained PILS pattern was processed using the Image-Pro-Plus 6.0 program. Red, green, and blue components were indicated for each pixel of a color picture. One of the 256 shades of red, or green, or blue was detected. After that, each color was counted by the amount of pixels possessing the exact shade of a color (*N*) and a dependence of amount of pixels on the color shade (I_{RGB}) were built. Then, the maxima on the graph were compared with the areas on the picture that correspond to these maxima.

3. Discussion

Figure 1 demonstrates dependences of the color shade (*I*) on the pixels amount (*N*) on the PILS pattern pictures of crystals $LiNbO_3$:Zn (0.018, 0.3, 0.53, 0.62, 0.88, 2.05wt.%), excited by a laser radiation power 110mW. Figure 2 demonstrates dependences *I*(*N*) for PILS of $LiNbO_{3stoich}$ and $LiNbO_{3stoich}K_2O$ excited by laser radiation powers 160mW, 35mW. A detailed analysis of curves revealed a number of regularities.

A red component of PILS patterns of $LiNbO_3$:Zn (0.018 – 0.88wt.%) contains two obvious maxima on *I*(*N*) dependences in the area $I_R = 10 – 70$ (Figure 1, R). The round scattering in $LiNbO_3$:Zn (2.05wt.%) also contains such maximum, but it occupies values of $I_R = 70 – 110$. A red component of PILS patterns of $LiNbO_{3stoich}$ and $LiNbO_{3stoich}K_2O$ crystals also has maxima in the range $I_R = 40 –100$, but they are much less pronounced (Figure 2, R). The maxima correspond to the edge of PILS pattern, i.e., to the layer furthest from the laser beam.

A green component of PILS patterns of $LiNbO_3$:Zn (0.018 – 0.88wt.%) contains blurred maxima on *I*(*N*) dependences in the area $I_G = 70 – 160$ (Figure 1, G). This maximum depends on a Zn concentration in the crystal: it is the greatest for $LiNbO_3$:Zn (0.52wt.%) and less visible for $LiNbO_3$:Zn (0.03wt.%). At the same time, round scattering of a crystal $LiNbO_3$:Zn (2.05wt.%) has a more pronounced maximum on an *I*(*N*) dependence at values $I_G = 100 – 140$ (Figure 2, R). A green component of PILS patterns of $LiNbO_{3stoich}$ and $LiNbO_{3stoich}K_2O$ crystals has significant noise, but one can find blurred low maxima on the *I*(*N*) dependences at approximately $I_G = 96 – 160$. The maxima on a green component of PILS patterns of all studied crystals correspond to an area inside the area with an excess of a red component on PILS patterns and are located approximately in the middle of scattered radiation.

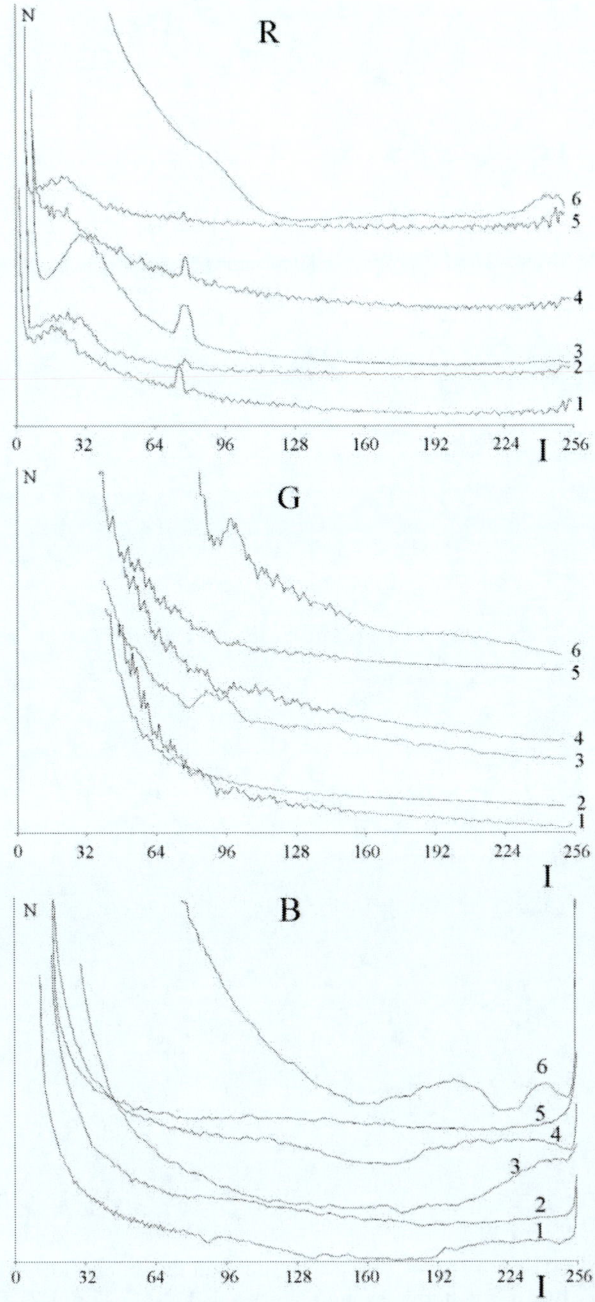

Figure 1. Dependence of shade of color (I, red, green, blue) on the amount of pixels (N) that correspond to these colors on images of PILS patterns, excited in crystals LiNbO3:Zn: 1 – 0.018wt.% ZnO, 2 – 0.03, 3 – 0.52, 4 – 0.62, 5 – 0.88, 6 – 2.05.

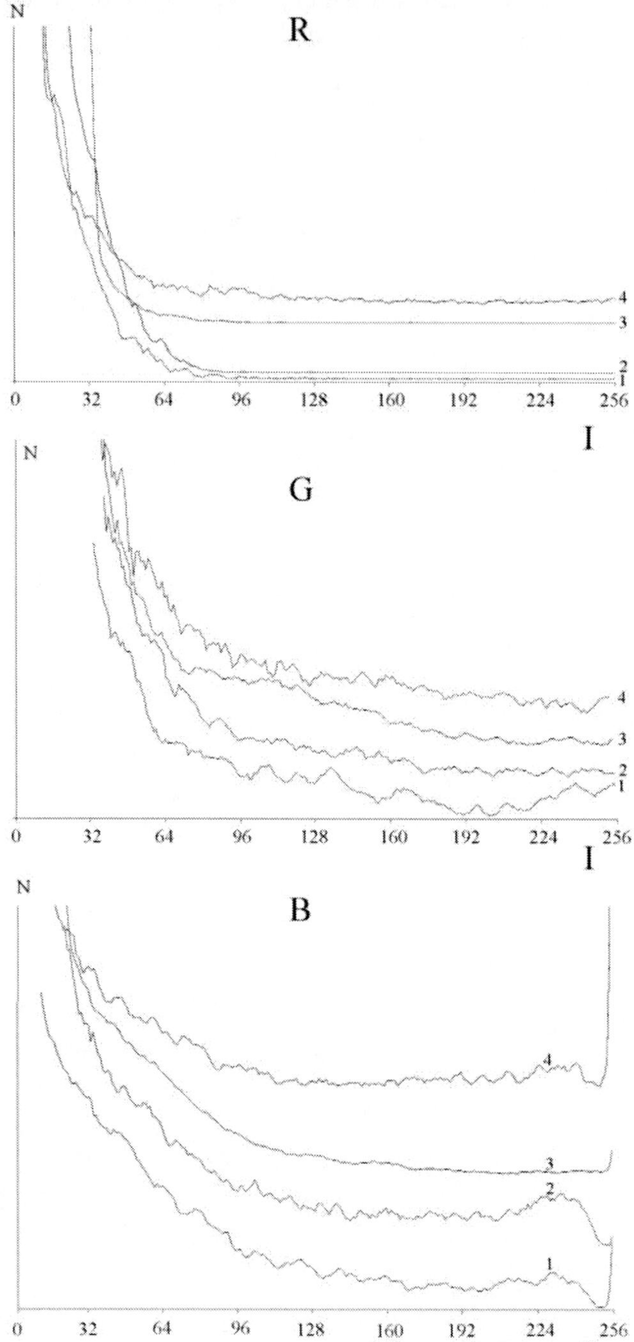

Figure 2. Dependence of shade of color (*I*, red, green, blue) on the amount of pixels (*N*) that correspond to these colors on images of PILS patterns, excited in crystals $LiNbO_{3stoich}$ (1,2) and $LiNbO_{3stoich}K_2O$ (3,4), excited by the laser radiation power 35mW (1,3) and 160mW (2,4).

A blue component of PILS patterns of $LiNbO_3$:Zn (0.018 – 0.88wt.%) sharply increases in the area $I_B = 250 - 255$ of $I(N)$ dependences (Figure 1, B). The amount of pixels N in this area surprisingly reaches the amount in the $I_B = 0 - 32$ area. The latter corresponds to a dark unlit edge of a picture, and it is obvious that the greatest amount of all pixels are expected in $I_{RGB} = 0 - 32$. The round scattering in $LiNbO_3$:Zn (2.05wt.%) also contains such a maximum, but it occupies little more values $I_R = 245 - 255$ and has approximately the same shape as the maxima of $LiNbO_3$:Zn (0.018 – 0.88wt.%) PILS patterns. The shades $I_B = 250 - 255$ correspond to the brightest and lightest shades and are located in the center of a PILS pattern in the area including the laser beam and some volume of the crystal around it. A blue component of PILS patterns of $LiNbO_{3stoich}$ and $LiNbO_{3stoich}K_2O$ crystals are more complex (Figure 2, B). Curves corresponding to $LiNbO_{3}$Stoich 35mW, $LiNbO_{3stoich}K_2O$ 35mW and $LiNbO_{3stoich}K_2O$ 160mW contain a maximum at $I_R = 250 - 255$. At the same time, a curve corresponding to $LiNbO_{3stoich}$ 160mW contains a minimum at $I_R = 250 - 255$. This is explained by the fact that to the 60's second of a laser irradiation, the central spot disappears from the PILS pattern picture (for more detail see [11]) of the crystal $LiNbO_{3stoich}$ 160mW. In this crystal, irradiated by a greater laser radiation power (160mW), such an amount of laser-induced defects appear to be enough to scatter all laser beam power and prevent it from reaching the exit face of the crystal.

CONCLUSION

The results of the experiment, demonstrated in this work, show that pictures of patterns that appear on the exit face of a photorefractive LN crystal, irradiated by the laser beam of a different power, show unambiguous regularities. The center of the PILS pattern picture close to a laser beam has an excess of a blue component, the area around it has an excess of a green component, and the edge of the PILS pattern picture has an excess of a red component.

The PILS pattern is actually monochromic because the only source of light in the experiment is a monochromic laser radiation. However, the digital camera distorts this picture and shifts colors due to particularities of image data record and storage. Thus, a human eye separates three layers according to the areas of PILS pattern pictures with excess of green, blue, or red, and the program takes into consideration only the illumination of pixels on pictures and separates 6 illumination levels.

REFERENCES

[1] Sidorov, N. V.; Yanichev, A. A.; Palatnikov, M. N. et al. *Opt. Spectrosc.,* 2014, *vol. 116(2)*, 281 - 290.

[2] Sidorov, N. V.; Yanichev, A. A.; Palatnikov, M. N. et al. *Opt. Spectrosc.,* 2014, *vol. 117(1)*, 72 - 81.

[3] Sidorov, N. V.; Palatnikov, M. N.; Teplyakova, N. A. et al. *Opt. Spectrosc.,* 2016, *vol. 120(4)*, 633 - 639.

[4] Chernaya, T. S.; Maksimov, B. A.; Volk, T. R., et al. *JETP Lett.*, 2001, *vol. 73(2)*, 103 - 106.

[5] Sidorov, N. V.; Palatnikov, M. N.; Gabrielyan, V. T. et al. *Inorg. Mater.*, 2007, *vol. 43(1)*, 60 - 67.

[6] Sidorov, N. V.; Syui, A. V.; Palatnikov, M. N. et al. *Dokl. Phys. Chem.*, 2011, *vol. 437(1)*, 47 - 49.

[7] Sidorov, N. V.; Palatnikov, M. N.; Yanichev, A. A. et al. *Dokl. Phys. Chem.*, 2013, *vol. 452(2)*, 243 - 246.

[8] Sidorov, N. V.; Kruk, A. A.; Yanichev, A. A. et al. *Dokl. Phys. Chem.*, 2014, *vol. 459(1)*, 173 - 176.

[9] Sidorov, N. V.; Teplyakova, N. A.; Yanichev, A. A. et al. *Inorg. Mater.*, 2017, *vol. 53(5)*, 489 - 495.

[10] Syuy, A. V.; Shtarev, D. S.; Antonycheva, E. A. et al. *J. Opt. Techn.*. 2015, *vol. 82(5)*, 319 - 322.

[11] Sidorov, N. V.; Manukovskaya, D. V.; Palatnikov, M. N. *Opt. Spectrosc.,* 2015, *vol. 118(6)*, 955 - 963.

[12] Manukovskaya, D. V.; Sidorov, N. V.; Starodub, O. R. et al. *Ferroelectrics*, 2014, *vol. 473(1)*, 100 - 109.

[13] Sidorov, N. V.; Manukovskaya, D. V.; Palatnikov, M. N. et al. *Bull. Rus. Academ. Sci. Phys.,* 2016, *vol. 80(11)*, 1404 - 1406.

[14] Manukovskaya, D. V. *Trudy KNC RAS.*, 2017, *vol.8(8)*, 88 - 94 (In Russian).

[15] Manukovskaya, D. V. *Trudy KNC RAS.*, 2018, *vol.9/2(9)*, 673 - 677 (In Russian).

[16] Manukovskaya, D. V.; Sidorov, N. V.; Palatnikov, M. N. In: *Proceedings of the 2017 International Conference on Physics and Mechanics of New Materials and Their Applications*. Ivan A. Parinov, Shun-Hsyung, Muaffaq A. Jani (Eds.). Nova Science Publishers, New York, 2018, 97 - 104.

[17] *Photorefractive Materials and Their Applications.* P. Günter, J.-P. Huidnard (Eds.), Springer, New York, 2007, pp. 1 - 256.

[18] Maksimenko, V. A.; Syui, A. V.; Karpets, Yu. M. *Photo-induced Processes in Lithium Niobate Crystals*, Fizmatlit, Moscow. 2008, pp. 1 - 256 (in Russian).

In: Proceedings of the 2018 International Conference … ISBN: 978-1-53615-862-5
Editors: Ivan A. Parinov et al.

Chapter 13

INTERRELATION BETWEEN OPTICAL AND RADIATION RESISTANCE OF LINBO3: ZNO SINGLE CRYSTALS

Mikhail N. Palatnikov[1,*], Nikolay V. Sidorov[1], Olga V. Makarova[1], Dmitriy V. Ivanenko[1], Sergey L. Panasyuk[2] and Igor' V. Judin[2]
[1]Tananaev Institute of Chemistry – Subdivision of the Federal Research Centre, Kola Science Centre of the Russian Academy of Sciences, Apatity, Murmansk Region, Russia
[2]St. Petersburg State Technological Institute, St. Petersburg, Russia

ABSTRACT

A radiation resistance and optical characteristics were studied for crystals $LiNbO_3$:ZnO ([ZnO] ~ 0.07 – 5.9 mol. % ZnO in a crystal) radiated by β- and γ-rays. Particularities of optical absorption (transmission) were detected for β- and γ-radiated $LiNbO_3$:ZnO crystals in dependence of concentration of a dopant, type and dose of radiation.

Optical and radiation resistances were shown to be connected and depend on threshold effects in $LiNbO_3$:ZnO crystals.

Keywords: crystal, lithium niobate, doping, β-radiation, γ-radiation, optical absorption, radiation resistance

---

* Corresponding Author's E-mail: palat_mn@chemy.kolasc.net.ru.

1. Introduction

A search of materials resistant to optical damage is demonstrated in publications dedicated to a study of $LiNbO_3$:ZnO crystals [1-3]. An influence of dopants on $LiNbO_3$:ZnO crystals properties has a jump-like character [1-3, 5-7], which is determined by the term "concentration threshold". $LiNbO_3$:ZnO crystals have more than one threshold [1, 3].

A number of applications of devices based on lithium niobate crystals involve their operation under ionizing radiation. At this, color centers contribute to optical absorption of crystals.

The centers appear as a result of charge carriers capture by antisite defects and impurities under laser radiation [4]. Ionizing radiation excites secondary electrons, which determines the type of appearing point defects.

Rearrangement of structure defects in lithium niobate crystals occurs mostly as a recharge of transition metals ions (for example, $Fe^{3+} \rightarrow Fe^{2+}$) and creation of polarons based on antisite defects Nb_{Li} [5].

Doping cations increase optical damage resistivity of $LiNbO_3$ crystals. This work studies influence of doping $LiNbO_3$ crystals by (Zn^{2+}) cations on creation and annealing of radiation defects during consequent accumulation of a radiation dose, absorbed by a crystal in dependence on radiation type and doze. Optical spectroscopy is used to study the radiation resistivity and optical transmission of β- and γ-radiated $LiNbO_3$:ZnO crystals (~ 0.07 – 5.9 mol. % ZnO in a crystal).

2. Method of Experiment

A charge synthesis, growth methods and an after growth treatment are described in detail in the work [3].

Samples for studying of transmission spectra were cut from $LiNbO_3$:ZnO crystals as parallelepipeds $6 \times 7 \times 8$ mm^3 with thoroughly polished faces.

$LiNbO_3$:ZnO crystals were irradiated at a room temperature by a source with radionuclides Sr-90 + Y-90 of a type BIS-20 with an activity 6.9×10^{10} Bq and by the equipment MPX-γ-20 with a source Co^{60} with a γ-radiation dose power ~ 0.5 Gy/s. Power of an adsorbed dose of a β-radiation source was detected using a ferrosulfate dosimeter.

Optical spectra were measured using a spectrometer Specord M40. Spectra of transition were recorded in the range 250 – 900 nm; after that absorption spectra were calculated.

Irradiation of samples by β-radiation and detection of optical spectra was carried out at the same face of a sample.

3. DISCUSSION

Papers [6, 7] improved mechanisms of creation of point and electron defects in $LiNbO_3$ crystals. The papers suggested a model, describing processes that takes place under γ-radiation, and revealed a contribution of oxygen sublattice defects to creation and annealing of color centers. The papers also revealed that doping of $LiNbO_3$ crystals could change their radiation resistivity. The latter depends on both dopant type and concentration, thus the radiation resistivity can possibly be much higher or much lower than in nominally pure $LiNbO_3$ crystals [7]. At this, an increase in an optical absorption after exposure to an ionizing radiation occurs in a wide area of wavelengths ~ 350 – 700 nm. The greatest change occurs around ~ 400 nm. We have researched a radiation resistance of $LiNbO_3$:ZnO crystals. Physic-chemical, structural and electrical characteristics of these crystals were previously studied in a wide range of dopant concentrations in [3]. An increase in an optical absorption in a wide range of wavelengths with a maximum around ~ 400 nm occurs at β- and γ-radiation of $LiNbO_3$:ZnO crystals, Figure 1. The similar increase was observed in $LiNbO_3$ crystals, doped by Gd, Y, Mg [6, 7].

Processes of a structure rearrangement in crystals $LiNbO_3$:ZnO that govern changes in optical and radiation resistances at doping are closely interconnected. At this, a decrease in an optical transmission (an increase in absorption) drastically and non-monotonously depends on a dopant concentration for β- and γ-radiated $LiNbO_3$:ZnO crystals, Table 1.

A study of optical characteristics of β- and γ-radiated $LiNbO_3$:ZnO crystals have revealed significant concentration anomalies of a radiation resistance, corresponding to concentration thresholds. A threshold effect in a low dopant concentration area (~1.0 – 1.17 mol. % ZnO in a crystal [3]) drastically changes saturation dose of a radiation color (SDRC) in $LiNbO_3$:ZnO crystals, Table 1, Figure 2.

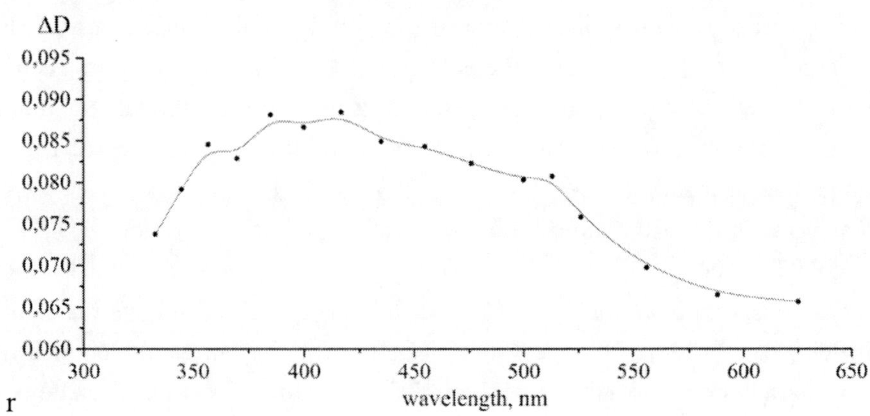

Figure 1. Absorption induced by an ionizing radiation (a difference between spectra of radiated and not radiated crystal No. 4 $LiNbO_3$:ZnO ([ZnO] ≈ 4.59 mol. %). A dose of β–radiation is ~ 237 Gy.

Table 1. Changes in optical transmission and SDRC of $LiNbO_3$:Zn crystals in dependence on type and dose of radiation and dopant concentration

No.	[ZnO] in crystals	Radiation type	SDRC	Initial transmission T	ΔT
1	0.07 mol. %	β	30 Gy	61%	8%
2	1.17 mol. %	β	3800 Gy	72%	27%
3	2.01 mol. %	β	85 Gy	68%	14%
4	3.98 mol. %	β	97 Gy	68%	2%
5	4.59 mol. %	β +β 3.6 kGy	120 Gy	74%	5%
6	4.66 mol. %	γ 20 kGy +γ 20 kGy	< 20 kGy	72%	13%
7	4.77 mol. %	β	4400 Gy	71%	13%
8	5.19 mol. %	β	147 Gy	71%	1%
9	5.89 mol. %	β 13.6 kGy +γ 20 kGy	0 Gy	73%	0%

SDRC is a dose at which crystal adsorption stops to increase under ionizing radiation.

For a crystal $LiNbO_3$:ZnO with a dopant concentration ~ 1.17 mol. % ZnO SDRC is two orders of magnitude greater than that for a pre-threshold crystal with a dopant concentration ~ 0.07 mol. % ZnO. Concentration anomalies of radiation effects in $LiNbO_3$:ZnO crystals clearly reveal the main threshold (~ 5.19 mol. % ZnO in a crystal or ~ 6.76 mol. % ZnO in the melt [3]), Figure 2. It should be noted, that an increase in the threshold area occurs not only in radiation resistance, but also in an optical resistance of $LiNbO_3$:ZnO crystals [3]. At this, sharp anomalies are also observed on a concentration dependence of absorption of $LiNbO_3$:ZnO crystals irradiated by ionizing (β- and γ-) radiation, Table 1, Figure 3.

A sharp increase in SDRC in the area of threshold dopant concentrations (Figure 2) shows a decrease in the amount of electron defects that are able to recharge and recombine in $LiNbO_3$:ZnO crystals during ionizing radiation treatment. Such a decrease is also a necessary condition for improving of optical damage resistance of $LiNbO_3$:ZnO crystals under the laser radiation.

One can note, that SDRC value for a crystal with zinc concentration ~ 5.19 mol. % (~ 147 Gy, Table 1) is comparable with that of a previously studied crystal $LiNbO_3$:Gd (~ 160 Gy) [7]. The crystals obviously possess an improved optical damage resistance.

An influence of a dopant on crystal properties at the threshold concentrations is jump-like. Characteristics of melts and crystals structure drastically change at these dopant concentrations [3].

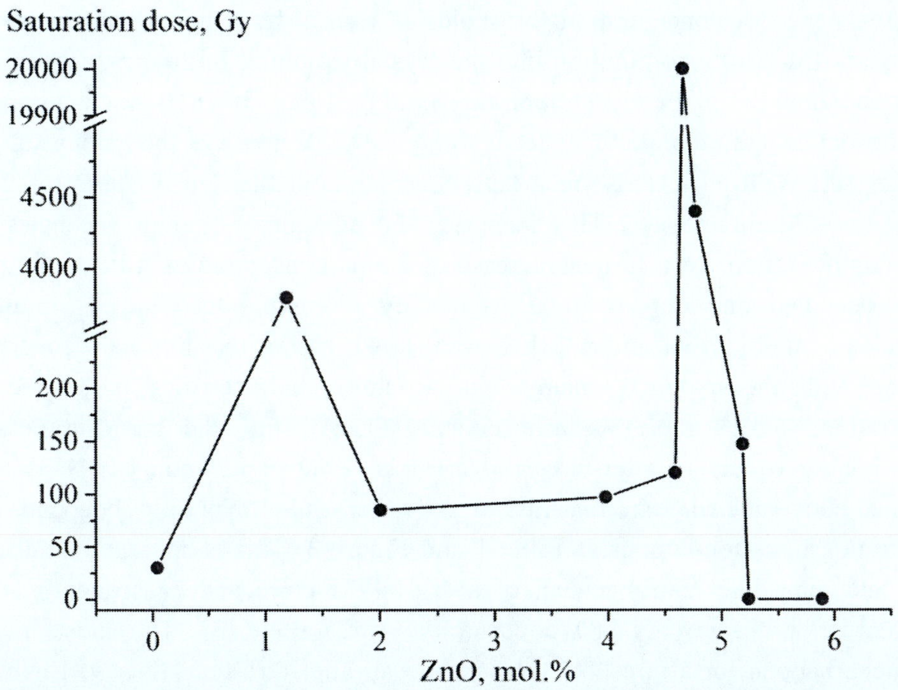

Figure 2. Dependence of SDRC on ZnO concentration in β- and γ-radiated $LiNbO_3$:ZnO crystals.

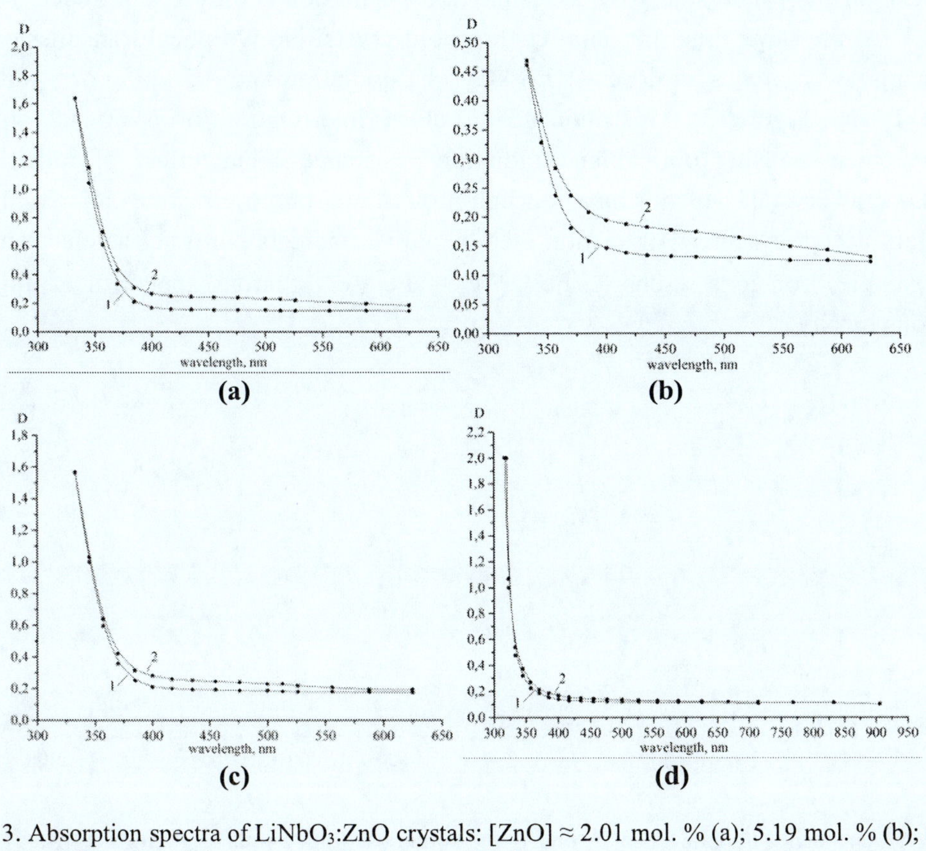

Figure 3. Absorption spectra of $LiNbO_3$:ZnO crystals: [ZnO] ≈ 2.01 mol. % (a); 5.19 mol. % (b); 0.07 mol. % (c); 5.89 mol. % (d); 1 – not radiated, 2 – radiated.

An existence of concentration thresholds is caused by a jump-like change of the mechanism of dopant incorporation into the crystal structure. In the first approximation, in the area below the first concentration threshold (~ 2 mol. % ZnO) zinc isomorphically substitutes an excess niobium. This leads to a general decrease of the amount of lithium vacancies [8]. With an increase of a zinc concentration, the dopant starts to substitute lithium ions in lithium sites. This increases the amount of lithium vacancies during charge compensation. At a further increase of a zinc concentration, all lithium sites are already occupied and dopant starts to occupy niobium sites with a simultaneous appearance of niobium vacancies [8]. It was shown earlier that the main concentration threshold with the maximal changes in structure characteristics is located at Zn concentration ~ 5.19 mol. % in a crystal [3]. At this, the best suppressing of photorefraction occurs in after-threshold crystals (ZnO > 5.2 mol. %) [1-3]. As was mentioned above, a change in transmission ΔT for crystals $LiNbO_3$:ZnO depends on zinc concentration non-monotonously, Table 1 and Figure 3. This is apparently caused by a drastic and jump-like rearrangement of a crystal structure and consequently, electron subsystem at a change of ZnO concentration in a melt [3]. The most interesting phenomena happen for a threshold (No. 8) and an after-threshold (No. 9) $LiNbO_3$:ZnO crystals. Due to data in [1-3], the effect of photorefraction is suppressed the best of all in these crystals. For a crystal No. 8 a change in transmission is only 1% at SDRC 147 Gy, Table 1. At the same time, for an after-threshold crystal No. 9 optical transmission does not change even after sum dose ~ 13.6 kGy of β-radiation and ~ 20 kGy of γ-radiation, Table 1 and Figure 3. This indicates a close interrelation of a crystal structure rearrangement causing optical and radiation resistance. Thus effect of radiation on $LiNbO_3$:ZnO crystals depends on a concentration of non-photorefractive cations Zn^{2+}.

Main acceptor centers (transition metals, antisite defects Nb_{Li}) act as electron traps. For example, iron ions re-charge $Fe^{3+} + e^{-} = Fe^{2+}$ or polarons appear at capturing of electrons by Nb_{Li}.

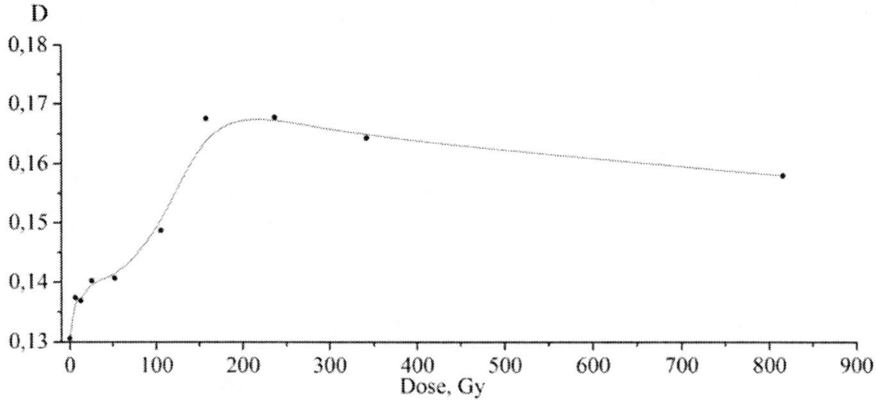

Figure 4. Dependence of optical density on a β–radiation dose for a crystal $LiNbO_3$:ZnO ([ZnO] ≈ 4.59 mol. %).

At after-threshold concentrations of the dopant an effect of ionizing radiation on $LiNbO_3$:ZnO crystals changes. For example, in such crystals Fe^{3+} cations localize in Nb sites [9]. It is obvious that centers $[Fe^{3+}_{Nb}]^{2'}$ with effective charge -2 cannot be an electron trap. At the same time, antisite defects Nb_{Li} are absent from after-threshold $LiNbO_3$:ZnO crystals [2, 3]. Figure 4 demonstrates that after an achievement of a maximum, optical density slightly decreases with an increase in a dose of β-radiation. The increase occurs due to a radiation annealing of defects.

Moreover, Figure 4 demonstrates that creation of radiation defects that contribute to optical absorption has a stepwise character. First, in the area of small doses (~50 Gy) occurs a rearrangement of the existing defect structure, and only after that new defects appear. The ongoing increase in absorption happens until the dose is ~ 150 – 200 Gy; after that (at a dose > ~ 200 Gy) radiation defects are annealed, Figure 4.

CONCLUSION

A radiation resistance and optical characteristics were studied for β- and γ-radiated $LiNbO_3$:ZnO crystals. A change in optical absorption (transmission) and saturation dose of a radiation color was shown to depend on a dopant concentration. The dependence is drastic and non-monotonous. This is caused by the influence of concentration threshold effects, which are in their turn caused by a structure rearrangement of $LiNbO_3$:ZnO crystals that happens at a change in the dopant concentration. Rearrangements of crystal structure that cause an increase in optical resistance and radiation resistance are closely interconnected. At this, a rearrangement in after-threshold ([ZnO] > 5.2 mol. % in a crystal) $LiNbO_3$:ZnO crystals leads to an absence of changes in optical absorption (transmission) even at very high doses of ionizing radiation.

REFERENCES

[1] Chernaya, T. S.; Volk, T. R.; Verin, I. A. et al. *Cryst. Rep.*, 2008, *vol. 53(4),* 573 - 578.

[2] Palatnikov, M. N.; Sidorov, N. V.; Manukovskaya, D. V. et al. *J. Am. Ceramic Soc.*, 2017, *vol. 100(8)*, P. 3703 - 3711.

[3] Palatnikov, M. N.; Sidorov, N. V.; Makarova, O. V.; Birjukova, I. V. *Fundamental Aspects of Technology of Strongly Doped Lithium Niobate Crystals.* Publishing of KCS RAS, Apatity, 2017, pp. 1 - 241 (in Russian).

[4] Parfinovich, I. A.; Penzina, E. E. *Electronic Color Centers in Ionic Crystals.* East-Siberian Book Press, Irkutsk, 1977, pp. 1 - 208 (in Russian).

[5] Volk, T. R.; Ivanov, M. A.; Mejl'man, M. L.; et al. *Fizika Tverdogo Tela*, 1987. *vol. 29(3),* 871 - 873 (in Russian).
[6] Palatnikov, M. N. *Inorg. Mater.*, 2008, *vol. 44(5)* 538 - 541.
[7] Palatnikov, M. N.; Efremov, I. N.; Sidorov, N. V. et al. *Inorg. Mater.*, 2013, *vol. 49(8)* 821 - 825.
[8] Simonov, V. I. *Priroda*, 2003, *vol. 11*, 4 - 100. (in Russian).
[9] Feng, H.; Wen, J.; Wang, H. et al. *J. Phys. Chem. Solids*, 1990, *vol. 51(5)* 397 - 400.

In: Proceedings of the 2018 International Conference … ISBN: 978-1-53615-862-5
Editors: Ivan A. Parinov et al.

Chapter 14

INFLUENCE OF DOPANTS ON THE PECULIARITIES OF FORMING OF COMPLEX DEFECTS IN A HOMOGENEOUSLY DOPED CRYSTAL $LiNbO_3$: MG (5.05 MOL%):FE (0.009 MOL%)

N. V. Sidorov, L. A. Bobreva*, S. M. Masloboeva and M. N. Palatnikov

Tananaev Institute of Chemistry - Subdivision of the Federal Research Centre
Kola Science Centre of the Russian Academy of Sciences,
Science Centre of Russian Academy of Sciences, Russia, Apatity

ABSTRACT

We investigated features of the incorporation of doping ions Mg^{2+} and Fe^{3+} into the structure of the crystal $LiNbO_3$: Mg (5.05 mol%) : Fe (0.009 mol%). We obtained this crystal using a homogeneously doped precursor, Nb_2O_5. We found in the IR spectrum of $LiNbO_3$:Mg (5.05 mol%):Fe (0.009 mol%), a significant shift ($\approx$ 50 cm^{-1}) to the high-frequency region of bands compared to a spectrum of a congruent crystal. This shift is due to the formation in the crystal's complex defects $Mg_{Li} - Fe_{Nb} - OH -$, $Mg_{Li} - Mg_{Nb} - OH$.

Keywords: single crystal, lithium niobate, co-doping, photorefractive effect, infrared absorption, stretching vibrations, OH groups

* Corresponding Author's E-mail: bobreva@chemy.kolasc.net.ru.

1. Introduction

Nonlinear optical single crystals of lithium niobate ($LiNbO_3$) are used in optical signal processing. The co-doping of crystals by "non-photorefractive" Mg^{2+} ions and "photorefractive" Fe^{3+} ions enables obtaining of materials with a fast photorefractive response and a high resistance to optical damage.

Despite the huge practical and scientific interest in a single crystal of $LiNbO_3$, some of its structural features have not been thoroughly investigated. The effect of these features on physical characteristics of the crystal is also not investigated. The presence in the structure of $LiNbO_3$ single crystals hydroxyl of OH groups affects the physical characteristics of the crystal: conductivity, the effect of photorefraction (optical damage), and the magnitude of the coercive field, etc. [1, 2]. The concentration of H^+ ions in the crystal ranges from 10^{18} to 10^{19} cm^{-3} [3]. The presence of H+ protons leads to the formation of complex defects in the crystal structure: V_{Li} – OH –, Me – OH –, Me – OH – Me. These defects affect a crystal's optical stability and photorefractive properties [1].

In this paper, we analyzed the incorporation of Mg^{2+} and Fe^{3+} ions to the structure and effect of these dopants on a homogeneously doped $LiNbO_3$:Mg (5.05 mol%):Fe (0.009 mol%) crystal. We performed a study of complex defects in $LiNbO_3$:Mg (5.05 mol%):Fe (0.009 mol%) on the base of IR absorption spectra in the valence vibrations region of OH groups.

2. Experimental Method

The crystal $LiNbO_3$:Mg (5.05 mol%):Fe (0.009 mol%) was obtained by the homogeneous doping method [4]. The precursor Nb_2O_5:Mg:Fe was isolated during the stage of extraction purification of fluoride-hydrochloric acid solutions from impurities. The precursor was obtained during the processing of single-crystal lithium niobate waste [5, 6]. A mixture of 35% dimethylamides of carboxylic acids of C_{10} – C_{13} fraction, 30% of octanol-1 and 35% of diluent Escide was used as an extractant. The high-purity re-extract (composition, g/l: Nb_2O_5-55, F-45) was obtained on an extractive cascade of the mixer-settler type. MgO was dissolved in this re-extract at a given concentration. Niobium hydroxide was precipitated with a 25% solution of NH_4OH at $pH > 11.5$ from the solution. The precipitate was washed with deionized water by repulpation 3 times at a ratio of $S : V_L = 3$. The precipitate was dried and calcined at 1000°C. The losses of magnesium with filtrate did not exceed 0.3 wt.%. The magnesium content was < 0.5 mg/l in filtrates and washings.

Under these conditions, magnesium almost completely converted to niobium hydroxide solution and then, after drying and calcination, it stayed with Nb_2O_5. The Mg content in Nb_2O_5 was determined by inductively coupled plasma mass spectrometry using the ELAN 9000 DRC-e. This value was 0.912 wt.%. The trace concentrations of cationic impurities were studied in a precursor Nb_2O_5 using a DFS-13 instrument. The impurities' concentrations were: Mo, Zr, Al, Ti, Co $< 5 \times 10^{-4}$; Mn, Ni, Cu, V $< 1 \times 10^{-4}$; Cr, Pb, Sn $< 3 \times 10^{-4}$; Fe 5×10^{-3}; Ca $< 1 \times 10^{-3}$; Si $< 1 \times 10^{-3}$ wt.%. We obtained a single-phase product using additional annealing of Nb_2O_5 : Mg at 1250°C. We used the precursor for synthesis of the granular charge of lithium niobate by the method described in [7].

The $LiNbO_3$:Mg:Fe crystal was grown in the <001> direction by the Czochralski method. The samples were cut in the form of rectangular parallelepipeds (dimensions ~ 8 × 7 × 6 mm^3). The edges of the parallelepipeds coincided with the crystal-physical axes *X*, *Y*, *Z* (*Z* is the polar axis of the crystal). Mg, Fe concentrations were controlled in the samples. The crystal was examined by IR spectroscopy. The concentration of Mg and Fe was determined by atomic emission analysis with inductively coupled plasma (AESM ICP). The results showed that in the crystal, the average concentration of Mg is 5.05 mol%, and Fe is 0.009 mol%. The difference between the concentrations in the conical and end parts of the crystalline boule does not exceed the error of determination. The IR spectra were recorded using an IFS 66 v/s spectrometer by Bruker.

3. Results and Discussion

We obtained a batch of a homogeneously doped precursor Nb_2O_5 : Mg : Fe. Mg was intentionally introduced into Nb_2O_5. Fe was detected in the lithium niobate crystal later. Dimethylamides of carboxylic acids of the $C_{10} - C_{13}$ fraction were used as an extractant in the extraction purification of niobium-containing fluoride-hydrochloric acid solutions.

These acids excellently extract iron [8]. Washing the extracts with subsequent stripping of the niobium with water does not purify the re-extracts from the impurity of Fe to concentrations $< 10^{-3}$ wt. %. Iron can form complexes with the organic phase. These complexes can get into reextracts due to the solubility of dimethylamides in aqueous solutions. It is believed that the solubility of these organic compounds is low enough when compared to other extractants. The solubility value is 0.05 kg/m^3 [9]. An admixture of Fe is also present in the re-extract at the ionic-molecular level. The introduction of magnesium homogenizes the resulting system by dissolving MgO in the re-extract.

The presence of organic inclusions affects the composition and structure of the ionic complexes in the melt. These complexes determine the composition and structure of the $LiNbO_3$ crystal. The presence of organic iron complexes affects the choice of sites by ions when the crystal grows. The type of connection in the Li site is ionic. There are many Li sites in a crystal. The Mg cation replaces Li in the Li site. The organic complex

Fe occupies Nb sites. Note that the type of bond is partially covalent in this site. Thus, the parameters of the crystal lattice change. It is known [10] that the addition of a small amount of Fe ions (1 mol%) leads to a greater increase in the parameters of the crystal lattice than the addition of 5 mol% of Mg ions. It is believed that a small Fe content (0.009 mol%) can lead to significant changes in the properties of a $LiNbO_3$ crystal [11]. Changes occur even at lower concentrations of Mg. Thus, the "threshold" concentration of Mg can significantly decrease in a crystal co-doped with Fe. Doping with Mg^{2+} cations reduces the coefficient of incorporated Fe^{3+} ions from 0.85 to 0.5 in $LiNbO_3$ [11]. Controlling the concentration of "non-photorefractive" ions reduces the electron-hole recombination and changes photorefractive properties in $LiNbO_3$ crystals co-doped with "photorefractive" ions, such as Fe, Cu, etc. [12]. The microscopic suppression of photorefraction occurs due to doping with magnesium, since a Fe ions position changes in the lattice of $LiNbO_3$:Mg:Fe [13].

IR absorption spectra are shown in the region of stretching vibrations of OH groups for a nominally pure congruent crystal ($LiNbO_{3cong}$) and a $LiNbO_3$:Mg (5.05):Fe (0.009 mol%) crystal grown from a congruent melt, Figure 1. All absorption bands have the same polarization (perpendicular to the ferroelectric *Z*-axis). The vibrations are dipole-active. The dipole moment of OH groups is also perpendicular to the polar axis of the crystal. The absorption band of stretching vibrations of OH groups in the spectrum of the $LiNbO_{3cong}$ crystal splits into three components with frequencies 3470, 3483 and 3486 cm^{-1}, Figure 1. These bands are shifted (≈ by 50 cm^{-1}) into the high-frequency region of stretching vibrations of OH groups in the spectrum of the $LiNbO_3$:Mg (5.05):Fe (0.009 mol %) crystal.

These absorption bands have frequencies 3506, 3526, 3535 cm^{-1} in this spectrum. The splitting of the absorption band into several components indicates different positions of the OH groups in the crystals. This indicates different values of the quasielastic constants of O – H bonds in vacant octahedral, octahedra occupied by the basic (Li^+, Nb^{5+}), and doping Mg^{2+}, Fe^{3+} cations. The number of proton sites is the same in the $LiNbO_3$: Mg (5.05) : Fe (0.009 mol%) crystal and $LiNbO_3$ crystal. Only an absorption band with a frequency of about 3467 cm^{-1} is observed in the IR absorption spectrum of the stoichiometric $LiNbO_3$ crystal. This absorption band indicates one position of H^+ ions in the crystal structure. It also indicates one value of the quasielastic constant hydrogen bond O – H [2].

A $LiNbO_3$ crystal of any composition has point defects of the cation sublattice, defect complexes, clusters, and a widely developed secondary structure. Vacancy split models help to describe the defective structure of the crystal and the formation of OH bonds [14 – 17]. According to the Li-vacancy compensation model, ~ 1 mol% of point defects Nb_{Li}^{5+} and ~ 4 mol% of point defects V_{Li}^- exist in the lattice of a congruent $LiNbO_3$ crystal [14, 15]. Negatively charged defects (V_{Li}^-) attract protons (H^+). H^+ are located on long O – O bonds (336 pm) and form new hydrogen bonds with oxygen. Thus, complex defects (V_{Li}) – OH^- are formed [18]. These defects correspond to absorption bands with frequencies 3483 and 3486 cm^{-1}, Figure 1, curve 1.

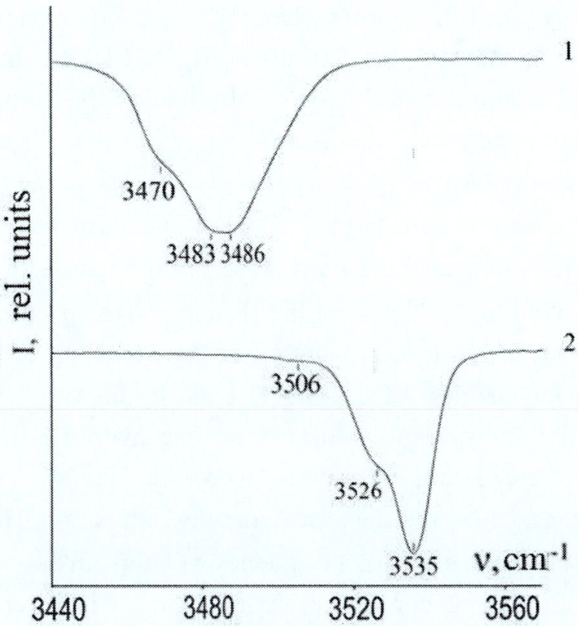

Figure 1. IR absorption spectra of $LiNbO_3$ single crystals in the valence vibrations region of OH groups: 1 – $LiNbO_{3cong}$, 2 – $LiNbO_3$:Mg (5.05):Fe (0.009 mol%).

Many properties of the crystal change with the transition of the dopant concentration through the threshold value. The reason for the change is that the mechanism of incorporation of the doping cation into the crystal structure changes. The main defects of the cation sublattice, Nb_{Li}, are completely displaced at this. Note that Nb^{5+} cations are located in lithium sites of a perfect stoichiometric composition. These defects are the deepest electronic traps. Displacement of Nb_{Li} leads to a decrease in the effect of photorefraction [14]. Doping cations do not affect the addition of a proton. Doping cations are positively charged and do not attract protons [19]. However, some doping cations occupy vacant octahedra and sites of niobium and lithium. The amount of such cations is small at low levels of doping. However, their amount becomes significant when their concentration exceeds the threshold value. Doping cations influence as negatively charged defects (for example, as defects of Mg_{Nb}^{3-}) at concentrations greater than the threshold in the $LiNbO_3$ crystal lattice. Such defects can attract protons and H^+ can localize on them, therefore forming new bonds with oxygen [20].

The shift of the IR absorption bands to the high-frequency region (Figure 1) can be explained as follows. When the Mg^{2+} cations concentration is below the first threshold (~ 3.0 mol% MgO [14]) in a $LiNbO_3$ crystal, most of the Mg^{2+} ions and Fe^{3+} ions displace Nb_{Li} defects, forming defect centers $(Mg_{Li})^{2+}$ and $(Fe_{Li})^{2+}$. Mg^{2+} ions displace part of Fe^{3+}

ions from lithium sites. Such a process occurs when the Mg^{2+} dopant concentration increases in the $LiNbO_3$:Mg (5.05):Fe (0.009 mol%) crystal. Thus, defects $(Fe^{3+}{}_{Nb})^{2-}$ appear. These defects occupy niobium sites. Virtually almost all structural defects of Nb_{Li} are replaced by Mg^{2+} cations. This occurs when the magnesium concentration approaches the second concentration threshold (~ 5.0 mol% MgO [14]). A further increase in the concentration of Mg^{2+} cations leads to a substitution of Li^+ and Nb^{5+} sites with an appearance of defective centers $Mg_{Li}{}^{2+}$ and $Mg_{Nb}{}^{3-}$. These defects form the self-compensating pairs $Mg_{Li}{}^{2+} - Mg_{Nb}{}^{3-}$. These pairs have a large attractive force for H^+. The proton will leave the complex defect $((V_{Li}) - OH^-)$ and form a complex defect $Mg_{Li} - Mg_{Nb} - OH^-$ with a self-compensating pair. Absorption bands with frequencies ~ 3526 and 3535 cm^{-1} correspond to this defect in the IR absorption spectrum. The formation of such defects leads to an increase in the energy of the hydrogen bonds. This increase is due to an increase in the stiffness of the O – H bond in the $Mg_{Li} - Mg_{Nb} - OH^-$ defect. The magnitude of the quasielastic constant of the hydrogen bond increases. The frequency of the corresponding vibrations increases, Figure 1. Defective centers $(Fe^{3+}{}_{Nb})^{2-}$ attract protons. Thus, the defective complex is $Fe_{Nb} - OH - Mg_{Li}$. This defect corresponds to an absorption band with a frequency of 3506 cm^{-1}.

It is believed that the second concentration threshold Me is lower in $LiNbO_3$: Me : Fe co-doped crystals than in single-doped $LiNbO_3$: Me crystals [21], since the defective complex $Fe_{Nb} - OH - Mg_{Li}$ is present in the structure of the $LiNbO_3$ crystal. The method of homogeneous doping allows us to introduce a large concentration of dopant into the crystal. The distribution coefficient of the dopant has the value of $K_d > 1$ [22]. Thus, it is possible to significantly reduce the photorefractive effect.

CONCLUSION

We have analyzed the incorporation and effect of Mg^{2+} and Fe^{3+} doping ions on a homogeneously doped $LiNbO_3$:Mg (5.05 mol%):Fe (0.009 mol%) crystal. It was established that the coefficient of incorporation of Fe^{3+} ions decreases in the result of doping the crystal with Mg^{2+} cations. Despite this fact, even a small amount of Fe leads to a change in the parameters of the $LiNbO_3$:Mg:Fe crystal lattice. These changes affect the properties of the crystal.

Three absorption bands in the valence vibration of the OH groups region appear in the IR absorption spectra of $LiNbO_3$ and $LiNbO_3$:Mg (5.05 mol%):Fe (0.009 mol%). Different positions of the OH groups in crystals indicate different values of the quasi-elastic bonds of hydrogen with the oxygen atom in vacant octahedra and octahedra occupied by the main (Li^+, Nb^{5+}) and doping cations Mg^{2+} and Fe^{3+}. We found that the IR absorption spectrum of the $LiNbO_3$:Mg (5.05):Fe (0.009 mol%) crystal is substantially (by 50 cm^{-1}) shifted into the high-frequency region compared to the IR spectrum of the $LiNbO_{3cong}$ crystal. This shift is related to the achievement of the second threshold concentration by the dopant Mg^{2+} in the $LiNbO_3$:Mg (5.05):Fe (0.009 mol%) crystal and the appearance of complex defects of the type $Fe_{Nb} - OH - Mg_{Li}$ and $Mg_{Li} - Mg_{Nb} - OH^-$.-

REFERENCES

[1] Cabrera, JM; Olivares, J; Carrascosa, M; Rams, J; Müller, R; Diéguez, E. *Adv. Physics*, 1996, vol. 45(5), 349-392.

[2] Lengyel, K; Peter, A; Kovacs, L; Corradi, G; Palfavi, L; Hebling, J; Unferdorben, M; Dravecz, G; Hajdara, I; Szaller, Zs; Polgar, K. *Appl. Phys. Rev.*, 2015, vol. 2, 040601-040628.

[3] Kong, Y; Xu, J; Zhang, W; Zhang, G. *J. Phys. Chem. Solids*, 2000, vol. 61, 1331-1335.

[4] Masloboeva, SM; Palatnikov, MN; Arytyanyan, LG; Ivanenko, DV. *News St. Petersburg State Technological University*, 2017, vol. 38 (64), 34-43.

[5] Masloboeva, SM; Sidorov, NV; Palatnikov, MN; Arutyunyan, LG; Chufyrev, PG. *Rus. J. Inorg. Chem.*, 2011, vol. 56 (8), 1194-1198.

[6] Palatnikov, MN; Masloboeva, SM; Biryukova, IV; Makarova, OV; Sidorov, NV; Efremov, VV. *Rus. J. Inorg. Chem.*, 2014, vol. 59 (3), 178-182.

[7] Palatnikov, MN; Sidorov, NV; Kalinnikov, VT. *Ferroelectric Solid Solutions Based on Nniobium and Tantalum Oxide Compounds*, St. Petersburg. Science, 2001 (In Russian).

[8] Skabichevskaya, GI; Balabanov, YuI; Kalinnikov, VT; et al. *The Method of Extracting Iron (III) from Aqueous Hydrochloric Acid Solutions*. USSR Patent 1655906 SU, 15.06.1991 (In Russian).

[9] Masloboeva, SM; Balabanov, YuI; Arutyunyan, LG; Masloboev, VA. *Proc. Sci. Conf. "Processing of Natural and Technogenic Raw Materials Containing Rare, Noble and Non-ferrous Metals".* Apatity, Kola Scientific Center of the Russian Academy of Sciences, 2003, 54-57 (In Russian).

[10] Prokohorov, AM; Kuz'minov, YS. *Physics and Chemistry of Crystalline Lithium Niobate*, Hilger, New York, 1990.

[11] Bae, SI; Ichikawa, J; Shimamura, K; Onodera, H; Fukuda, T. *J. Cryst. Growth*, 1997, vol. 94, 94-100.

[12] Yanlu, Li; Lili, Li; Xiufeng, Cheng; Xian, Zhao. *Phys. Chem. Solids*, 2017, vol. 121, 8968−8975.

[13] Jianjnn, Lin; Wanlin, Zhang; Cuangyin, Zhang. *Sol. State Comm.*, 1996, vol. 98 (6), 523-526.

[14] Iyi, N; Kitamura, K; Izumi, F; Yamamoto, JK; Hayashi, T; Asano, H; Kimura, S. *J. Solid State Chem.*, 1992, vol. 101, 340-352.

[15] Sidorov, NV; Volk, TP; Mavrin, BN; Kalinnikov, VT. *Lithium Niobate: Defects, Photorefraction, Vibrational Spectrum, Polaritons*, Moscow, Sience, 2003, 255 (In Russian).

[16] Donnerberg, HJ; Tomlinson, SM; Catlow, CRA. *J. Phys. Chem. Solids*, 1991, vol. 52(1), 201-210.

[17] Zotov, N; Boysen, H; Frey, F. *J. Phys. Chem. Solids*, 1994, vol. 55 (2), 145-152.

[18] Kong, Y; Wanlin, Z; Xiaojun, C; Jingjun, X; Guangyin, Z. *J. Phys.: Cond. Matter*, 1999, vol. 11, 2139-2145.

[19] Feng, Xi-Qi; Tang, Tong B. *J. Phys.: Cond. Matter*, 1993, vol. 5, 2423-2430.

[20] Arizmendi, L; Emilio Ambite, J; Plaza Jose, L. *Opt. Mater.*, 2013, vol. 35, 2411–2413.

[21] Meng, Q; Suhua, L; Xiudong, S. *Proc. SPIE*, 2005, vol. 5646, 378-382.

[22] Palatnikov, MN; Sidorov, NV; Biryukova, IV; Makarova, OV. *Fundamental Aspects of the Technology of Heavily Doped Lithium Niobate Crystals*, Apatity, KSC RAS, 2017, 241 (In Russian).

In: Proceedings of the 2018 International Conference … ISBN: 978-1-53615-862-5
Editors: Ivan A. Parinov et al.

Chapter 15

THE INFLUENCE OF HYPER-HARDENING ON THE STRUCTURE AND PROPERTIES OF CERAMIC SOLID SOLUTIONS $LI_{0.12}NA_{0.88}TA_{Y}NB_{1-Y}O_3$

O. B. Shcherbina*[1,*]*, V. V. Efremov*[1]*, L. A. Aleshina*[2]*,
***A. V. Kadetova*[2]*, O. V. Sidorova*[2] *and M. N. Palatnikov*[1]**
[1]I. V. Tananaev Institute of Chemistry, Subdivision of the Federal Research Centre Kola Science Centre of the Russian Academy of Sciences, Apatity, Murmansk Region, Russia
[2]Petrozavodsk State University, Petrozavodsk, Russia

ABSTRACT

An influence of hyper-fast hardening on the structure and mechanical properties of ceramic solid solutions $Li_{0.12}Na_{0.88}Ta_yNb_{1-y}O_3$ (LNTN), was researched by XRD and scanning probe microscopy. The co-existence of phases with a rhombohedral R3cH and orthorhombic $P2_1ma$ unit cells was determined for LNTN with $y = 0.05$; 0.25; 0.4. A hyper-fast hardening of a sample $Li_{0.12}Na_{0.88}Ta_yNb_{1-y}O_3$ ($y = 0.25$) causes a structure transition of the main phase from a rhombohedral R3cH to an orthorhombic $P2_1ma$ one. The rest samples with $y = 0.05$; 0.4; 0.5; 0.7; 0.8; 1 show the structural transition with the change of a space group from $P2_1ma$ to Pbnm. Hardened ceramics $Li_{0.12}Na_{0.88}Ta_yNb_{1-y}O_3$ are characterized by a complex texture and high values of micro-distortions. A hyper-fast hardening was demonstrated to increase micro-hardness and surface strength of ceramic samples.

---

*Corresponding Author's E-mail: shcerbina@chemy.kolasc.net.ru.

Table 1. Unit cell periods and space symmetry groups (SPGRs) of initial SSs of $Li_{0.12}Na_{0.88}Ta_yNb_{1-y}O_3$.

	$NaNbO_3$ [10]	$NaTaO_3$ [11]	$Li_{0.12}Na_{0.88}Ta_yNb_{1-y}O_3$ initial ceramic samples						
			y= 0.05		*y*= 0.25		*y*= 0.4		*y** = 1
SPGR	Pbcm	Pbmn	P2₁ma	R3cH	P2₁ma	R3cH	P2₁ma	R3cH	Pbmn
C,%	-	-	85.1	13.8	15.4	84.0	81.3	1.5	100
a, Å	5.507(1)	5.477(1)	5.5624(2)	5.4639(7)	5.5552(6)	5.4703(1)	5.5067(1)	5.492(2)	5.483(3)
b, Å	5.569(8)	5.521(1)	7.7514(6)	13.718(3)	7.7475(7)	13.720(2)	7.7514(1)	13.688(4)	5.488(1)
c, Å	15.524(5)	7.789(1)	5.4943(2)		5.4895(6)		5.4828(1)		7.748(7)
R_{wp}(%)	-	-	23.3		14.8		12.5		10.9
R_p(%)	-	-	17.4		11.2		9.4		7.8

*Results from paper [9].

Table 2. Unit cell periods and space symmetry groups (SPGRs) of hyper-hardened SSs of $Li_{0.12}Na_{0.88}Ta_yNb_{1-y}O_3$

	$Li_{0.12}Na_{0.88}Ta_yNb_{1-y}O_3$ hyper-hardened samples									
	y = 0.05		*y* = 0.25		*y* = 0.4		*y** = 0.5	*y** = 0.7	*y** = 0.8	*y** = 1
SPGR	Pbnm	R3cH	Pbnm	R3cH	Pbnm	R3cH	Pbmn	Pbmn	Pbmn	Pbmn
C,%	66.4	32.1	84.9	14.8	94.7	1.5	100	100	100	100
a, Å	5.4974(2)	5.4795(3)	5.4863(2)	5.488(5)	5.4709(1)	5.492(2)	5.466(1)	5.460(1)	5.458(1)	5.449(1)
b, Å	5.5633(3)	13.680(1)	5.5497(2)	13.747(2)	5.5323(1)	13.688(4)	5.527(1)	5.524(1)	5.516(1)	5.515(1)
c, Å	7.7510(4)		7.7691(3)		7.7710(2)		7.757(1)	7.774(1)	7.747(1)	7.764(1)
R_{wp}(%)	18.2		22.6		15.7		10.6	13.4	14.77	7.13
R_p (%)	13.7		17.2		11.3		8.2	14.9	11.00	7.69

*Results from paper[9].

A sample with tantalum content $y = 0.25$ dislocated near the special concentration point of a morphotropic phase transition [2]. The main phase in this sample (84%) is a rhombohedral one with a structure of a type $Li_{0.12}Na_{0.88}NbO_3$ and a space group R3cH. The concentration of tantalum in this phase corresponds to the value of *y*, demonstrated in Table 1. An orthorhombic ferroelectric phase with a space group $P2_1ma$ and four formula units per a unit cell is a minor phase (15.4%) for this compound. This phase was characterized in detail in paper [11].

A Rietveld's method has demonstrated that LNTN ceramics with $y = 0.5 - 1$ has a rhombic lattice with a space group Pbnm and four formula units per a unit cell [9]. Constants *a* and *b* of unhardened $Li_{0.12}Na_{0.88}TaO_3$ ceramics differ only in the third decimal place, i.e., the structure is close to a tetragonal one.

The ratio between unit cell constants *a* and *b* of hyper-hardened $Li_{0.12}Na_{0.88}TaO_3$ is close to the values of the rhombic phase of $NaTaO_3$.

The main phase in hyper-hardened LNTN compounds with $0.05 \leq {\sim}y \leq 1$ is a rhombic one. The phase corresponds to a structure type of $NaTaO_3$, with SPGR Pbnm. The amount of this phase increases with an increase in a tantalum content; it reaches

100% at $y = 0.5$ (Table 2). The structure of this phase is characterized in papers [9, 12]. An increase in tantalum content actually affects LNTN structure at an increase in temperature (compare data in Tables 1 and 2). Perhaps isovalent substitution by tantalum ions in a B-sublattice causes a "straightening" of octahedra and leads to destruction of the superstructure. The second phase of hyper-hardened samples $y = 0.05 - 0.4$ is a rhombohedral one with a structure of type $Li_{0.12}Na_{0.88}NbO_3$ [12, 13]. However, a tantalum concentration in the phase corresponds to values from Table 2.

A structure transition with a change in a space group from $P2_1ma$ to Pbnm occurs in SSs of $Li_{0.12}Na_{0.88}Ta_yNb_{1-y}O_3$ with a tantalum content $y = 0.05$; 0.4; 0.5; 0.7; 0.8; 1 as in the result of hyper-hardening in conditions of extreme heating-cooling. A partial substitution of sodium by lithium in $NaTaO_3$, with a rhombic lattice (a space group Pbnm), leads to three consequences: (i) a decrease in constants of a distorted perovskite sublattice; (ii) a change in octahedra TaO_6 orientation towards axes (a Ta – O – Ta bond angle increases towards *c*-axis and decreases in a *ab*-plane of a perovskite sublattice); (iii) a redistribution of Na – O bonds lengths. Hyper-fast hardening of the initial ceramics leads to an increase in constants of a perovskite sublattice; a sharp increase in a bond angle between TaO_6 octahedra towards *ab*-plane and a decrease in the angle towards *c*-axis.

In a case of SSs $Li_{0.12}Na_{0.88}Ta_yNb_{1-y}O_3$ with $y = 0.25$, the main phase transits from a rhombic phase R3cH to an ortorhombic phase Pbnm. Moreover, we have figured out that all hyper-hardened LNTN samples possess textures with axes [110] and [112]. The ratio between them depends on the composition of the ceramic. Analysis of the reflections widths, carried out by the Gauss approximation, revealed that micro-tensions cause the main contribution to the widening of reflections on X-ray patterns of hardened LNTN.

The microhardness of LNTN was studied by sclerometry at loads 25 – 30 mN because scratches did not appear at smaller loads. Young's modulus of samples was detected by a probe method based on a mechanical action of an indenter on the material surface. Properties of hyper-hardened $Li_{0.12}Na_{0.88}Ta_yNb_{1-y}O_3$ samples were compared to the ones of ceramic LNTN samples obtained due to CCT. Measurements of hardness and Young's modulus are demonstrated in Table 3.

It should be noted that strength characteristics (*E*) and microhardness (*H*) considerably increase for hyper-hardened $Li_{0.12}Na_{0.88}Ta_yNb_{1-y}O_3$ samples.

Table 3. Microhardness and Young's modulus of $Li_{0.12}Na_{0.88}Ta_yNb_{1-y}O_3$

$Li_{0.12}Na_{0.88}Ta_yNb_{1-y}O_3$ ceramic type	Microhardness, *H*, GPa, an average value		Young's modulus, *E*, GPa	
	CCT	Hyper-hardened	CCT	Hyper-hardened
y=0.05	9.64±0.87	28.5±3.6	112.9±1.5	277.4±4.0
y= 0.25	6.35±0.52	29.98±4.3	109.4±1.3	319.6±7.2
y= 0.4	7.15±0.6	23.89±3.0	105.8±2.2	264.2±3.4
y=1	5.34±0.84	29.3±1.86	96.6±1.4	186.9±2.9

CONCLUSION

Microstructure and mechanical characteristics were researched for hyper-hardened LNTN samples. A high-energy treatment was shown to increase surface hardness and strength of ceramic samples. A Rietveld's method was used to refine structure characteristics and to reveal changes in a perovskite cell of hyper-hardened $Li_{0.12}Na_{0.88}Ta_1O_3$. A structure transition of a main phase from a rhombohedral R3cH to an orthorhombic $P2_1ma$ phase was detected for a hyper-hardened SSs of $Li_{0.12}Na_{0.88}Ta_yNb_{1-y}O_3$, $y = 0.25$. The rest of the row with $y = 0.05$; 0.4; 0.5; 0.7; 0.8; 1 demonstrated a change of a space group from $P2_1ma$ to Pbnm. A hyper-fast hardening of ceramic solid solutions $Li_{0.12}Na_{0.88}Ta_yNb_{1-y}O_3$ in tantalum concentration range ≥ 0.5 leads to a creation of single-phase materials with an orthorhombic unit cell.

ACKNOWLEDGMENT

The reported study was funded by RFBR according to the research project No. 18-33-00099\18.

REFERENCES

[1] Palatnikov, M.; Sidorov, N.V.; Kalinnikov, V.T. Ferroelectric Solid Solutions Based on Niobium and Tantalum Oxide Compounds: Synthesis, Investigation of Structural Ordering and Physical Characteristics. *Nauka,* St.-Petersburg, 2001, pp. 1 – 350 (In Russian).

[2] Sidorov, N.V.; Palatnikov, M.N.; Teplyakova, N. A.; Kalinnikov, V. T. Ferroelectric Solid Solutions $Li_xNa_{1-x}Ta_yNb_{1-y}O_3$. Synthesis, Structure, Properties.*Nauka,* Moscow. 2015, pp. 1 – 231 (In Russian).

[3] Kalinichenko, A.S. *Casting and Metallurgy*, 2010, vol.58(4), 49-53 (InRussian).

[4] Traverse, J. P. *These de docteur es science sphysiques* [*PhD thesis in physical sciences*]. 1071, 2009. Nd'enregistrementau CNRS: A.O. 5.879.

[5] Useinov, A.S. *Devices and Experimental Equipment*, 2004, No.1, 134 – 138. (InRussian).

[6] Useinov, S.S.; Soloviev, V.V.; Gogolinsky, K.V.; Useinov, A.S.; Lvov, N.A. *Nanoindustry.Scientific and Technical Journal*, 2010, No.2, 30 – 35 (InRussian).

[7] Allmann, R.; Hinek, R. *Acta Cryst,* 2007, vol. A63, 412 – 417.

[8] Zlokazov, V. B.; Chernyshev, V. V. *J. Appl. Crystallogr*, 1992, vol.25, 447 – 451.

[9] Aleshina, L. A.; Palatnikov, M. N.;. Shcherbanich, Ya. I; Feklistova, E. P.; Shcherbina, O. B.. *Technical Physics*, 2017, vol. 62(3), 424 – 430.

[10] Shiratori, Y.; Magrez, A.; Dornseiffer, J.*J. Phys. Chem. B*, 2005, vol. 109(43), 20122 – 20130.

[11] Yuzyuk, Yu. I.; Gagarina, E.; Simon, P.; Reznitchenko, L.A;, Hennet, L.; Thiaudiere, D. *Physical Review*, Serie 3. *B - Condensed Matter*, 2004, vol. 69, 144105-1 – 144105-7

[12] Kennedy, Brendan J.; Prodjosantoso, A. K.; Howard, Christopher J. *J. Phys.:Condens. Matter*. 1999, vol. 11,6319 – 6327.

[13] Woodward, P. M. *Acta Crystal,* 1997, vol. B53, 44 – 66.

In: Proceedings of the 2018 International Conference … ISBN: 978-1-53615-862-5
Editors: Ivan A. Parinov et al.

Chapter 16

INVESTIGATION OF MECHANICAL STRESSES IN THIN CHROMIUM FILMS ON SAPPHIRE SUBSTRATE

Yu. V. Klunnikova[1,*], S. P. Malyukov[1], A. V. Sayenko[1] and Yu. N. Biyatenko[1]
[1]Southern Federal University, Taganrog, Russia

ABSTRACT

In this work, we investigate the mechanical stresses in chromium films on a sapphire substrate, which occur during their magnetron sputtering. We experimentally studied the influence of thermal processes on mechanical stresses in chromium films. The obtained experimental results correspond to theoretical calculations of mechanical stresses in a chromium film.

Keywords: sapphire substrate, film, magnetron sputtering, mechanical stresses, micromechanical devices

1. INTRODUCTION

The possibility of thin chromium films formation on dielectric and semiconductor substrates gives wide possibilities for constructing micromechanical devices. Chromium films of various thicknesses are used in the construction of micromechanical sensors of

* Correponding Author's E-mail: yvklunnikova@sfedu.ru.

the membrane type to form reflection from the membrane (when the signal is removed by interference methods) and as a universal sublayer when other metals (such as nickel) are used in high-sensitivity sensors [1, 2]. The mechanical properties of the layers that make up the membrane determine the mechanical characteristics of the device, as well as its purpose.

2. Technological Basis of Films Formation on Sapphire Substrates

In order to produce sensitive sensors, mechanical stresses in chromium films have to provide the minimal overall level of mechanical stresses in the membrane structure.

The aim of this work is to study the mechanical stresses formation in thin chromium films on sapphire substrates.

We used the magnetron deposition method [2, 3], which is one of the promising methods of these coatings formation, to obtain chromium films on a sapphire substrate (0.43mm thick, 50mm diameter). The method of magnetron deposition allows forming a perfect nanocomposite structure of a film with a high surface smoothness with a low friction coefficient and effective protection against wear and corrosion at increased temperatures. We obtained chromium films on a sapphire substrate with a vacuum deposition device VSE-PVD-DESK-PRO (the target is 99.95% chromium) in argon atmosphere (99.998%) with the process parameters shown in Table 1.

We carried out the experimental studies to measure mechanical stresses in thin chromium films with Tencor FLX-2320 (Japan) at "Piezopribor" research center of Southern Federal University (Rostov-on-Don, Russia). This study is based on estimating the change in the radius of curvature of the substrate caused by the influence of a thin film formed on it [6-10] in order to measure surface stresses of thin films. Figure 1 shows the internal stresses in thin chromium films of different thicknesses on a sapphire substrate.

Table 1. Parameters of chromium films formation on a sapphire substrate

Film thickness, nm	Magnetron discharge currents, mA	Magnetron discharge voltages, V	Working pressure, mbar	Deposition time, min
50	420	400	5×10^{-3}	5
100	572	400	7×10^{-3}	10
120	508	400	5×10^{-3}	12

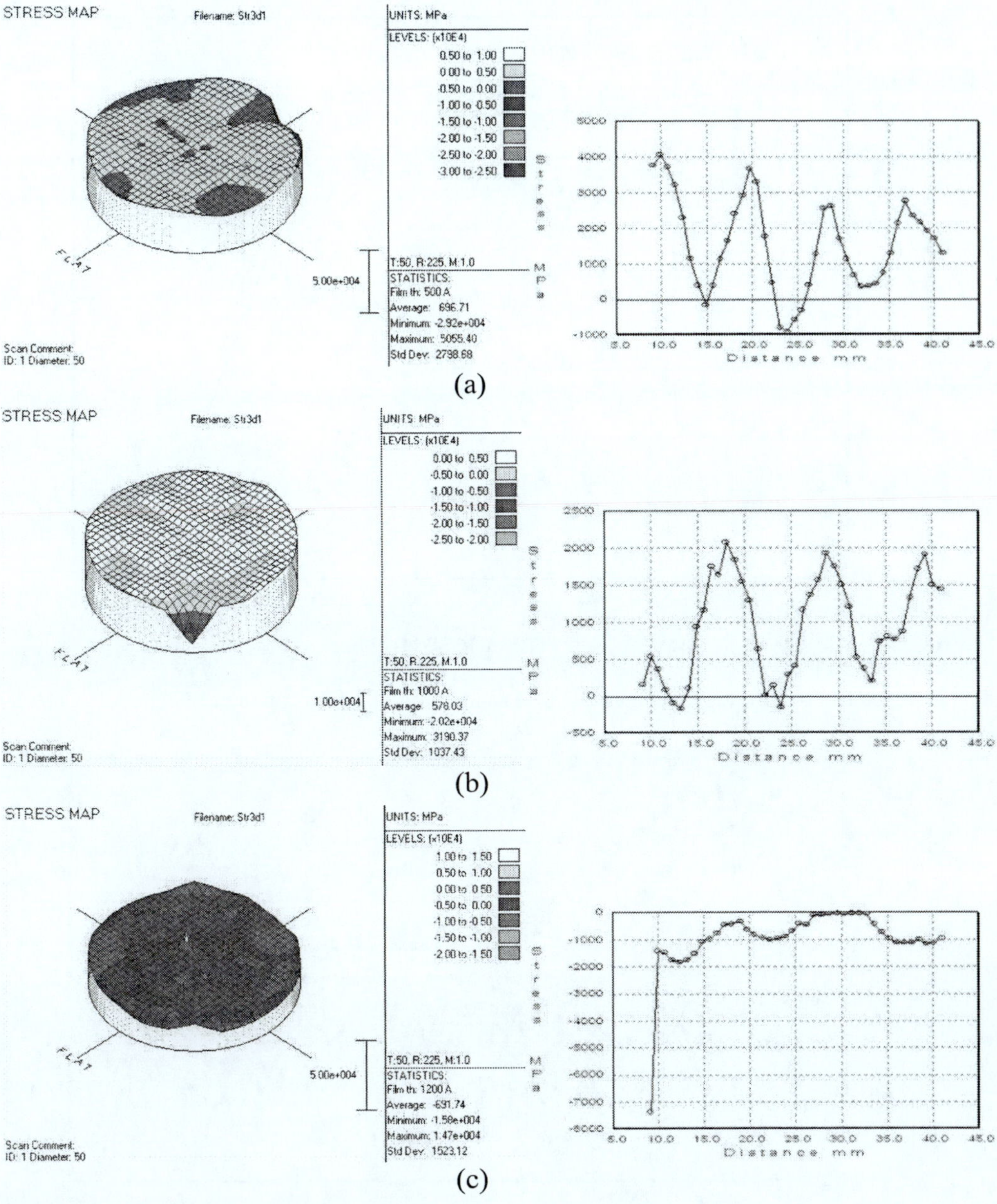

Figure 1. Internal stresses in thin chromium films on a sapphire substrate with a thickness of a chromium film: (a) 50nm; (b) 100nm; (c) 120nm.

Figure 2 shows the experimental dependence of the average mechanical stresses in chromium films on their thickness.

The stresses are tensile (positive) for chromium films on sapphire up to about 100nm thick, obtained by magnetron deposition. The stresses value depends on the parameters of the deposition process, which were selected experimentally. At chromium film thickness of more than 100nm, the level of tensile mechanical stresses decreases and the compressive mechanical stresses begin to predominate.

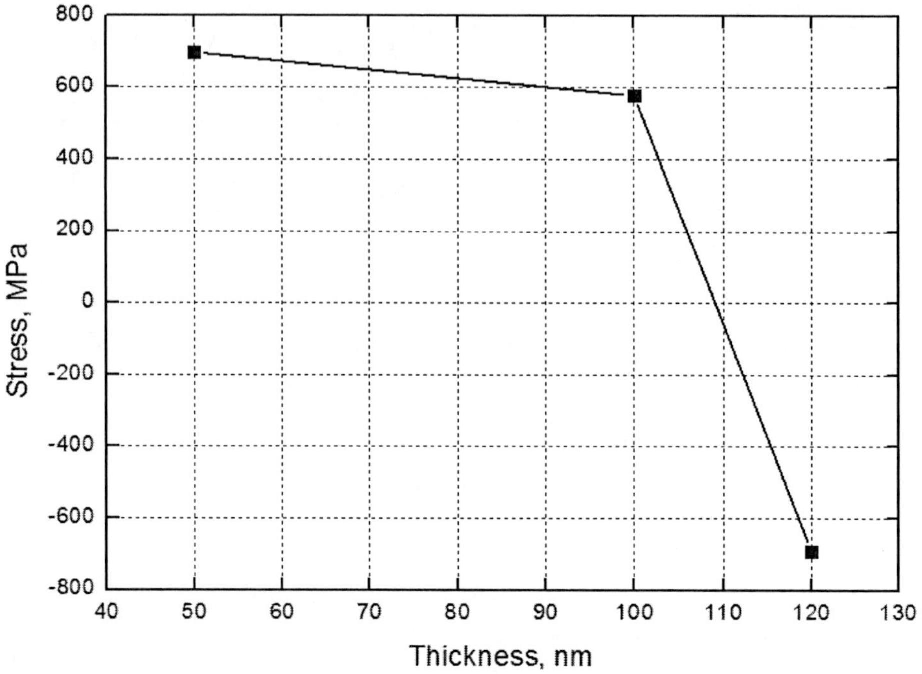

Figure 2. Dependence of average mechanical stresses in chromium films on their thickness.

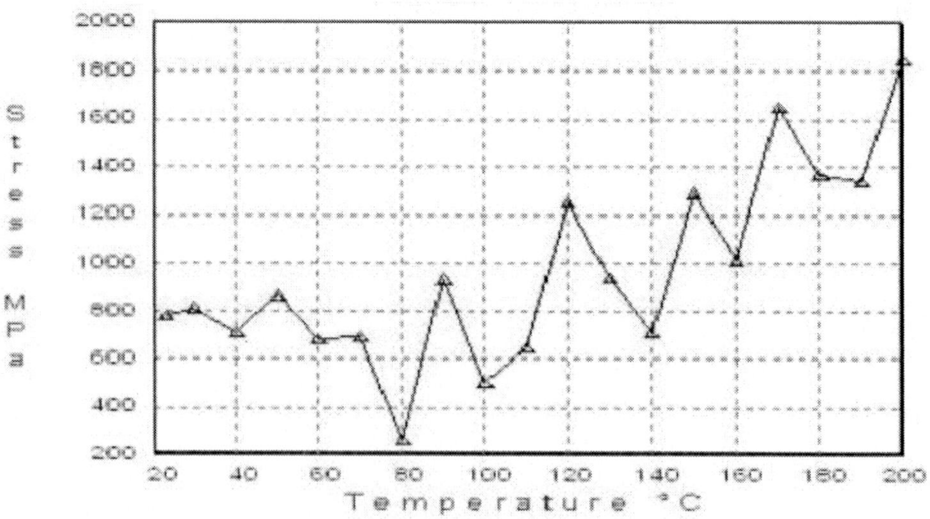

Figure 3. Influence of the temperature on the level of thermoelastic stresses in chromium films on sapphire substrate.

Investigations of the temperature influence on mechanical stresses in chromium films on a sapphire substrate have shown that when we heat the structure to 200°C, the value of mechanical stress varies from 400 – 600MPa to 1600 – 1800MPa (Figure 3). Further temperature increase leads to the growth of mechanical stresses in chromium films on a sapphire substrate, which may have a negative effect on the characteristics of micromechanical devices based on these films.

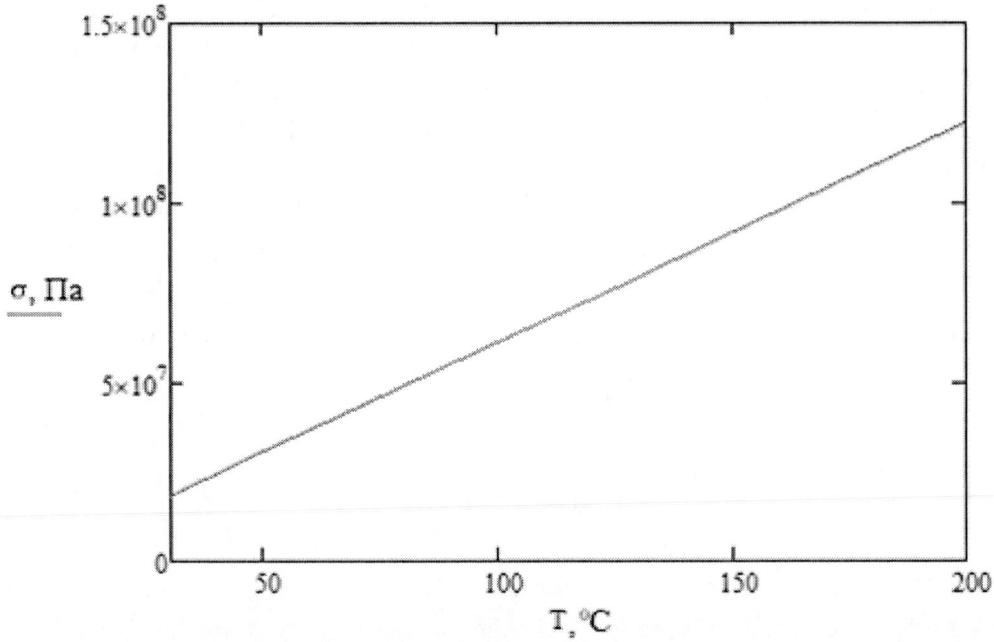

Figure 4. Theoretical dependence of mechanical stresses on the temperature in chromium films.

In order to confirm the experimental results of the study of mechanical stresses, we made theoretical calculations based on thermoelasticity theory using the following equation [3, 11-17]:

$$\sigma_T = \frac{E_f}{1-v_f}(a_f - a_s) \cdot \Delta T, \tag{1}$$

where E_f is the Young's modulus of film; v_f is the Poisson's ratios of film; α_f, α_s are the coefficients of linear expansion of the film and substrate; and ΔT is the temperature change.

The results of theoretical calculating the mechanical stresses in chromium films on a sapphire substrate are shown in Figure 4.

We can see that the theoretical values of mechanical stresses in chromium films on a sapphire substrate correspond to the experimental data (Figures 3 and 4).

CONCLUSION

In this paper, we have described the chromium thin films on a sapphire substrate obtained by magnetron deposition, which can find wide application for the construction of the micromechanical sensors of membrane type. We investigate mechanical stresses in thin chromium films on a sapphire substrate that occur because of the difference in the coefficients of the thermal expansion of the film and the substrate. Theoretical and experimental studies have shown that when the chromium film is formed with a thickness of about 100nm by magnetron deposition, the stresses are tensile (positive) [15]. On the other hand, when the thickness of the chromium film exceeds 100nm, the compressive nature of the mechanical stresses begins to predominate. We study also the effect of temperature on the mechanical stresses in chromium films. We determine that when we heat the structure to 200°C, the value of mechanical stress varies from 400 – 600MPa to 1600 – 1800MPa.

The experimental studies results of mechanical stresses in chromium films on a sapphire substrate correspond to the theoretical calculations.

Acknowledgments

This work is supported by the Ministry of Education and Science of the Russian Federation and German Academic Exchange Service (DAAD) in the framework of Michail-Lomonosov-Program.

References

[1] Astashenkova, O. N.; Korlyakov, A. V. *Modern Science: Researches, Ideas, Results, Technologies*, 2014, *vol. 2(15)*, 57 - 61 (in Russian).

[2] Astashenkova, O. N.; Korlyakov, A. V. *Izvestiya of the Kabardino-Balkar State University*, 2011, *No. 3*, 18 - 20 (in Russian).

[3] Druzhkov, S. S. *Bulletin of UGATU*, 2013, *No. 8*, 137 - 142 (in Russian).

[4] Astashenkova, O. N.; Korlyakov, A. V. *Nano- and Microsystem Technology*, 2013, *No. 2*, 24 - 29 (in Russian).

[5] Danilin, A. S. Magnetron Sputtering Systems. *Radio and Communication*, Moscow. 1982, pp. 1 - 75 (in Russian).

[6] Tamulevicius, S. *Vacuum*, 1998, *No. 2*, 127 - 139.

[7] Coman, B. P.; Yuzevich, V. N. *Solid State Physics*, 2012, *No. 7*, 1335 - 1341 (in Russian).

[8] Klunnikova, Yu. V.; Gusev, E. Yu. *Izvestiya of SFedU. Engineering Sciences*, 2018, *No. 2*, 16 - 25 (in Russian).

[9] Haider, J.; et al. *Journal of Materials Processing Technology*, 2005, *No. 1*, 36 - 41.

[10] Kovalenko, D. A.; Petrov, V. V. *Journal of Nano-Electron. Phys.,* 2015, *No. 3*, 03036 - 1 (in Russian).

[11] Dawber, M.; Rabe, K. M.; Scott, J. F. *Review of Modem Physics*, 2005, *vol. 77*, 1083 - 1130.

[12] Pichugin, V. F. *Material Science of Surface and Thin Films.* TPU Press, Tomsk. 2008, pp. 1 - 173 (in Russian).

[13] Feldman, F.; Mayer, D. Fundamentals of Surface Analysis and Thin Films. *Mir,* Moscow. 1989, 1 - 344 (in Russian).

[14] Bifano, T. G. *Journal of Micromechanical Systems*, 2002, *No. 5*, 592 - 597.

[15] Astashenkova, O. N. *Materials of the IV International Scientific and Technical Conference "Micro- and Nanotechnologies in Electronics"*, Nal'chik. 2011, 153 - 154 (in Russian).

[16] Klunnikova, Yu. V.; Malyukov, S. P.; Sayenko, A. V.; Sarychev, D. A.; Kitaev, V. V. *Technical Physics Letters*, 2018, *No. 6*, 535 - 537.

[17] Malyukov, S. P.; Klunnikova, Yu. V.; Sayenko, A. V. *Journal of Russian Laser Research*, 2015. *vol. 36(3)*, 276 - 280.

In: Proceedings of the 2018 International Conference … ISBN: 978-1-53615-862-5
Editors: Ivan A. Parinov et al.

Chapter 17

UHF Radiation Properties of PCR-13 Piezoceramics

E. N. Sidorenko[1,*], G. I. Prihodko[1], I. I. Natkhin[1], T. I. Debelova[1] and D. V. Kolodko[1]

[1]Southern Federal University, Rostov-on-Don, Russia

Abstract

The experimental study results of the electromagnetic field radiation by the microstrip line (MSL) with various ferroelectric resonators are presented. The resonators in the disc form of the same dimensions are made of PKR-13 type piezoceramics. The energy emission spectrum of the MSL-piezoelectric system different heights regarding the resonator for piezoelectric with various crystallites dimensions were plotted. The intensity distribution pictures of the electric field in the resonator vicinity were obtained for the different orientations regarding its axis.

Measurements were made in the frequency range from 3 to 12 GHz using whip or folded antenna.

The mechanical twins available at the resonator crystallites domainstructure oscillating walls field energy emission and absorption processes contribution is detected in the ferroelectric resonator electric field intensity angular distribution.

Keywords: piezoceramics, microstrip line (MSL), dielectric resonator, radiation spectra, field intensity distribution, domain structure elements

* Corresponding Author's E-mail: si-do-re@mail.ru.

1. Introduction

Ferroelectric materials have been successfully used for many years to construct various signal processing devices in the UHF range. They have a wide application field due to such properties as dielectric constant high values, a relatively low level of the UHF dielectric losses of ferroelectrics in the paraelectric phase, as well as the UHF devices characteristics operating possibility with the control voltage help. Ferroelectrics are used in such classical UHF devices as low-noise parametric amplifiers and phase shifters, tunable filters and resonators on the single-crystal ferroelectrics, planar capacitors with the electrically controlled capacitance, and the UHF energy transmission lines with electrically controlled phase velocity [1, 2]. The ferroelectric materials have a number of technical and economic advantages over other identical materials. So UHF devices based on ferroelectrics can be used in a wide range of power level values. The ferroelectrics in the control circuits can provide high speed work at low energy consumption due to a short tuning time of the operating character (~ 10^{-8}s). Ferroelectrics have lager electrical and radiation resistance than semiconductors. Ferroelectrics are much more economical in energy consumption than ferrites. The ferroelectric materials films manufacturing technology is relatively simple and cheap. The useage of the ferroelectric films on a dielectric substrate ensures the UHF microcircuit realization in a fully integrated design without manual mounting operation of the active elements. This allows to reduce the cost of devices based on ferroelectrics significantly [3].

All previously mentioned information is a cause of increased interest to the work on the piezoelectric ceramics useage in the UHF range. However, a large number of theoretical developments have not found proper practical confirmation. In our opinion, it is due to the fact that there is no sufficient number of practical studies of the ferroelectric ceramics types in the UHF.

2. Objective of the Work

The present work develops the earlier studies of the electromagnetic (EM) energy UHF absorption of the radio absorbing materials: crystals, individual compounds, and multicomponent solid solutions related to the perovskite oxides [4-6]. It was established that the mechanical twin oscillating walls, which are form the ceramics and crystals – ferroics domain structure basis, can be electromagnetic UHF energy absorption source [7].

The main idea is to determine the mechanicalhe twin walls contribution to the MSL system total radiation with a surface placed ferroelectric resonators with the help of the generator signal frequency, the average piezoceramic grains dimensions and the cavity samples orientations depending on their angular electric field intensity axes.

3. OBJECTS AND RESEARCH METHOD

The hard ferroelectric (with the domain structure stable to external influences) piezoceramics PCR-13 obtained by the hot pressing method was chosen as the research object. PCR is the Rostov-on-Don piezoceramics. It is a four-component solid solution with a perovskite-type structure, which has a tetragonal symmetry. PCR-13 has a phase transition temperature T_C = 335^{0}C, a spontaneous polarization P_0=0.32 C•m^{-2}, coercive field E_C=2.5•10^6 V•m^{-1} and a homogeneous strain parameter high value δ =0.016 Å^3 at T = 20 C [8]. There were investigated three types of the piezoceramics with average crystallites sizes of 3, 7 and 19μm. All the samples were disk-shaped with a diameter of 10mm and a thickness of 1 mm.

The setup for measurements consisted of a voltage standard wage ratio (VSWR) and attenuation indicator, three interchangeable swept-frequency generators covering the 3.2 – 12 GHz frequency band and the MSL. The dielectric constant of the MSL substrate ε ≈ 10. The MSL tape type central conductor width is 1mm (Figure 1). A 50-ohm optimum load was connected to the MSL output, so the line is operating in the traveling wave mode. The installation was tuned and operated in the reflection mode with the VSWR measuring possibility in necessity. The EM field measurement power was performed using a thermistor bridge.

Figure 1. MSL in the waveguide transmission line, the resonator and the whip antenna appearance.

The value proportional to the electric field strength E* and the magnetic field induction B* in the resonator vicinity on the MSL was determined using a whip or folded antenna.

The signal from the antenna, detected by the UHF diode, was fed to a digital voltmeter for the DC voltage component measuring. The occurring currents in the antenna circuit over weak signals measuring are low.

Consequently, the diode current-voltage characteristic used region can be approximated as a function with a quadratic dependence I = f (U). Therefore, the electric field intensity or the magnetic field induction were considered as values proportional to the square root of digital voltmeter U constant voltage component. The voltage was measured in Volts.

4. Results and Discussion

The electric and magnetic fields emission spectra were obtained at the different heights h from the sample surface.

Figure 2 shows the B*(f) dependencies measured with a folded antenna of 0.8cm diameter. B* is a value proportional to the magnetic field induction modulus here. There was a background over the microstrip line - a weak radiation B* = 0.005-0.01 without a sample.

The obtained B*(f) dependence is far from the ideal resonance curve. For all the samples, regardless the crystallites size, a sufficiently spike radiation maximum at a frequency of ~5 GHz in the transmission band in width of~ 0.5 GHz at h = 0 was observed in the emission spectra. With h was increased its amplitude was decreased.

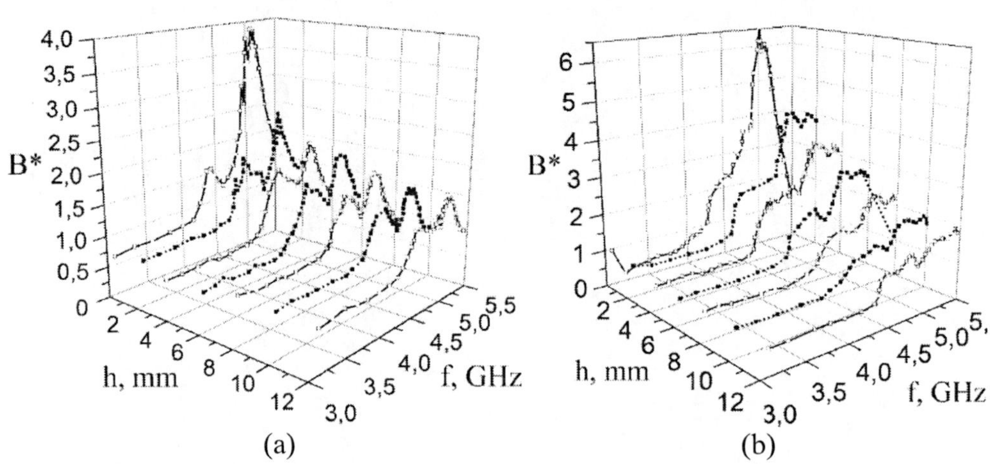

Figure 2. Integrated radiation spectra of magnetic field induction for piezoceramics with the crystallite sizes: (a) 7μm and (b) 19μm.

There is the second maximum with a smaller amplitude in spectra at lower frequencies (~ 4.4 GHz), explicitly (Figure 2a) or implicitly (Figure 2b), its value depends little on h.

Similar characteristics were obtained for the electric field. In the region of higher frequencies, the emitted signal was weak, apparently because of the inconsistency of the UHF diode used for this frequency range.

Large sharp energy absorption peaks were observed earlier [9] in the PCR-13 absorption spectra in the frequency range 3-12 GHz that were due to the domain structure peculiarity, and more specifically to the mechanical twins "jittering" in the EM field. We suppose that the obtained radiation spectra (Figure 2) was formed as two simultaneously existing resonant processes with close frequencies overall result. First, it is the resonator-ferroelectric disk radiation. Secondly, it is the EM field energy selection by mechanical twin resonant ensembles.

The electric field radiated by resonators with different d values for the different orientations on MSL intensity angular distribution patterns was obtained. Measurements were made at points near the lateral sample side surface and at the distance of 5mm from it in the plane located at the disk resonator height.

In the figure 3 the EM wave propagation is directed in the MSL from the generator to the load from right to left: $180^{\circ} \leftarrow 0^{\circ}$.

The electric field intensity angular distribution images are elongated along the MSL conductive strip, and their centers are displaced to the generator. The first, apparently, is the result of the MSL central conductor EM field emission, stimulated by the resonator radiation. And the second is due to the wave propagation direction in the MSL substrate.

Figure 3 shows that the distribution patterns are differ in different α and d. But the presence of the individual points with inflated E* values presence in the distribution patterns is the most interesting in our opinion.

It is watched on the figures: 3,2 - along the ray 210° for $\alpha = 180°$, 3,3 - on the ray 225° for $\alpha = 90°$, 3,4 -on the ray 60° for $\alpha = 90°$, 3,5 - on the ray 240° for $\alpha = 0°$, 3,6 - on the ray 135° for $\alpha = 0°$. On the other hand, there are also points with underestimate E* values in the distribution.

Apparently, contribution to the studied field of the resonator disk of the resonance nature absorption of the field energy by the domain structureelements is the reason for the points with low E* values presence in the electric field intensity distribution. On the other hand, the resonator domain structure various elements ensembles, oscillating in the MSL EM field, are EM energy radiation sources. The contribution of the energy radiated by them to the piezoceramic resonator field energy exceed the electric field individual points intensity.

The generator signal frequency in the observed frequency range change affects the electric field intensity distribution, where the points with anomalous intensity values are also observed.

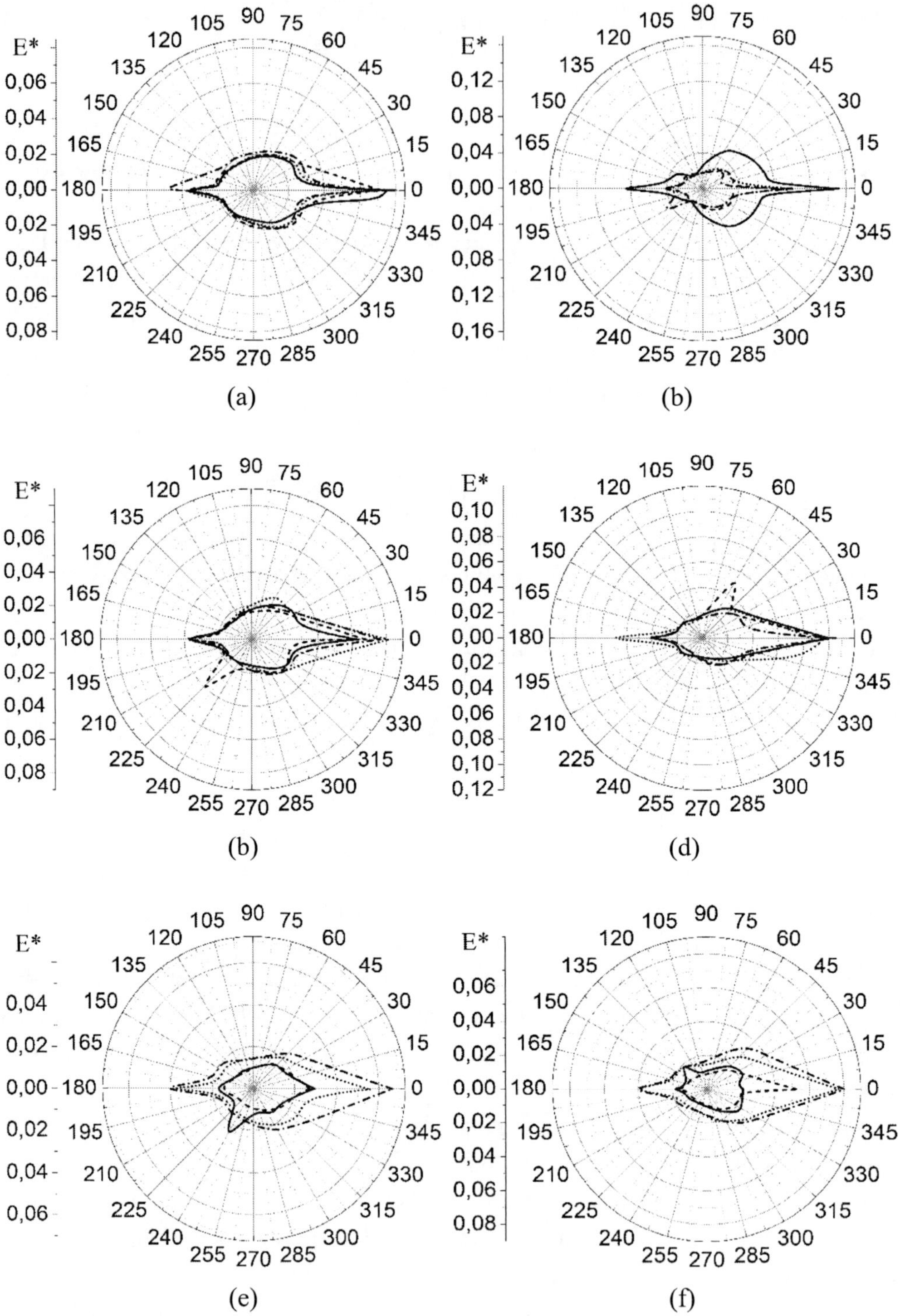

Figure 3. Angular distribution of resonator electric field intensity for its four positions on the MSL with different rotation angles α around th disk axis: 0º- —; 90º- ----; 180º- —•—; 270º - ••••••; (a), (b) d = 3 μm, (c), (d) d = 7 μm, (e), (f) d = 19 μm; (a), (c), (e) at the lateral resonator surface; (b), (d), (f) at a distance of 5 mm from the resonator; f = 4.5 GHz.

CONCLUSION

As a result of the investigation it was established that there is a contribution of the resonance absorption and a radiation of oscillating walls of mechanical twin in the electric field of the piezoceramic resonator located on the MSL surface.

REFERENCES

[1] Karin, M.; Rabe, K.M.; Charles, H.; Ahn, C.H.; Triscone, J.-M. *Physics of Ferroelectrics: A Modern Perspective*, Springer, 2010, pp. 1 - 222.

[2] Antonov, N.N.; Buzin, I.M.; Vendik, O.G.; et al. Ferroelectrics in Microwave Technology.Moscow. *Sovetskoe Radio*, 1979,pp. 1 -271. (in Russian).

[3] Rabe, K.M.; Ana, C.G.; Triskoni, Zh.M.; Binom, M. Physics of Ferroelectrics. A Modern Perspective.*Laboratory of Knowledge*, 2011,pp. 1 - 440 (in Russian).

[4] Sidorenko, E.N.; Turik, A.V.; Natkhin, I.I.; Andreev, I.S.*Ferroelectrics,*2003. *vol. 286,* 131 - 140.

[5] Gershenovich, V.V.; Sidorenko, E.N.; Reznichenko, L.A.; Natkhin, I.I. *Izv. RAN, Ser. Neorg. Matter,*2007, *vol. 43(10)*, 1236 - 1238 (in Russian).

[6] Sidorenko, E.N.; Gavrilyatchenko, V.G.; Turik, A.V.; Semenchev, A.F.; Natkhin, I.I. *Electromagnetic Waves and Electronic Systems,*2013, *vol.18(9)*,51 - 54(in Russian).

[7] Sidorenko, E.N.; Gavrilyatchenko, V.G.; Smotrakov, V.G.; Eremkin, V.V.; Natkhin, I.I.; Borisova, A.V.; Akbaeva G.M. In: *Proceedings of the 2016 International Conference on Physics and Mechanics of New Materials and Their Applications*, Ivan A. Parinov, Shun-Hsyung Chang, Muaffaq A. Jani (Eds.). Nova Science Publishers.New York. 2017, 361 - 368.

[8] Fesenko, E. G.; Danziger, L. A.; Razumovskaya, O. N. *The New Piezoceramic Materials,*Rostov State University Press,Rostov-on-Don. 1983, pp. 1 - 110. (in Russian).

[9] Sidorenko, E.N.; Gavrilyatchenko, V.G.; Natkhin, I.I.; Debelova, T.I.; Kolodko, D.V. In: *Abstracts and Schedule of the 2017 International Conference on "Physics and Mechanics of New Materials and Their Applications",*(14-16 October, 2017, Jabalpur, India), Vijay K. Gupta, Ivan A. Parinov, Shun-Hsyung Chang (Eds.). PDPM Press, Jabalpur. 2017, 235 - 236.

In: Proceedings of the 2018 International Conference … ISBN: 978-1-53615-862-5
Editors: Ivan A. Parinov et al.

Chapter 18

ELECTROMAGNETIC FIELD OF PIEZOCERAMIC RESONATOR ON MICROSTRIP LINE WITH THE GAP

E. N. Sidorenko*[*]*, I. I. Natkhin, G. I. Prihodko, D. V. Kolodko and T. I. Debelova

Southern Federal University, Rostov-on-Don, 344006, Russia

ABSTRACT

The electromagnetic (EM) wave transmission and emission in a waveguide structure, consisting of an asymmetrical microstrip line (MSL) with a PCR-13 piezoceramics resonator, have been experimentally studied. Piezoceramic crystallites, resonator orientation regarding its axis, electromagnetic field frequency, resonator with the MSL coupling coefficient influence on the medium-sized resonator, and MSL emission EM field intensity angular distribution emission spectra were investigated. It was found that at the resonant frequency, the resonator can emit approximately a tenth of the generator signal energy.

Keywords: microstrip line (MSL), electromagnetic (EM) wave, piezoresonator, dielectric resonator antenna (DRA)

1. INTRODUCTION

Cylindrical form dielectric resonators made of anisotropic dielectrics today are waveguide transmission line elements, well studied both theoretically and experimentally [1, 2]. The combined usage of dielectric resonators and a microstrip line has led to the

[*] Corresponding Author E-mail: si-do-re@mail.ru.

appearance of elements of waveguiding structures on a new base. Thus, the dielectric resonators usage allows an obtaining of UHF filters both in waveguide and microstrip form [3]. Adjustable UHF energy couplers were created due to the existing possibility to change the coupling coefficient of the dielectric resonator with the transmission line. It is possible to obtain elements with electromechanical adjustment of their resonant frequency, bandwidth, and quality factor (Q-factor) by resonator material selecting. The combined usage of dielectric resonators of various shapes and microstrip lines of various types made it possible to create small-sized perspective dielectric resonator antennas (DRAs) [4].

This work is the continuation of the ferroelectric ceramics and crystals properties studyin the UHF field by the authors [5].

2. Problem Formulation

The resonators are made of dielectrics with low dielectric losses and low dielectric constant ε in DRA. The dielectrics with increased ε values usage allow us to reduce the resonators geometric dimensions. It is known that high ε values are characteristic of ferroelectric materials [6]. However, the UHF dielectric dispersion phenomenon conditioned by relaxation and resonance processes is observed in polycrystalline ferroelectrics in the frequency range 2 – 10 GHz. The dielectric losses overestimated values in such ferroelectrics are the consequence of the above. However, the ferroelectric resonators usage in the DRA is attractive in relation with the resonator parameters electromagnetic control possibility [7].

The microstrip lines (MSLs) of various constructions are used for resonator excitation with less energy losses [8]. The electromagnetic wave transmission and emission possibility in the waveguide structures containing an asymmetric microstrip line (MSL), with a resonator above the microstrip gap on a dielectric substrate, was experimentally investigated in this work. The resonator is a disk made of PCR-13 type ferroelectric ceramics. An attempt to determine the influence of the frequency, position, disk orientation on the MSL surface, and PCR-13 piezoceramics domain structure elements on the electric field emitted by the resonator angular distribution intensity was made.

3. Research Methods

The installation consists of three replaceable oscillating frequency generators covering the bandwidth of 3.2 – 12 GHz, the VSWR, and attenuation indicator, as well as an asymmetrical microstrip line with a piezoelectric resonator. The MSL can operate both in the traveling mode and in the reflected wave mode. A nonreflecting 50 Ω load extremely minimizing the wave reflection process in the waveguide path is connected to the MSL output while operating in the traveling wave mode.

A brass plate is the base or outer conductor of the MSL (Figure 1). A polycore plate with a corundum ceramics polished surface with a size of 12 × 30 mm^2 is used as a dielectric (substrate). Its relative dielectric constant ε is equal to 9.7 ± 0.25 at a frequency of 10 GHz and the dielectric loss angle tangent $\tan\delta = 10^{-4}$.

The MSL inner conductor consists of two open microstrip L-shaped conductors.

The distance between them is 8 mm in the gap area. The ratio *W*/*h*' = 0.98, where *W* is the microstrip conductor width, *h*' = 0.5 mm, is the substrate thickness. The piezoelectric ceramics investigated sample is placed on the parallel open conductors ends at the UHF field electrical component antinode *E*.

A loop or a probe was located above the sample or at its lateral surface as a measuring probe of the radiated EM field of the studied structure. The modulated UHF signal, guided in it, was detected by a UHF diode and fed to the high-resistance digital voltmeter input. The voltage constant component value of digital voltmeter was taken proportional to the magnetic field inductance *B**, or the constant component value square of electric field density *E**, due to the quadratic dependence of UHF diode voltage-current characteristics.

Figure 1. Photo of the piezoelectric resonator on the MSL surface in the waveguide path and whip antenna.

4. Results and Discussion

The MSL-resonator system's absorbing properties in the 3 – 12 GHz range with a symmetrical and asymmetrical resonator arrangement in the open microstrip conductors region were determined in the traveling wave mode. A MSL without a resonator does not transmit an EM wave with the exception of a narrow frequency band in the (4.0 – 4.2) GHz region. There is a maximum energy absorption L, equal to (45 – 50) dB, corresponding to the VSWR and the noise limit measurements of attenuation indicator in the entire frequency range under study, $L = -(17\ldots 25)$ dB in the (4.0...4.2) GHz frequency range.

The energy absorption spectra for MSL with different resonators, symmetrically placed over the open microstrip, are shown in Figure 2a. Resonators piezoceramics differ in average crystallite sizes of 3, 7 and 19 μm. It can be concluded that zones with energy absorption less than $L \approx -20$ dB appeared due to the resonator excitation in the researched frequency range where the traveling wave mode is realized. These zones alternate with zones, where $L = -(45\ldots50)$ dB, due to high-energy absorption or reflection resulting in connection between the MSL and the resonator's non-existence.

VSWR measurements were made for a system with a piezo-ceramic resonator of different types in the (3.2 – 5.6) GHz frequency range, while the installation operated in reflection mode. It is established that the VSWR is large in the low frequency range (3.2 – 4) GHz and equals to 4 – 6, which corresponds to a strong energy reflection. The VSWR nonlinearly decreases until 1.43. which indicates strong energy absorption with the frequency increasing until 5.6 GHz. However, the minimum VSWR of 1.2 and the maximum absorption are observed in the frequency range (5.2 – 5.3) GHz where the energy is reemitted. The radiation power is about a tenth of the generator signal power.

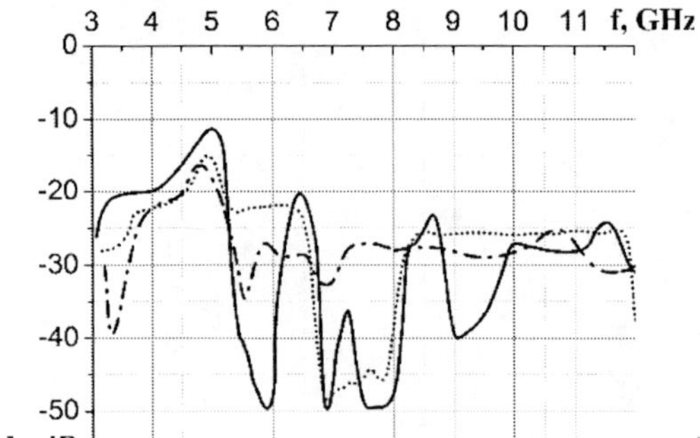

(a)

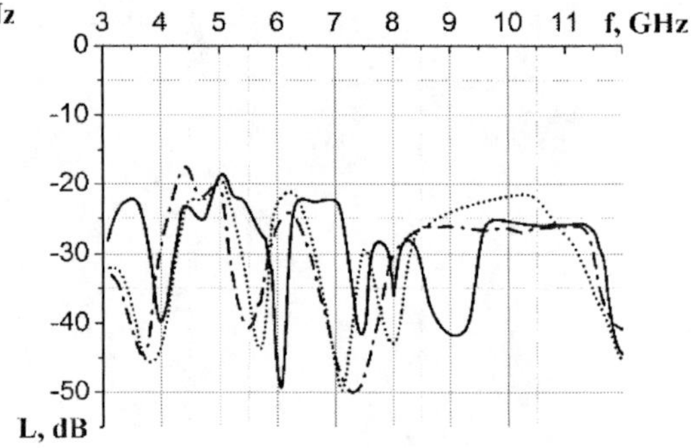

(b)

Figure 2. Energy absorption spectra of the MSL-piezoceramic resonator system with different crystallite sizes d: —— 19 μm; ------ 7 μm; ····· 3 μm; the resonator was located on the MSL surface between the open central MSL conductors in the electric field loop (a) or shifted right (b).

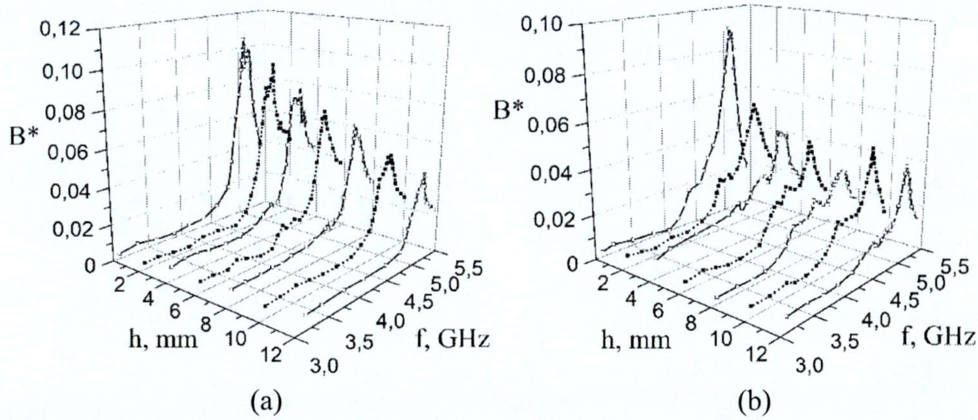

Figure 3. Induction spectra of the integral radiation of magnetic field at different heights *h* from the piezoresonator upper surface with the ferrite-ceramic crystallites sizes: (a) 19 μm and (b) 7 μm; the resonator is placed on the MSL surface between two open microstrips in the electric field antinodes; *B** is the proportional value to magnetic field inductance.

The integral magnetic field measurements above the resonator showed that the emission spectra of piezoresonators with different crystallite sizes have a similar appearance and close resonant frequencies f_0 = (5.1 – 5.2) GHz. The bandwidth is Δf = (60 –75) MHz and the *Q*-factor is Q = 70–90 for resonance curves, obtained by measuring the field at a distance of 1 mm above the piezoresonator.

The cylindrical resonator HEM_{11} wave resonant frequency calculation was made using the formula and taking into account the symmetric addition of an over-the-screen antenna by its imaginary pattern [8]:

$$f_0 = \frac{C}{2\pi R\varepsilon^{0,42}} \cdot (1{,}6 + 0{,}513k + 1{,}392k^2 - 0{,}575k^3 + 0{,}088k^4)$$

Here $k = R/2H$, where *R* and *H* are the resonator radius and height expressed in mm. ε, is the UHF dielectric constant of piezoceramics. It was established [8] through the permittivity UHF dispersion coefficient $K \leq 0.5$ in ferroelectric ceramics low-frequency and UHF ε values measurements. Therefore, the ε value is equal to 400 in resonant frequency calculations. The theoretically calculated f_0 value is 4.9 GHz, which satisfactorily coincides with the experiment.

Unfortunately, it was not possible to compare the experimentally obtained *Q* value with the calculation due to the $Q = f(k)$ nonlinear relationship's lack for a resonator with permittivity of 400.

The electric field intensity's distribution was measured using the probe directly at the lateral surface of the resonator disk and at a distance of 5 mm from it. It follows that, from Figure 4, the same order resonator radiation intensity does not depend on the piezoceramics crystallites' size. The field intensity angular distribution picture does not have a clear radiation direction, as it is observed in a MSL with a central conductor without a gap, where it has an elongated shape along the microstrip.

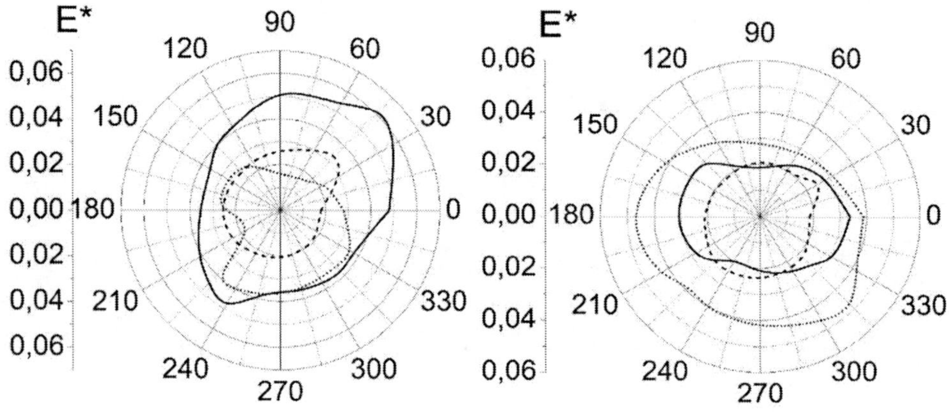

(a) (b)

Figure 4. Angular distribution of electric field intensity for a MSL and a piezoresonator with different average crystallite sizes values *d*: —— 19 μm; ------ 7 μm; ······ 3 μm near the lateral surface (a) and at a distance of 5 mm from it (b) with a symmetric resonator arrangement; the electromagnetic field frequency is 5,2 GHz, *E** is the proportional value to electric field inductance.

The field pattern changes to a greater or lesser extent depending on the frequency, size *d* and distance to the resonator, when the resonator disk orientation changes regarding its axis (Figure 5). Apparently, it is due to the radiation's contribution of the structure elements of ferroceramics domain to the resonator radiation.

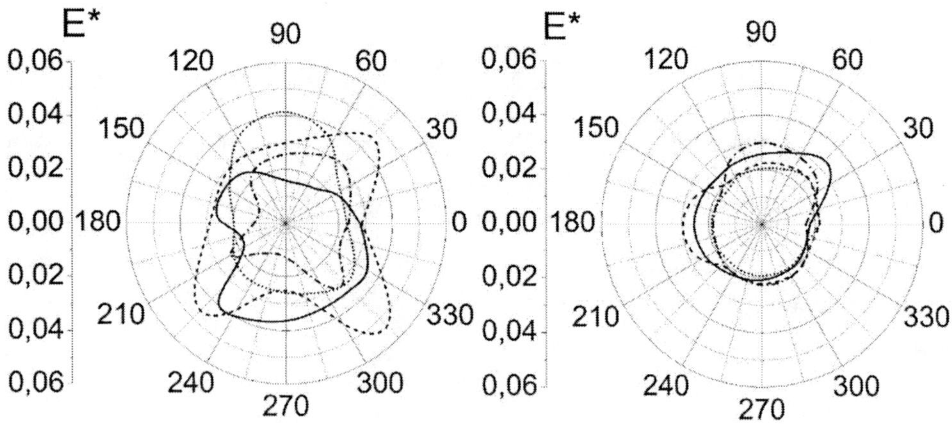

(a) (b)

Figure 5. Angular distribution of electric field intensity at the resonator lateral surface for its four positions on the MSL with different rotation angles α around the disk axis:0° ———; 90° ---------;180° -•-•-•-•-•-; 270° • • • • • • •. (a) $d = 3$ μm, (b) $d = 7$ μm; $F = 5.2$ GHz.

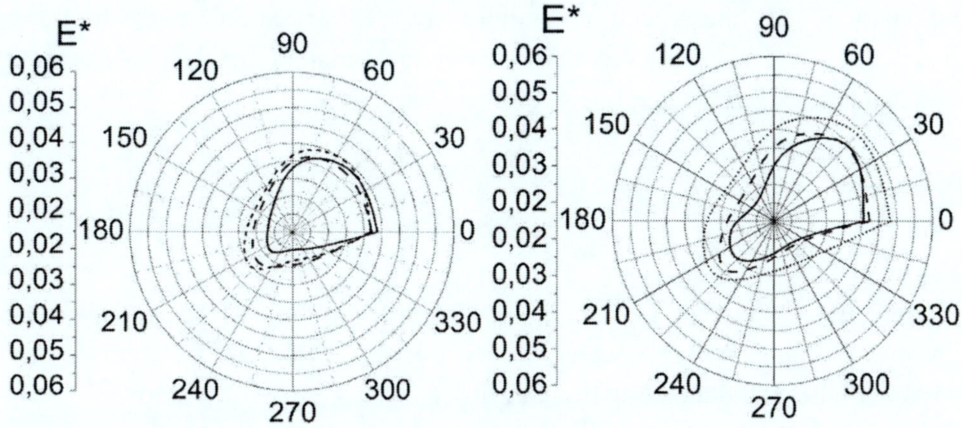

Figure 6. Angular distribution of electric field intensity for a MSL and piezoresonator with different crystallite sizes *d*: —— 19 μm; ------ 7 μm; ⋯⋯ 3 μm near the lateral surface (a) and at the distance of 5 mm from it (b) with an asymmetrical resonator arrangement; electromagnetic field frequency is 5,2 GHz.

The coupling coefficient between the MSL and the resonator changes, and the EM field distribution asymmetry, appears when the resonator is shifted to the right until its lateral surface touches the left microstrip (Figure 6). The radiation zone becomes directional regardless of the resonator piezoceramics*d*. The picture of field intensity distribution has a displaced petal form regarding the resonator center and directed at an angle of (30 – 45)°. A similar resonator's displacement to the left over the MSL surface causes the radiation zone in the petal form rotation to be counter clockwise through (60 – 90)° depending on the *d* size.

It is clear (Figures 4 and 5) that the field radiation areas are somewhat different when *E** is measured at the resonator lateral surface and at a distance of 5 mm from it. The field intensity measurements in the maximum radiation direction showed that *E** increases in the near zone and decreases in the far one. The zones' boundary is at a distance of 1 – 6 mm from the resonator surface and depends on the chosen radiation direction and *d* value. The maximum distance from the source was equal to 20 – 50 mm, where the field radiation was still observed in the far zone.

Conclusion

The emitting asymmetric microstrip line, together with the properties of PCR-13 piezoceramics resonators, has been experimentally studied. It was found that a resonator at a resonant frequency can emit about a tenth of the generator signal energy. The MSL and the angular distribution of the resonator electric field intensity depend on various factors, such as the piezoelectric ceramics average crystallite size, the resonator orientation regarding its axis, and the electromagnetic field frequency. The radiation direction can be changed by coupling coefficient varying

REFERENCES

[1] Jones, EM; Matthaei, G; Young, L. *Microwave Filters, Impedance-Matching Networks, and Coupling Structures*. Artech House., 1980, pp. 1 – 1096.

[2] Volman, VI. Handbook for the Calculation and Design of Microwave Strip-line Devices. *Radio and Communication*, Moscow., 1982, pp. 1 – 326 (in Russian).

[3] Groll, H; Nedkov, I. *Microwave Physics and Techniques*. Springer Science & Business Media., 2012, pp. 1 – 457.

[4] Constattine, AB. *Antenna Theory: Analysis and Design*, 3rd Edition, John Wiley &Sons., 2005, pp. 1 – 1138.

[5] Sidorenko, EN; Gavrilyatchenko, VG; Smotrakov, VG; et al. In: *Proceedings of the 2016 International Conference on Physics and Mechanics of New Materials and Their Applications*, Ivan A. Parinov, Shun-Hsyung Chang, Muaffaq A. Jani (Eds.). Nova Science Publishers.New York., 2017, 361 – 368.

[6] Rabe, KM; Ana, CG; Triskoni, ZhM; Binom, M. *Physics of Ferroelectrics. A Modern Perspective. Laboratory of Knowledge*., 2011, pp. 1 – 440 (in Russian).

[7] Kalinin, AS; Kovtunov, DA; Macinko, VY; Rudy, JB; Smirnova, AS.*Electronics and Microelectronics*, *Microwave.* Saint-Petersburg State Electrotechnical University Press, St. Petersburg., 2015, vol.2(3), 60 – 62 (In Russian).

[8] Makimoto, M; Yamashita, S. *Microwave Resonators and Filters for Wireless Communication.* Matsushita Research Institute, Tokyo., 2013, pp. 1 – 172.

[9] Slyusar, VI. *Electronics: Science, Technology,Business*., 2007, No. 2, 28 – 37 (In Russian).

[10] Turik, AV; Sidorenko, EN; Zhestkov, VF; Komarov, VD. *Izv. AN SSSR.,Ser. Fiz.*, 1970, vol. 34(12), 2590 – 2593 (in Russian).

In: Proceedings of the 2018 International Conference … ISBN: 978-1-53615-862-5
Editors: Ivan A. Parinov et al.

Chapter 19

FEATURES OF THE MICROSTRUCTURE OF FERROELECTRIC-SOFT MATERIALS BASED ON PZT

***A. V. Skrylev*[1,*], *G. M. Akbaeva*[2], *E. V. Barabanova*[3], *O. V. Malyshkina*[3] *and E. M. Panchenko*[2]**

[1]Institute of High Technologies and Piezotechnics, Southern Federal University, Rostov-on-Don, Russia
[2]Research Institute of Physics, Southern Federal University, Rostov-on-Don, Russia
[3]Tver State University, Tver, Russia

ABSTRACT

Ferroelectric ceramics, based on lead zirconate-titanate (PZT), have found a wide application in the world. In order to change its properties for a specific practical use, multicomponent systems based on it are created and various modifying additives are used, which makes it possible to obtain materials with various electrophysical parameters, including those possessing relaxor properties. The ferroelectric relaxors are widely studied today. This is due to their improved properties in comparison with classical ferroelectrics. In particular, this refers to the electromechanical properties of relaxors: large values of piezomodules and electrostriction coefficients. They find application in actuators, piezomotors, acoustoelectric and mechanoelectric converters [1-3]. A number of ferroelectrics-relaxors have high values of dielectric permittivity, electro-optical, and pyroelectric characteristics [4]. The maximum temperature of these and other characteristics depend strongly on the frequency of the external field, reaching tens and hundreds of degrees in some materials [5].

* Corresponding Author's E-mail: skrylyov@sfedu.ru.

The purpose of this work is to determine the characteristics of the microstructure of materials with a low coercive field depending on the electrophysical parameters.

Keywords: multicomponent systems, ferroelectrics-relaxors, electrophysical and mechanical properties, atomic force microscope (AFM)

1. Methods and Samples

In this work, the samples of two compositions of multicomponent systems, based on PZT, were studied:

(1) $0.40PbTiO_3 - 0.53PbZrO_3 - PbW_{1/2}Cd_{1/2}O_3 - PbW_{3/5}Li_{2/5}O_3 - 0.005Yb_2O_3$,
(2) $0.36PbTiO_3 - 0.33PbZrO_3 - 0.17Pb(MgZn)_{1/3}Nb_{2/3}O_3 - BaTiO_3 - SrZrO_3$.

The ceramics of composition (i) were obtained by the hot pressing method and of composition (ii) by conventional technology. The components were selected to produce ceramics with small coercive fields E_c [6]. Table 1 shows the main electrophysical characteristics of the compositions.

The electrical capacitance, resonance, and antiresonance frequencies of polarized piezoceramic samples were measured using the precision impedance meter "Wayne Kerr Electronics WK 6510B." The coefficient of electromechanical coupling k_p, relative permittivity $\varepsilon_{33}^{T}/\varepsilon_0$, and piezomodule d_{31} were calculated according to the current industry standard [7].

The domain structure was studied by using an atomic force microscope (AFM) with the help of the piezoelectric force microscopy (PFM) method. The PFM method is based on the inverse piezoelectric effect, which is characteristic for all ferroelectrics. When the sample is locally exposed to the electric field, applied to the conducting AFM probe, the sample under the probe deforms.

The deformation depends on the direction of polarization at a given point. If the field and polarization coincide in direction, then the sample stretches. If they are oppositely directed, the sample compresses.

Table 1. Main electrophysical characteristics of the samples

No.	T_c, °C	E_c, kV/cm	$\varepsilon_{33}^{T}/\varepsilon_0$	k_p	d_{31}, 10^{-12}C/N
(1)	320	5.5	570	0.52	69
(2)	150	6.0	4900	0.66	325

In the PFM method, an alternating voltage $V_{tip} = V_{ac}\cos(\omega t)$ is applied to the probe, and the amplitude (V_{ac}) is smaller than the coercive voltage. This causes the sample to oscillate in thickness, which in turn leads to vibration of the cantilever. In the case under consideration, the deformation of the sample is proportional to the applied voltage:

$$\Delta z = \pm d_{zz} V_{tip} \quad (1)$$

The proportionality coefficient is the piezomodule d_{zz}. The $\pm$ sign is determined by the mutual orientation of the polarization vectors and the field voltage.

The cantilever oscillations are fixed by the optical system. The signal of the piezoelectric response (electromechanical response, EMR) from the sample measured by the optical system is $A = A_1\cos(\omega t + \varphi)$, where A_1 is the amplitude of the first harmonic and φ, the phase shift between the applied voltage V_{tip}, and the strain-induced deformation of the material [8, 9].

To increase the magnitude of the signal (the amplitude of the oscillations), the frequency of the applied alternating field is chosen close to the frequency of the resonance oscillations of the cantilever-surface system of the sample [10]. The resonance frequency depends on the cantilever type and the local surface properties.

To study the local hysteresis loops in the spectroscopy mode, the variable V_{ac} and the constant voltage V_{dc}, $V_{tip} = V_{dc} + V_{ac}\cos(\omega t)$ are applied to the probe. The measurements were made in a pulsed mode, where the recording of the electromechanical response was registered at an alternating voltage after the DC voltage pulse was applied [11]. This method allows us to minimize the capacitive interaction of the probe with the sample by measuring EMR [12].

Studies of the domain structure and local piezoelectric loops were carried out at room temperature on a SolverNano (NT-MDT) instrument. The FMG01 probe was supplied with an alternating voltage of 3V amplitude and a frequency close to resonant. When measuring loops, the DC voltage varied from −10V to 10V and back in steps of 20mV. The pulse application time was 18ms.

2. Results and Discussion

An investigation of the temperature dependence on the dielectric constant showed a weak diffused character of the phase transition in composition 1 with a maximum temperature of dielectric permittivity $T_m = 320°C$.

The position of the maxima ε' does not change with the frequency of the external alternating field (Figure 1). Consequently, it exhibits the properties of the classical ferroelectric.

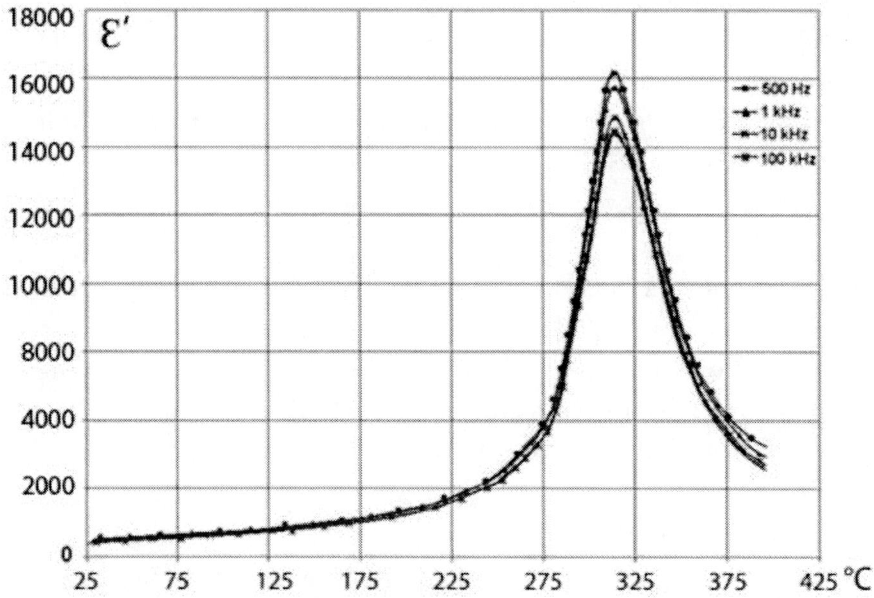

Figure 1. Temperature dependence of the dielectric constant at different frequencies for composition (1).

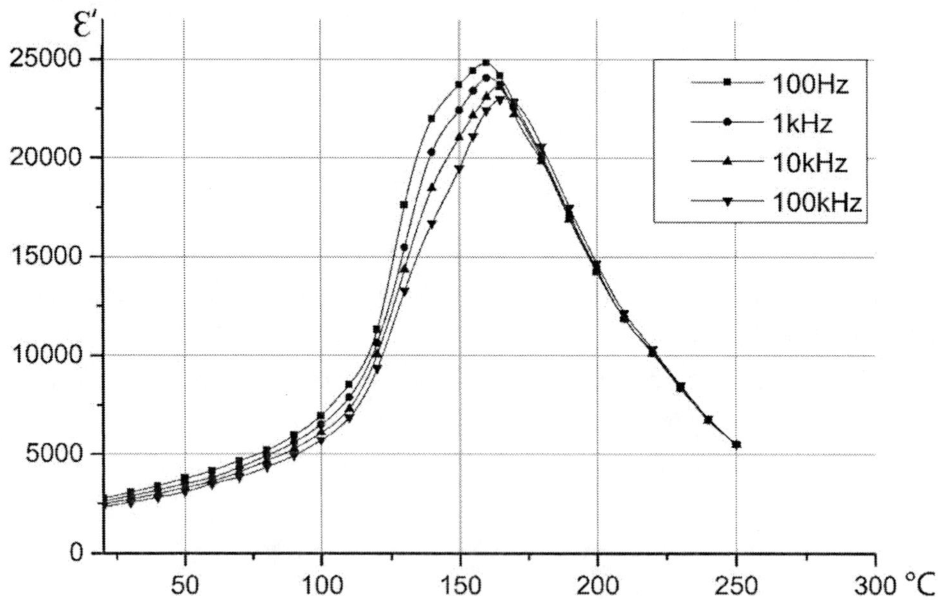

Figure 2. Temperature dependence of the dielectric constant at different frequencies for composition (2).

Composition 2 with $T_m = 160°C$ is a ferroelectric-relaxor. The maximum temperature dependence of the dielectric constant shifts to high temperatures with increasing field frequency (Figure 2) and the value of ε' decreases in this case. The temperature interval in which relaxation is observed for the frequency range 10^2 – 10^5Hz is 50°C.

The reason for the appearance of relaxor properties is structural disordering and composition fluctuations. The presence of polar nanoregions in the material are observed in unpolarized samples at room temperature and appear in polarized samples that are heated above the temperature T_m [13, 14]. In some relaxors, such as in strontium barium niobate (SBN), it manifests itself as a nanodomain structure formed in the ferroelectric phase upon cooling without the application of an external field [15, 16]. In this connection, for both compositions the microstructure and the domain structure were investigated. On the untreated surface of the ceramic, it is not possible to reveal the grain and domain structures of the materials due to a surface layer, different in structure from bulk, and forms during the production process. Therefore, a cleavage of a ceramic sample made along the polarization direction was studied on the AFM. The grains of both compounds have the form of polyhedra (Figure 3a, b) on which grown layers along one direction are clearly visible.

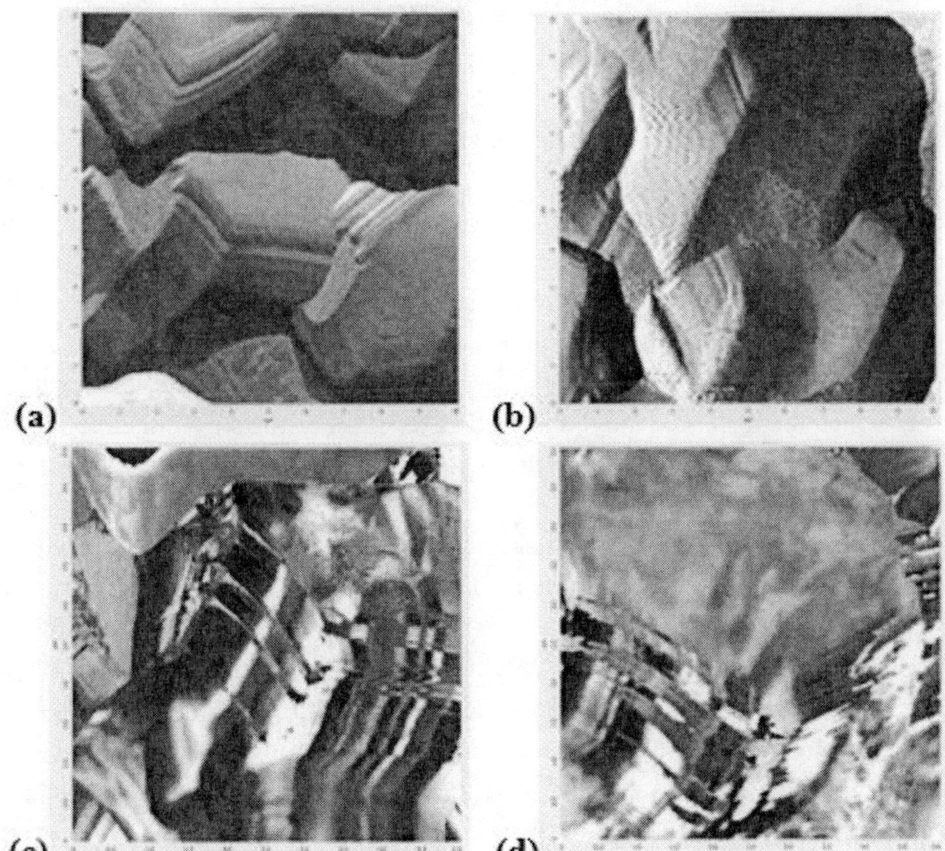

Figure 3. Topography (a, b) and domain structure (c, d) for composition 1 – (a, c) and composition 2 – (b, d); scale 10×10 μm² and (a, b) 5×5 μm² (c, d).

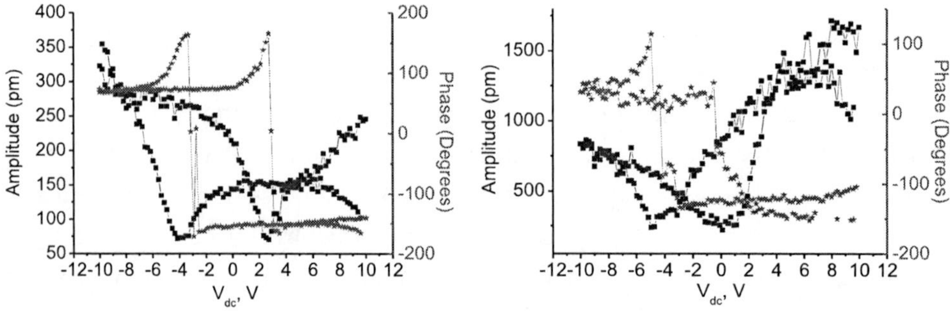

Figure 4. Dependence of the amplitude and phase of the electromechanical response on the direct voltage for (a) compositions 1 and (b) composition 2.

The grown layers are formed during the sintering of ceramics during the process of its recrystallization and are characteristic for ceramics of the PZT system [17]. The grown layers reflect the direction of recrystallization determined by the structure of the embryo. The grain size of the composition with a large value of ε' and a smaller T_m is twice as large as in ceramics obtained in other various ways. In both cases, the domain boundaries (Figure 3c, d) are perpendicular to the grown layers, therefore forming a striped domain structure. For a composition with a smaller T_m, the width of the domains is larger, and on the surface of the grain they have an irregular shape, whereas in the composition with a larger T_m, stripe, domains are seen from the grain end.

For both compositions, local loops of piezoelectric hysteresis were investigated (Figure 4). The loops were calculated from the average value of the piezoelectric response and measured at three different points on the surface of the grain end face.

For composition 1, the dependence of the amplitude and phase of the piezoelectric response on the applied DC voltage is shown in Figure 4a. For composition 2 (Figure 4b), along with a piezoelectric response similar to that shown in Figure 4a, loops can be observed, which are displaced along the voltage axis. As follows from the literature data, the displacement of local loops of piezoelectric hysteresis is characteristic for relaxor ferroelectrics, for example, SBN [18].

The presence of loops in different shapes at different points of the sample is due to the heterogeneity of the relaxor's structure. The magnitude of the signal amplitude of the electromechanical response in the relaxor is several times greater than the amplitude of the classical ferroelectric. Since the EMR is determined by the strain arising in the sample as a result of the inverse piezoelectric effect, according to (1) with the same field applied to the cantilever, the piezoelectric modulus of the ferroelectric relaxor will be larger, which agrees with the theory and the obtained electrophysical characteristics (Table 1).

According to the loops, obtained for the phase of piezo-response, in both compositions at the microlevel the coercive field voltages are small and practically identical (2.5 – 3V), as well as for the macrolevel.

Conclusion

The conducted studies have shown that the presence of relaxor properties do not have a significant effect on the value of the coercive field that characterizes the ferroelectric softness of piezoelectric ceramics. The difference in the value of the coercive field does not exceed 10%, both for macroscopic measurements and for measurements at the microlevel. At the same time, in composition 2, which is a ferroelectric-relaxor, in addition to stripe domains, the domains of irregular shape are observed. The latter is a distinguishing feature of ferroelectric crystals possessing relaxor properties [15].

References

[1] *Piezoelectric Materials, Global Technology Developments*, www.frost.com.

[2] Gupta, B. L.; Abraham, T. *ET-102 Piezoelectric Operated Actuators and Motors – A Global Industry and Market Analysis*, Innovative Research and Products, Inc., Stamford, 2006, pp. 1 - 111.

[3] Golovnin, V. A.; Kaplunov, I. A.; Malyshkina, O. V.; Ped'ko, B. B.; Movchikova, A. A. Physical Base, *Research Methods and Practical Application of Piezomaterials*, Tehnosfera Publ., 2013, pp. 1 - 272 (In Russian).

[4] Peräntie, J. *Electric-field-induced Dielectric and Caloric Effects in Relaxor Ferroelectrics*, University of Oulu, 2014, pp. 1 - 126.

[5] Smolenskij, G. A.; Krajnik, N. N. *Ferroelektrika und Antiferroelektrika*, Teubner Leipzig, 1972 (In Germanian).

[6] Akbaeva, G. M.; Dantsiger A. Ya.; Razumovskaya O. N. *Ferroelectrics,* 1994, vol. 154, 1 - 3.

[7] *Russian OST 11 0444-87*. Electrostandard, Moscow, 1987, pp. 1 - 141 (In Russian).

[8] Kalinin, S. V.; Morozovska, A. N.; Chen, L. Q.; Rodriguez, B. J. *Rep. Prog. Phys.,* 2010, vol. 73, 67.

[9] Kholkin, A. L.; Kalinin, S. V.; Roelofs, A.; Gruverman, A. *Scanning Probe Microscopy,* 2007, 173 - 214.

[10] Soergel, E. *J. Phys. D: Appl. Phys.,* 2011, vol. 44, 17.

[11] Honga, S.; Woo, J.; Shin, H.; Jeon, J. U.; Pak, Y. E.; Colla, E. L.; Setter, N.; Kim, E.; No, K. *J. Appl. Phys.,* 2001, vol. 89(2), 1377 - 1386.

[12] Ankudinov, A. V.; Titkov, A. N. *Physics of the Solid State,* 2005, vol. 47(6), 1148 - 1155.

[13] Isupov, V. A. *Physics of the Solid State,* 2003, vol. 45(6), 1056 - 1060.

[14] Samara, G. A. *J. Phys.: Condens. Matter.,* 2003, vol. 15, 367 0 411.

[15] Shvartsman, V. V.; Kleemann, W. *Physical review B,* 2008, vol. 77, 054105.

[16] Shaganov, A. P.; Vakhrushev, S. B.; Filimonov, A. V.; Ohwada, K. *St. Petersburg State Polytechnical University Journal, Computer Science,* 2013, vol. 3(177), 9 - 17.

[17] Malyshkina, O. V.; Eliseev, A. Yu.; Golovnin, V. A.; Daineko, A. V.; Ivanova, A. I.; Barabanova, E. V. *Journal of Surface Investigation, X-ray, Synchrotron and Neutron Techniques*, 2015, vol. 9(5), 899 - 903.

[18] Gainutdinov, R. V.; Volk, T. R.; Lysova, O. A.; Razgonov, I. I.; Tolstikhina, A. L.; Ivleva, L. I. *Appl. Phys. B,* 2009, vol. 95, 505 - 512.

In: Proceedings of the 2018 International Conference … ISBN: 978-1-53615-862-5
Editors: Ivan A. Parinov et al.

Chapter 20

MATERIALS FOR SURFACE ACOUSTIC WAVE DEVICES: ASPECTS FOR DEVELOPMENT

A. S. Bagdasaryan[1], V. G. Dneprovski[2,*], G. Y. Karapetyan[2] and I. A. Parinov[2]

[1]Kotel'nikov Institute of Radio Engineering and Electronics (IRE) of Russian Academy of Sciences, Moscow, Russia

[2]I. I. Vorovich Mathematics, Mechanics and Computer Science Institute, Southern Federal University, Rostov-on-Don, Russia

ABSTRACT

The range of piezomaterials, which are used to manufacture the devices on surface acoustic waves (SAWs), is considered. In particular, we consider single crystal piezoelectric substrates, thin films, multilayer film structures, and piezoceramic substrates to discuss their properties. The substrate determines the passband of the SAWs device, its thermal stability, durability, attenuation per unit length, frequency characteristics, dimensions, and topology of the interdigital transducers (IDTs). At present, single crystal substrates of various cuts of quartz, lithium niobate, and lithium tantalate are most widely used. Crystals of the langasite family and multilayer film structures are the new piezoelectric materials, which have already found wide application, in particular, for the realization of resonators, filters, etc.

Keywords: surface acoustic waves (SAWs) devices, substrate, interdigital transducers (IDTs), piezoceramics, thin piezoelectric films, monocrystalline quartz, lithium niobate

* Corresponding author: E-mail: valdnepr@ya.ru.

1. Introduction

In the process of developing surface acoustic waves (SAWs) devices, it is quite important to have a choice of the substrate material. In particular, the substrate determines the bandwidth of the SAWs device, such as its thermal stability, durability, attenuation per unit length, frequency characteristics, dimensions, and topology of the interdigital transducers (IDTs). Therefore, by designing the SAWs devices, it is very important to make the right choice of the substrate and its crystallographic cut if it is a single crystal. Currently, the monocrystalline substrates of various cuts of quartz, lithium niobate, and lithium tantalate are used most widely. At the same time, polycrystalline materials, such as piezoceramics and thin piezoelectric films, are used too. During the half-century period of acoustoelectronic development, a lot of materials for the substrates of SAW devices were proposed and many of them are now used for various applications. The characteristics of piezoelectric substrates that were widely used before the beginning of the 2000s are widely represented in the books [1-3]. Below we shall consider the further research in this field.

2. Monocrystalline Piezoelectric Substrates

The single crystals are distinguished into perfection of the structure, which provides small losses for the SAW propagation (about 0.1 - 0.5 dB/cm at frequencies up to 2.5 GHz). In addition, they are stable in time. Single-crystal quartz, lithium niobate, and lithium tantalate are widely used as piezomaterials for SAWs devices [4]. In the narrow-band filters with a bandwidth of up to 5 - 6%, quartz of various cuts are most widely used. In addition, quartz is characterized by high-temperature stability, especially the ST-cut. For the acoustic lines of filters with a bandwidth of up to 50 - 60%, lithium niobate $LiNbO_3$ is mainly used. Among monocrystalline materials, bismuth germanite $Bi_{12}GeO_{20}$, lithium iodate, bismuth silicate, paratelluride TeO_2, berlinite $AlPO_4$, gallium arsenide GaAs, selenium Se, as well as ZnO films, AlN films on sapphire, and some others can be classified as single-crystal materials for use in SAWs devices. Some of them have been successfully used for the development of SAWs devices. Recently, interest in materials, such as lithium tetraborate $Li_2B_4O_7$, langasite $La_3Ga_5SiO_{14}$, langatate $La_3Ga_{5.5}Ta_{0.5}O_{14}$, gallium orthophosphate $GaPO_4$, and catangasite CTGS - ($Ca_3TaGa_3Si_2O_{14}$), has been increasing [5]. Lithium tantalum $LiTaO_3$ is perhaps the only material in which high piezoelectric activity is so well combined with good thermal stability. Bismuth germanium is a suitable material for the delay lines of long duration due to the very low SAW propagation velocity and large dimensions of the produced crystals. Unfortunately, one of its drawbacks is its tendency to thermal shock, which makes it difficult to form the device based on it. The Curie temperature of lithium niobate (1165°C) is large and, at first glance, this crystal can be used for high-temperature measurements. However, for the use of lithium niobate in high-temperature sensors, there is one significant limitation: it has a significant intrinsic conductivity at the temperatures above 600°C [6].

Crystals of the langasite family are new piezoelectric materials that have found wide application, in particular, for the realization of resonators, filters, and sensors of various types, since they can successfully replace traditionally used ceramics and quartz. The crystals have high acoustic Q-factor, zero temperature coefficients of the first-order frequency (TFC), and high coupling coefficients [7].

Recently, big size high quality piezoelectric single crystals of $Ca_3TaGa_3Si_2O_{14}$ (CTGS - catangasite) have been grown [8]. This temperature durable crystal is very promising for future application in the different devices based on surface acoustic waves due to its high electromechanical coupling compared to quartz. In contrast to the popular langasite, CTGS is a more ordered compound and therefore stands for better characteristics at both high temperatures and high frequencies. During the last decade, several sets of material constant for CTGS were published and extracted from bulk acoustic wave measurements on small samples, but these results are strongly different from each other. Moreover, all published constant sets lead to unacceptable numerical differences for calculated SAW velocities when compared to recent measurements of SAW velocities on large CTGS wafers. In [8], a new set of room- temperature CTGS material constants extracted by a combination of high-precision SAW phase velocity measurements with proprietary numerical procedures developed in-house, are present.

For a functionally graded piezoelectric substrate with exponential variation, the existence and propagation of transverse surface waves were studied by the analytical technique [9]. The dispersion equations for the existence of the transverse surface waves with respect to phase velocity are obtained for both electrically open and short conditions. A detailed investigation of the effect of gradient coefficient on dispersion relation, phase velocity, group velocity, and electromechanical coupling factor is carried out. Because of the negligible initial stress in functionally graded piezoelectric materials, this model could serve as an excellent substitution for the typical layered piezoelectric structures used in SAWs devices, thus providing a theoretical foundation for designing SAWs devices with high performance.

In contrast to quartz, which has a low temperature of αβ-phase transition (350°C), the operation temperature of catangasite is limited only by the melting point of the material 1350°C [10]. At the same time, catangasite is more chemically stable than langasite, which has an advantage with the viewpoint of technology. Investigations of the electrophysical properties of the material have shown that the most promising use in SAWs techniques are (0°, 90°, 40° – 50°)-cuts of catangasite, which provide a coupling coefficient of 0.1% – 0.15% and have a fairly-good temperature stability, which makes it possible to treat this material as an alternative material to silica SiO_2 and langasite

$La_3Ga_5SiO_{14}$. The article [10] presents the results of development and research of SAW filters for nominal frequencies of 125 MHz and 140 MHz realized on the base of two cuts of catangasite.

3. THIN FILMS AND MULTILAYER FILM STRUCTURES

The piezoelectric zinc oxide films and aluminum nitride films on the sapphire make it possible to use a non-piezoelectric material (sapphire) for both the acoustic waveguides of SAW devices and the substrates forming the structure of amplifying cascades [4]. Films of AlN are promising for high-frequency filters and resonators due to the high velocity of the SAW in this material. Using piezoelectric films on non-piezoelectric substrates can significantly reduce the cost of SAW devices.

In [11] the devices on surface acoustic wave, based on the layered structures comprising PFD and piezoelectric film, AlN are present. Polycluster films of diamond (PFD) were formed from the gas phase hydrogen and methane, activated arc or microwave discharge, and heat (the method of "hot-wire") on Si- and W-substrates. According to experts, the most promising area of PFD application in electronic engineering is the microwave acoustoelectronics, since PFD have a high SAW velocity of Rayleigh type (VR ≈ 10 km/s). PFD, because of their phenomenal elastic characteristics, are promising for use in SAW devices, operating in the frequency range > 1 GHz, in which only single crystals were used to reduce losses.

To create high-frequency SAW devices as a piezoelectric, AlN is a promising material with the nearest diamond velocity (VR ≈ 6 km/s) and strong piezoelectric effect (coefficient of electromechanical coupling k2 ≈ 0.8%). The use of layered substrate/AlN structures, where the substrate is a material with a high SAW velocity, and AlN as a piezoelectric layer, allows one to increase the upper limit of the frequency range of SAW devices. Currently, with the use of PFD, SAW filters have been made which operate at a frequency > 3 GHz [11]. Figure 1 shows the SAW filter on the layered structure of PFD/AlN/Si(W).

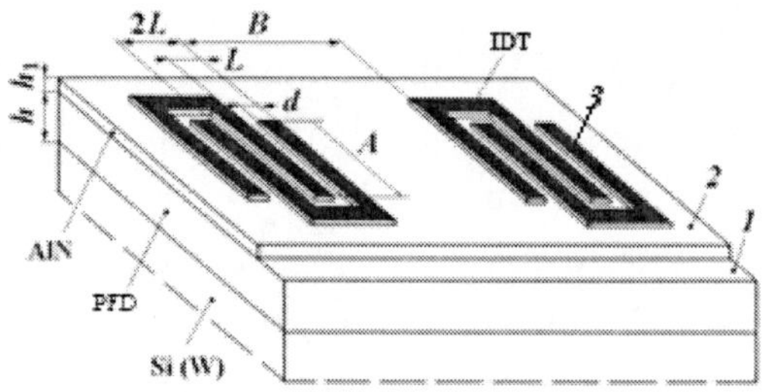

Figure 1. SAW filter on the layered structure of PFD/AlN (1 – polished surface of PFD, 2 – growth surface of AlN, 3 – interdigital transducer, IDT) [11].

Until recently, the effective method for the implementation of thermocompensated middle band and wideband SAW devices was to use $SiO_2/LiNbO_3$ layered structures. However, additional insertion loss (from 2 to 5 dB), inherent in this method, and complex manufacturing technology significantly limits the scope of its application [12].

Further studies in this direction have made it possible to create a new type of thermally compensated layered material [13], which provides a high coupling coefficient in combination with low TFC. As a carrier substrate, a material with a low coefficient of thermal expansion (silicon or sapphire) is used, which is connected by diffusion or by means of a thin layer of glue with a piezoelectric plate, which is then ground to the required thickness. Thus, the plates of NANO'LN (China) manufactured have a thickness of the piezoelectric film that varies between 300 and 700 nm for lithium niobate and 5 – 50 μm for lithium tantalate. In [12], comparative studies of SAW filter samples at a nominal frequency of 800 MHz with a bandwidth of 4.5%, made on a 42° $LiTaO_3$ & Si layered plate with a 42°-cut $LiTaO_3$ single crystal film of thickness 20 μm, and a 42°-cut $LiTaO_3$ standard plate, were conducted. Investigations of the temperature stability of filters in the temperature range from minus 60°C to plus 85°C showed more than a twofold improvement in the TFC for the filters realized on the layered plate.

4. Piezo-Ceramic Substrates

The principal shortcoming of piezoceramics in comparison with the single crystals is the considerable attenuation of SAW propagation, which increases sharply with frequency. From all the technological methods of the manufacture of piezoceramic goods, only hot pressing increases the density of the samples to 90 - 95% of the theoretical; wherein the scattering of the values of the electrophysical parameters of the material decreases. However, even when using hot pressing, the reproducibility of piezoceramic properties are low in comparison with single crystals. This circumstance makes it possible to use piezoceramics for the manufacture of acoustic lines that are only broadband SAW devices with non-rigid performance characteristics, where the values of temporary instability and reproducibility of properties have an insignificant value. In particular, piezoceramic TV SAW filters were successfully implemented in the serial production of enterprises "Horizont" (Minsk, Belarus), "Murata" (Japan) et al. [1, 14].

SAW filters for the tract of IF TV receivers were manufactured and tested on the substrates from hot-pressed piezoceramics, which had a reduced attenuation of SAW at the frequencies of the IF path due to a reduction in the number of pores. There were two types of piezoceramics: PCR-53 (V_{SAW} = 2440 m/s, k^2_{eff} = 2.2%, SAW attenuation per unit length α_p = 3.2 dB/cm) and PCR-55 (V_{SAW} = 2600 m/s, k^2_{eff} = 2.2%, the attenuation per unit length of the SAW α_p = 2.1 dB/cm). These filters had insertion attenuations of 25

[15] Karapetyan, GY; Panich, AE; Rybyanyets, AN; Khasabova, GI. *Conf. Proc. "Acoustoelectronic Devices for Processing Information on Surface Acoustic Waves".* Cherkassy, Ukraine. 1990: 133-134 (in Russian).

In: Proceedings of the 2018 International Conference … ISBN: 978-1-53615-862-5
Editors: Ivan A. Parinov et al.

Chapter 21

MICROSTRUCTURE ANALYSIS OF HYPOEUTECTIC AL-CU ALLOY

***Supardi*[1,*] *and Djoko Sulistiyono*[1]**
[1]Department of Mechanical Engineering, University of 17 Agustus 1945 Surabaya, Jawa Timur, Indonesia

ABSTRACT

This study analyzed an effect of temperatures of casting and mold on the existing structure of the material. The hypoeutectic Al-Cu is an aluminum and copper alloy with eutectic phase, which offers advantages of melting and pouring (especially around the eutectic region) at a temperature of 548°C. Aluminum, one of reactive metals, is easily oxidized and form a layer of aluminum oxide, named alumina (Al_2O_3), thus has good corrosion resistance. When added with Fe, Cu, and Ni, however, it can reduce the corrosion-resistant properties as the alumina content decreases. Copper can form proper eutectic phase with aluminum at fraction of 33%. Copper alloys (Al/Cu), containing 10 – 20% of Al, indicate exceptional strength, for hardness and wear, adjusting some steel alloys. By using gravity, casting method in a permanent mold will be easily dismantled. The material composition was made of 3.375 grams of aluminum and 1.125 grams of copper.

The pouring temperature used was 600°C, 650°C, and 700°C, while the mold temperature used was room temperature at 100°C, and 200°C. The specimen was divided into A', B, and A'' codes at the time of cutting using lathe and cutter. Pouring temperature and mold temperature will affect the microstructure formed. The higher the pouring temperature, the clearer the structure visible at the pouring temperature of 700°C. The smaller the mold temperature, the rougher the shape of the structure and the more apparent the outward appearance; in other words, a dendrite structure is formed. Codification B is the core or midpoint of the mold of the farthest point of the cooling rate process. The structure formed in the B-code is clear, thick and coincident.

[*] Corresponding Author's E-mail: supardis@gmail.com.

Keywords: hypoeutectic Al-Cu, dendrite, microstructure, mold temperature, pouring temperature

1. Introduction

The Hypoeutectic Al-Cu alloy area is called hypoeutectic, that is, if the alloy contains Cu < 32.7%, spade, etc. to extract (dredge) mud [1]. Aluminum is a lightweight metal weighing 2.7g/cm^3 after magnesium (1.7 g/cm^3) and beryllium (1.85g/cm^3), consisting of 1/3 of the specific gravity of iron or copper. It has 60% of electrical conductivity higher than copper; thus used for electrical equipment. Aluminum has a good property of heat conductor, therefore it is used on machine components, and so on [2].

Additionally, aluminum is a reactive metal that is easily oxidized with oxygen to form a layer of aluminum oxide, alumina (Al_2O_3), thus has good corrosion resistance. Moreover, adding Fe, Cu, and Ni will help reduce the corrosion-resistant properties, for the alumina content decreases [3]. This does not apply to the addition of Mg, Mn, which will not affect its corrosion resistance properties. Aluminum is also ductile, easily machined and formed with tensile strength for pure aluminum around 4 – 5kgf/mm^2. The mechanical strength of aluminum can be optimized by adding combination elements such as Cu, Mg, Zn, Mn, and Ni. The element Cu in Al alloys, particularly, can improve the mechanical properties of namely hardness and tensile strength yet reduce its ability to cast. Further integration with silicon is able to improve the level of liquidity (fluidity) and can reduce shrinkage defects, that may positively affect castability (weldability).The following are properties of aluminum used as a technical material.

(i) resistant to corrosion;
(ii) good heat and electricity delivery;
(iii) has light specific gravity (only 2.7g/cm^3, while iron has ± 8.1g/cm^3);
(iv) easy to form (fabricate).

Aluminum can increase hardness and strength, and the machining capability at about 12% of Cu content. Al-Cu is too brittle for the technique above this mixture level. At conditions of equilibrium, up to 5.65% of Cu can dissolve in aluminum at binary Al-Cu alloy temperatures. This solid solubility will drop to less than 0.1% at room temperature and reduced solid dissolution can explain good deposition for heat treatment. 4.5% of Cu in copper-aluminum alloy creates mechanical properties and allows good machining but leads to poor cast. Copper-silicium aluminum alloys can be created by adding 4 - 5% of silicium to copper aluminum alloys to improve their castability. Additionally, this alloy can be used for parts of motorcycles, cars, gauges and main parts of valves [4].

Metal casting is a manufacturing process in which liquid metal and molds are used to produce shapes which resemble the final geometric shape of the finished product. Liquid metal is pressed into a mold with a cavity created following a desired shape or design. After the molten metal meets the print cavity, the mold is removed and the cast results can be used for secondary processes.

In the metal casting process, several steps are to be taken to produce metal molded products. The metal material to be melted is heated in a heating kitchen with a certain temperature until it melts. The casting process includes the following stages: (i) preparing tools and materials; (ii) heating the melting furnace to a temperature of 400°C; (iii) entering 1.125kg of aluminum into the melting furnace; (iv) reheating the melting furnace to a temperature of 600°C, 650°C, 700°C (according to the Hypoeutectic Al-Cu temperature variation); (v) inserting Cu according to a variation of 25% from 500 grams, that is 125 grams of Cu per temperature variation pouring; (vi) stirring by using a stirrer; (vii) preparing a mold, heated according to the temperature variation of the mold, that is, a room temperature, 100°C, 200°C; (viii) taking the molten metal from the melting furnace by using cawing pour for the pouring process into the mold; (ix) starting the process of pouring the liquid metal into the mold; and finally, (x) after the casting metal has hardened, the metal mold is unloaded and the casting material is taken off [5].

Permanent printing is a mold intended for continuous uses. This mold consists of parts that are combined with screws, plates, claims and other tools that can be removed after the metal hardens. As the finishing, this permanent mold is coated with a heat and soot resistant material to keep it from sticking. Microstructure illustrates numerous phases that are observable through metallographic techniques. The microstructure of metals can be observed with the help of an optical microscope. Before observing with a microscope, the metal surface must be cleaned.

2. EXPERIMENT

2.1. Metal Casting

The main stages of metal casting include:

(i) preparing the tools and materials;
(ii) heating the melting furnace;
(iii) inserting all aluminum into the melting furnace;
(iv) reheating the furnace to a temperature of 600°C, 650°C, 700°C (according to the hypoeutectic Al-Cu temperature variation);
(v) incorporating Cu according to 25% variation from 500 grams, that is 125 grams of Cu for each variation of casting temperature, followed by mixing with stirrer;

(vi) preparing the heated mold according to the temperature variation of the mold, that is at room temperature, 100°C and 200°C;
(vii) taking off the molten metal from the melting furnace by using casting cast;
(viii)pouring the molten metal into the mold;
(ix) unloading the mold and grabbing the foundry.

2.2. Metallography

The main stages of metallography include:

(i) preparing a specimen with the codification A', B, A";
(ii) rubbing the specimen by using scouring paper with the level of highest order 5000 to the lowest 300;
(iii) if it looks shiny and the fiber from aluminum is not visible, the specimen is rubbed with bludruh fabric;
(iv) applying Autosol to make the surface tested (polished) shiny;
(v) preparing etching by measuring and pouring with the cup and applying Etsa using Aquades (45ml), HCl (1.5ml of HNO_3) (25ml), and Hf (1ml);
(vi) performing the etching by dipping for 8 – 16 seconds the surface of specimen that has been rubbed and polished, and drying the specimen;
(vii) observing the microstructure, using an optical microscope;
(viii)analyzing the microstructure image according to the variation of the mold temperature and pouring temperature.

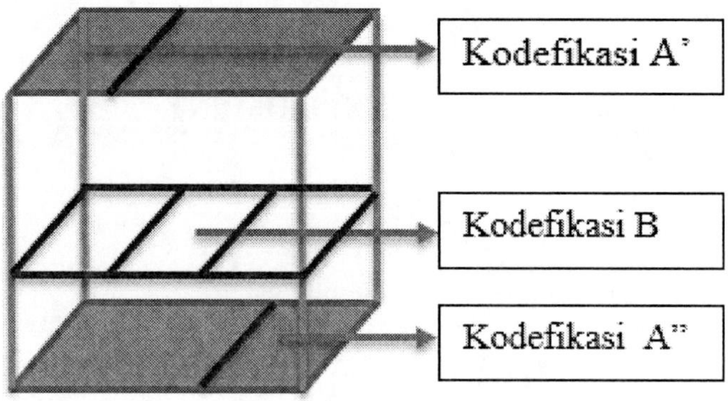

Figure 1. Coding of the specimen

The specimen is divided into A', B, and A" codes (see Figure 1) at the time of cutting using lathe and cutter.

3. Results and Discussion

At the mold temperature of room temperature (codification A"), the dendrite structure begins to form at the pouring temperature of 700°C (see Figure 2) as illustrated by rough and round shape. While at the mold temperatures of 100°C and 200°C, the structures formed are almost the same. The difference consists in that at 200°C where the mold temperature is clearer and longer due to the slow cooling rate.

At 200°C the mold temperature and 650°C pouring temperature (codification B), the formed structure looks clearer and smoother than that at the mold temperature of room temperature and of 100°C due to slow cooling rate (see Figures 3, 4). Also, the distance between structures looks denser.

At the temperature of casting 600°C (codification B) and the mold temperature of room temperature, the same dendrites are formed with elongated structure.

The distance between structures is denser due to the slow cooling rate. The difference is in that at a pouring temperature of 600°C, 650°C and 700°C, where the structure is clearer (see Figure 3).

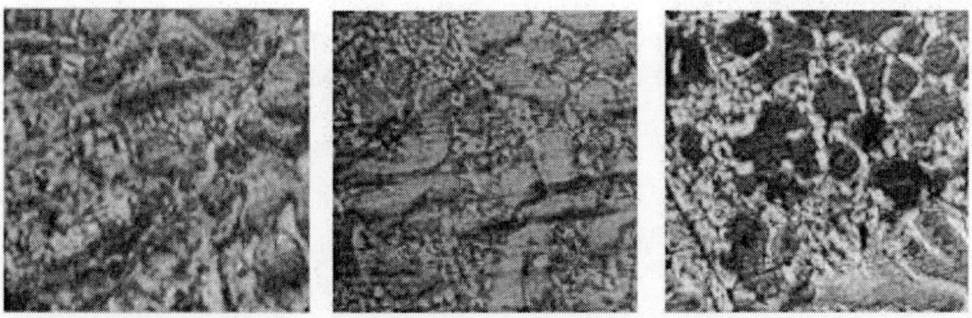

Figure 2. Microstructure of surface view (code A) with 600, 650, and 700°C pouring temperature (from left to right), at room temperature molds.

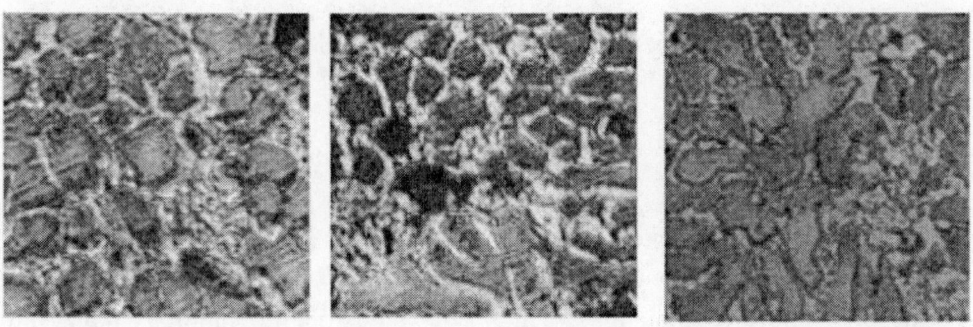

Figure 3. Microstructure of middle view (code B) with 600, 650, and 700°C pouring temperature (from left to right), at room temperature molds.

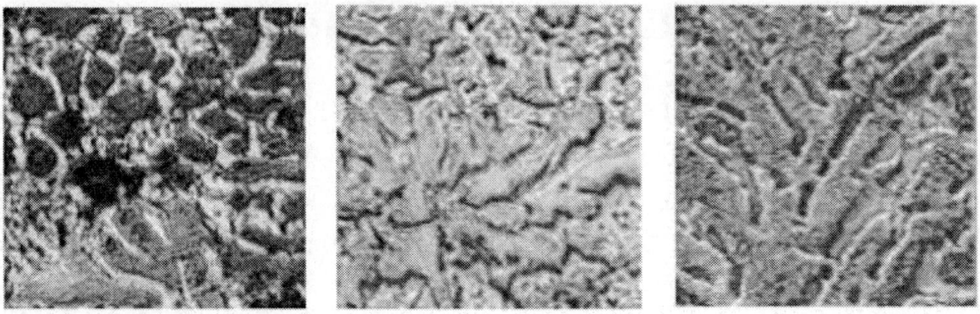

Figure 4. Microstructure of middle view (code B) with 600°C pouring temperature, at 100, 150, and 200°C mold temperature (from left to right).

Conclusion

(1) Higher pouring temperature and mold temperature will result in columnar structure.
(2) Greater cooling rate from the mold wall will likely gradually change microstructure.
(3) The structures of hypoeutectic Al-Cu and hypereutectic Al-Cu are not very different; only that the structure of hypoeutectic Al-Cu tends to be columnar and regular.

Acknowledgments

The authors thank The Ministry of Research, Technology, and Higher Education of Republic of Indonesia for funding this research.

References

[1] Surdia, T., Chijiwa, K. (1980). *Teknik Pengecoran Logam* [*Metal Casting Techniques*]. Jakarta: Pradya Paramita.

[2] Budiyanto. (2008). *Temperatur Penuangan Paduan Al – Si (Seri 4032) Terhadap Hasil Pengecoran* [*Temperatures for Casting Al-Si Compund (4032 Series) towards Casting Products*]. Malang: Faculty of Industrial Technology of National Technology Institute of Malang.

[3] Wijaya, M. T., Zubaidi, Wijoyo. (2017). *Temperatur Tuang terhadap Ketangguhan Impak dan Struktur Mikro pada Pengecoran Aluminium* [*Casting Temperatures on*

Impact Resilience and Microstructures in Aluminium Casting]. Surakarta: Faculty of Industrial Technology of Surakarta University.

[4] Raharjo, T. Sujana, W. (2010). *Analisis Pengaruh Temperatur penuangan dan Temperatur Cetakan terhadap Sifat Mekanis bahan paduan Al-Zn* [*An Analysis of Effects of Casting and Mold Temperature on Mechanical Properties of Al-Zn Compound*]. Malang: Faculty of Industrial Technology of National Technology Institute of Malang.

[5] Triajati, L. (2008). *Mekanisme Fading* [*Fading Mechanism*]. Jakarta: Faculty of Engineering of University of Indonesia.

In: Proceedings of the 2018 International Conference … ISBN: 978-1-53615-862-5
Editors: Ivan A. Parinov et al.

Chapter 22

EXPERIMENTAL STUDY OF OPTICAL INTERFERENCE METER FOR SMALL DISPLACEMENTS

I. P. Miroshnichenko[1,*]***, I. A. Parinov***[2] ***and V. P. Sizov***[1]
[1]Don State Technical University, Rostov-on-Don, Russia
[2]I. I. Vorovich Mathematics, Mechanics and Computer Sciences Institute, Southern Federal University, Rostov-on-Don, Russia

ABSTRACT

The paper describes in detail the results of an experimental study of an optical interference meter for small displacements of surfaces of test objects, based on a two-way laser interferometer with combined branches. The analysis of the obtained measurement results is performed. Technical proposals for improving the quality of measurement results when solving practical measurement problems are formulated.

Keywords: laser interferometer, interference pattern, displacement meter, thin-layer hologram, small displacements

1. INTRODUCTION

In works [1 – 4], an optical interference meter for small displacements of the control object surfaces, based on a two-way laser interferometer with combined branches, was proposed and grounded. Experimental studies of the noted meter were carried out. The objectives of the research were experimental confirmation of the technical advantages of

* Corresponding Author's E-mail: ipmir@rambler.ru.

using thin-layer holograms as a beam splitter and the development and justification of technical proposals for improving the quality of measurement results in solving practical measurement problems.

2. Experimental Setup and Experimental Method

An experimental setup was used for the studies, the scheme of which is shown in Figure 1. This setup includes a radiation source 1 (He-Ne laser, wavelength $\lambda = 0.63$ μm), a collecting lens 2, a spatial filter aperture 3, a beam splitter 4, a reflector 5, a screen 6 and a photodetector 7. The reflector 5 is fixed to the end surface of the cylindrical piezoelectric transducer 8, which is electrically connected to a source of stabilized voltage 9 and a universal digital voltmeter 10, but a photodetector 7 is electrically connected to a device for recording and processing results 11.

The setup works as follows. The laser radiation 1, after passing through the lens 2, is converted into a diverging beam. The beam splitter 4 carries out splitting the beam according to amplitude: one part is reflected from its surface (direction $S - S$), and the other from the surface of the reflector 5 (direction $S^* - S^*$). The beams thus formed are coherent. Regulators of the fastening mechanism 12 of the piezotransducer 8 achieve spatial alignment of the object and reference beams in the plane of the screen 6, on which the interference pattern is formed. By controlling the voltage on the piezoelectric transducer 8, we can set the movement of the reflector 5, which leads to a change in the parameters of the interference pattern, which is recorded by the photodetector 7 and processed in the device 11.

The experimental procedure consisted in the step-by-step statement of moving the reflector 5 in the range from 0 to λ with step $\lambda/30$. The signal from the photodetector 7 was recorded at each step of moving the reflector 5. The digital images of the interference pattern (interferogram) were transmitted to the device 11 in the form of standard graphic image formats.

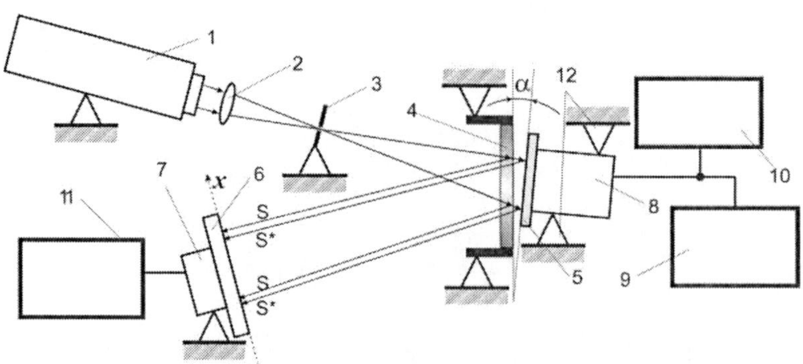

Figure 1. Scheme of experimental setup.

3. Experimental Results and Their Discussion

The result of the observation was an interferogram, which in digital form is an image, discretized in rows i ($i = 1, \ldots, k_Z$) and columns j ($j = 1, \ldots, k_X$). The values of the image elements are quantized in the interval $[0, \ldots, (2^K - 1)]$, $K = 8$, i.e., 8-bit level quantization is used.

During the experiment, two types of beam splitters made from a single holographic HFG-03M plate were investigated: a beam splitter 1 in the form of a plate with a thin-layer hologram, exposed in colliding beams, and a beam splitter 2 in the form of a plate, exposed in the same conditions, but without a holographic scheme.

The process of making a hologram is quite laborious and multistage. At each stage, errors are accumulated, which introduce distortions in the interference pattern. Therefore, all beam splitters underwent further processing by using the same special method simultaneously.

A typical interferogram for the beam splitter 1 is shown in Figure 2a, and for the beam splitter 2, in Figure 2b.

For greater clarity of the results, one-dimensional intensity distributions in the cross-section of the interferogram were analyzed, passing through its center, shown in Figure 3 for two positions of the reflector differing by $\lambda/4$, while the phase of the object beam changed on π. The amplitude values of the measuring signal (dynamic range in intensity) were determined in accordance with the expression:

$$\Delta I_j = \left| I_j^0 - I_j^{180} \right|, \tag{1}$$

where I_j^0, I_j^{180} are the intensity values at the specified positions of the reflector (in Figure 3, curves 1 and 2); j is the X-axis image element number in pixels.

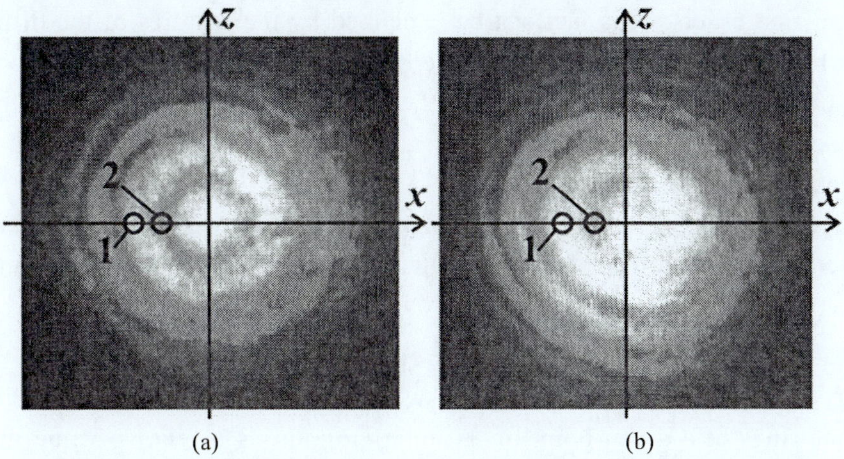

Figure 2. Typical interferogram: (a) beam-splitter 1; (b) beam-splitter 2.

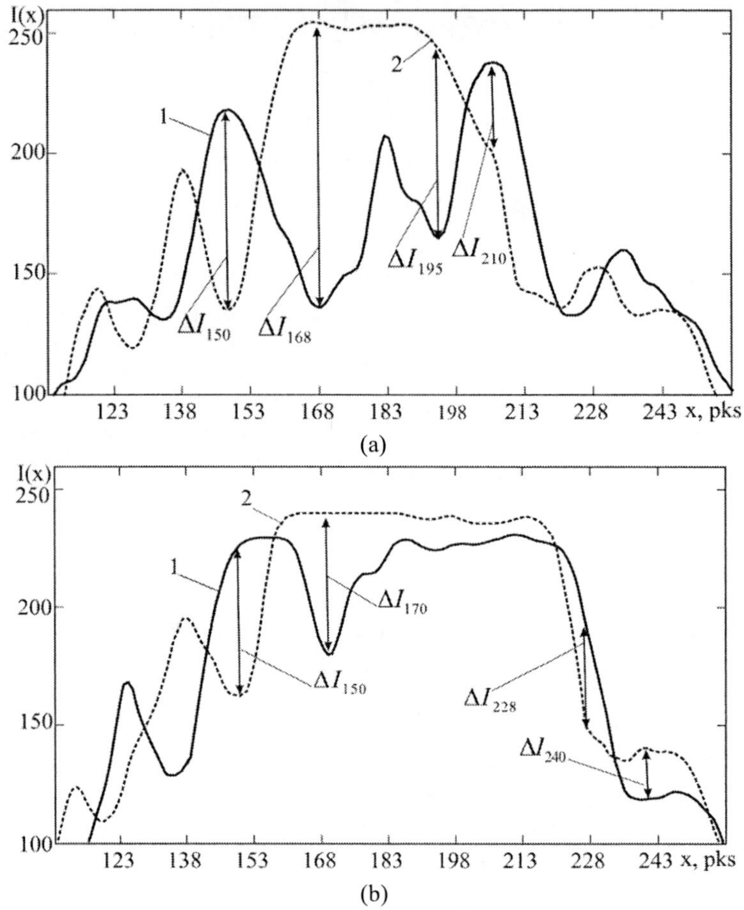

Figure 3. Intensity distribution in the horizontal section of the interferogram (along *X*-axis, see Figure 2): (a) beam-splitter 1; (b) beam-splitter 2

From the analysis of Figure 3 and the calculated results, it follows that the greatest changes in intensity when the reflector moves on $\lambda/4$ take place in the interference rings of zero and first orders. This fact can be explained by the features of the diffraction of light on a hologram as a result of which the bulk of the optical radiation is concentrated in the central part of the image. Regarding the beam splitter 2, it should be noted that the dynamic range of the interferogram is lower, especially in the central region, as can be clearly seen from Figure 3b.

The dynamic range in the interference rings of the same order for different regions of the interferogram and different beam splitters has different values. For example, it is fooled from Figure 3a that $\Delta I_{150} > \Delta I_{210}$; $\Delta I_{168} > \Delta I_{195}$. Values of I_{max} and I_{min} for rings of different orders and areas of the interferogram also have different values. Therefore following [5], for comparative analysis, we take the measure of the definition of interference rings at a given point of the interferogram as a relative value where their "visibility" is defined as:

$$\gamma = \frac{I_{max} - I_{min}}{I_{max} + I_{min}}, \tag{2}$$

where I_{max} and I_{min} are the maximum and minimum intensity values in the immediate vicinity of a given point.

The distribution of the "visibility" γ_i of the interference rings in the section under consideration was determined in accordance with the expression:

$$\gamma_j = \left| \frac{I_j^0 - I_j^{180}}{I_j^0 + I_j^{180}} \right|. \tag{3}$$

Such a calculation is valid, since when the phase difference between the reference and object beams changes on π, the intensity of the interference rings completely changes.

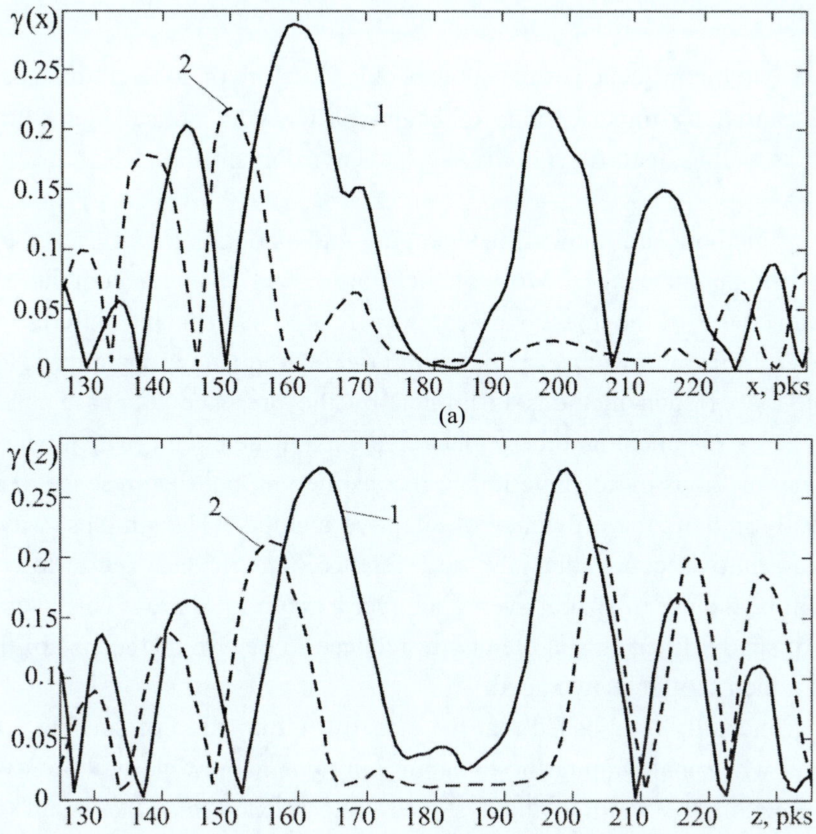

Figure 4. Values of the "visibility" of interference rings (curve 1 – beam-splitter 1, curve 2 – beam-splitter 2): (a) along the *X*-axis; b) on the *Z*-axis.

From the analysis of the distributions (Figure 4), determined by the formula (3), it follows that the values of "visibility" for interference rings of the same order, calculated on the left and right of the *Z*-axis (Figure 2) are significantly different. This feature has a systematic manifestation, is characteristic of both types of beam splitters, and can be explained by the influence of the angle α between the reflector and the beam splitter. As a result, in the area farther from the rib of wedge, formed by the beam splitter and reflector planes, it is taken place a reducing contrast and blurring interference rings [5]. To verify this assumption, a series of similar experiments were carried out when the beam splitter was turned on $\pi/2$. In this case, the pointed feature was preserved. However, in the study of the distributions in the orthogonal cross-section of the interferogram (along the *Z*-axis), the "visibility" in the rings of the corresponding orders had approximately the same values, regardless of the position and type of beam splitter.

The result of measurement during the experiment is the dependence of the intensity of the optical field at a given point or fragment of the interference pattern on the movement of the reflector. In this case, the measurement result is influenced by the following factors, associated with the formation, registration, and processing of interferograms:

- errors due to the influence of the spot speckle structure of coherent images;
- errors, caused by imperfections of beam-splitters and optical elements of the circuit, as well as distortions in the external environment.

A number of methods are known for averaging and smoothing speckle noise during the formation of coherent images. Most of them are based on the accumulation of an ensemble of realizations of the image's object uncorrelated in a speckle pattern, followed by averaging. With this aim, we use spatial, temporal or frequency separation. However, the implementation of such methods is limited by the presence of phase distortions, introduced by the environment and the optical elements of the circuit, which necessitates the use of special methods of adaptive image reconstruction, but the presence of speckle noise significantly impairs the efficiency of adaptive methods. The simplest way in this situation, taking into account that the interference pattern has quite low spatial frequencies, will be the use of spatial averaging over a given local area. This averaging is a special case of spatial filtering and the task is reduced to determine the size of the local image area, on which should be averaged.

It was experimentally established that the effective elimination of speckle noise and phase distortion, while maintaining the dynamic range, is achievable with an averaging window not exceeding the width of the first-order interference ring. The result obtained explains the dependence of change in the intensity of the optical field on the movement of the reflector, shown in Figure 5. Figure 5a shows changes in intensity, determined in each of the three adjacent image elements: curve 1 corresponds to an element (i, j), located in the middle of a dark first-order ring on the *X*-axis; curve 2 corresponds to an element $(i + 1, j)$; curve 3 corresponds to an element $(i + 2, j)$. In this case, the error in determining the displacement from the error of setting the measurement point (for example, Δ1 – Δ4) has a significant value with a random manifestation. Figure 5b shows the change in intensity, calculated as an average over the local areas of the size $n \times n$ elements with centers at the points used to construct the curves 1, 2 and 3 (Figure 5a). Respectively, here the error is less and has systematic character.

The analysis of the results shows that the dynamic range of intensity at a quantization level of 8 bits includes approximately 100 intensity gradations (Figure 5), which corresponds to the displacement of the reflector at a distance of $\Delta h = \lambda/4 \approx 1.583 \times 10^{-7}$ m. In this case, we can talk about the sensitivity threshold of the meter's order of 1.6×10^{-9} m.

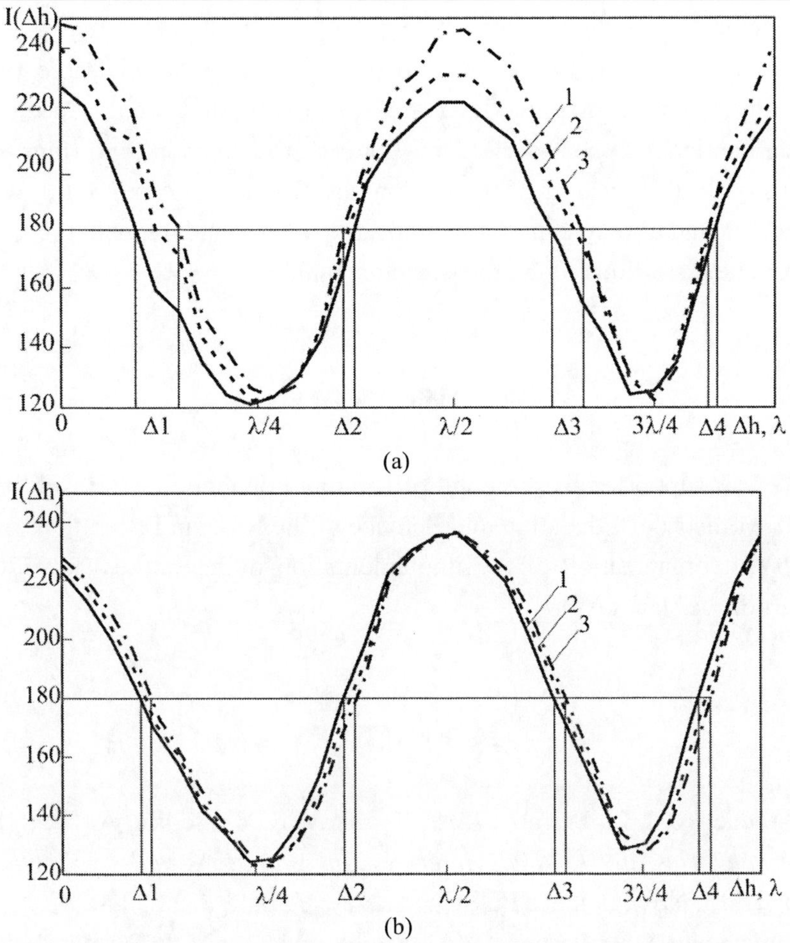

Figure 5. Dependences of the intensity on the displacement of the reflector: (a) defined in each of the three adjacent elements to image; (b) average over the local area with size of $n \times n$, where (n = 5).

Conclusion

The analysis of the results of the experimental study described allows us to formulate the following proposals for improving the quality of measurement results through solving practical measurement problems:

- it is advisable to use thin-layer holograms as the beam-splitters, which allows for a relatively higher contrast and dynamic range of the interference pattern;
- during measurements, photodetectors should be placed in the rings of the zero and first orders, using the area of the interferogram with the highest dynamic range;
- by using the digital registration of interferogram, for example, using CCD photodetectors, it is necessary to apply spatial averaging of intensity over the local image area not exceeding the width of the interference ring;
- to increase the signal-to-noise ratio of the meter, it is advisable to record the intensity, averaged over two local areas, that is located at a distance of half of the spatial period of the interference pattern, i.e., in adjacent rings of opposite intensity. In this case, the information signal is the intensity ratio, which allows us to expand the dynamic range of the meter and reduce the sensitivity to amplitude distortions of the measuring signal.

Acknowledgments

The work was supported by the grant of Russian Foundation for Basic Research, No. 16-08-00740, Ministry of Education and Science of the Russian Federation: projects Nos. 3.5378.2017/VU (organization of the implementation of researches), 9.9770.2017/BCh (the basic part of the state task).

References

[1] Miroshnichenko, I. P.; Parinov, I. A.; Rozhkov, E. V.; Serkin, A. G.; P. Sizov, V. P. In: *Piezoelectrics and Related Materials: Investigations and Applications*. Ivan A. Parinov (Ed.), Nova Science Publishers, New York, 2012, 239 – 282.

[2] Miroshnichenko, I. P.; Parinov, I. A.; Rozhkov, E. V.; V. P. Sizov; Shevtsov, V. A.. In: *Physics and Mechanics of New Materials and Their Applications*. Ivan A. Parinov, Shun-Hsyung Chang (Eds.), Nova Science Publishers, New York, 2013, 145 – 154.

[3] Miroshnichenko, I. P.; Parinov, I. A.; Chang, S. H. In: *Advanced Materials. Studies and Applications*. Ivan A. Parinov, Shun-Hsyung Chang, Somnuk Theerakulpisut (Eds.), Nova Science Publishers, New York, 2015, 437 – 451.

[4] Miroshnichenko, I. P.; Parinov, I. A.; Rozhkov, E. V.; Chang, S.-H. In: *Advanced Materials Manufacturing, Physics, Mechanics and Applications*, Ivan A. Parinov, Shun-Hsyung Chang, Vitaly Yu. Topolov (Eds.), Springer *Proceedings in Physics,* Springer Cham, Heidelberg New York Dordrecht London, 2016, *vol. 175*, 341 – 356.

[5] Born, M.; Volf, E. *Fundamentals of Optics*. Nauka, Moscow, 1979, pp. 1 – 719.

PART 3. MECHANICS

In: Proceedings of the 2018 International Conference …
Editors: Ivan A. Parinov et al.
ISBN: 978-1-53615-862-5

Chapter 23

INFLUENCE OF VISCOELASTIC PROPERTIES OF A POLYMERIC MATRIX ON THE AMPLITUDE-FREQUENCY CHARACTERISTICS OF COMPOSITE CYLINDRICAL SHELL

*V. G. Safronenko**
I. I. Vorovich Mathematics, Mechanics and Computer Sciences Institute, Southern Federal University, Rostov-on-Don, Russia

ABSTRACT

The development and updating of mathematical models and describing dynamics of stationary composite shells with the complex physical and mechanical properties takes one of the central places in the mechanics of thin-walled structures. The problem of reducing vibration levels and improving the vibroacoustic and dissipative characteristics of shell structures is crucial for a wide range of applications in advanced technologies. Mathematical and computer models for the stationary dynamics of multilayered composite shells of revolution under vibration load are proposed. The Timoshenko model is used, which takes into account the deformation at transverse displacement and rotation inertia. Constitutive equations of polymeric binding are applied, taking into account nonlinear and non-monotone dependence of physical and mechanical properties on both load frequency and temperature. For the numerical solution to the problem, the Fourier factorization method is used. The resulting system of equations of the normal type is solved with the convolution method. The forced vibration of cylindrical shells made of a polymer fiber composite material is considered. The influence of viscoelastic, instant-elastic and long-elastic properties of a polymeric matrix on the amplitude-frequency characteristics of a shell are studied numerically.

* Corresponding Author's E-mail: safron@math.sfedu.ru.

Keywords: cylindrical shell, composite, polymer, forced oscillations, Fourier factorization method

1. Introduction

It is well known that the anisotropy of physical and mechanical properties, increasing compliance to transverse shear and high nonlinear dependence of a polymer binding's thermoviscoelastic properties on the frequency, temperature, and load is typical for polymer composites. The methods of mathematical modeling and the experiment are the main methods of investigation in both an efficient and economically feasible way in the design of advanced thin-walled structures. In some cases, the calculation of thin-walled structures made of polymer composite is possible within the theory of elasticity. In this regard, the influence of viscoelastic properties of the polymer matrix of the composite on the amplitude-frequency characteristics of the shell is of interest.

2. Methods

As the base for further calculations, we shall use the Timoshenko type theory according to the kinematic and strain relationships, recorded for the rotation shells in conventional notation, have the form [1, 2]:

$U(\alpha_1,\alpha_2,z)=u(\alpha_1,\alpha_2)+z\varphi_1(\alpha_1,\alpha_2),\ V(\alpha_1,\alpha_2, z)=v(\alpha_1,\alpha_2)+z\varphi_2(\alpha_1,\alpha_2);$

$W(\alpha_1,\alpha_2,z)=w(\alpha_1,\alpha_2);$

$\varepsilon_{11}(\alpha_1,\alpha_2,z)=E_{11}(\alpha_1,\alpha_2)+zK_{11}(\alpha_1,\alpha_2);\ (1\leftrightarrow 2)$

$\varepsilon_{13}(\alpha_1,\alpha_2,z)=E_{13}(\alpha_1,\alpha_2);\ (1\leftrightarrow 2)$

$$\varepsilon_{12}(\alpha_1,\alpha_2,z)=E_{12}(\alpha_1,\alpha_2)+zK_{12}(\alpha_1,\alpha_2); \qquad (1)$$

$$\varepsilon_{33}(\alpha_1,\alpha_2,z)=0.\ E_{11}=\frac{1}{A_1}\frac{\partial u}{\partial\alpha_1}+k_1 w;\ E_{22}=\frac{1}{A_2}\frac{\partial v}{\partial\alpha_2}+\psi u+k_2 w;$$

$$E_{13}=\varphi_1-\theta_1;\ E_{23}=\varphi_2-\theta_2;\ E_{12}=\frac{1}{A_1}\frac{\partial v}{\partial\alpha_1}+\frac{1}{A_2}\frac{\partial u}{\partial\alpha_2}-\psi v;$$

$$K_{11}=\frac{1}{A_1}\frac{\partial\varphi_1}{\partial\alpha_1};$$

$$K_{22}=\frac{1}{A_2}\frac{\partial\varphi_2}{\partial\alpha_2}+\psi\varphi_1;\ K_{12}=\left(\frac{1}{A_1}\frac{\partial\varphi_2}{\partial\alpha_1}\right)+k_1\left(\frac{1}{A_2}\frac{\partial u}{\partial\alpha_2}-\psi v\right)+$$

$$+\left(\frac{1}{A_2}\frac{\partial\varphi_1}{\partial\alpha_2}-\psi\varphi_2\right)+k_2\left(\frac{1}{A_1}\frac{\partial v}{\partial\alpha_1}\right);\ \theta_1=k_1u-\frac{1}{A_1}\frac{\partial w}{\partial\alpha_1};\ \theta_2=k_2v-\frac{1}{A_2}\frac{\partial w}{\partial\alpha_2}$$

where $\psi=\frac{1}{A_1A_2}\frac{\partial A_2}{\partial\alpha_1}$.

Equations of harmonic oscillations, obtained from the Hamilton variational principle in conventional notation, have the following forms:

$$\frac{\partial T_{11}}{A_1\partial\alpha_1}+\psi(T_{11}-T_{22})+\frac{\partial S}{A_2\partial\alpha_2}+k_1Q_{11}+\Omega^2\rho_{11}u+q_1=0;$$

$$\frac{\partial S}{A_1\partial\alpha_1}+2\psi S+\frac{\partial T_{22}}{A_2\partial\alpha_2}+k_2Q_{22}+\Omega^2\rho_{11}v+q_2=0;$$

$$\frac{\partial Q_{11}}{A_1\partial\alpha_1}+\psi Q_{11}+\frac{\partial Q_{22}}{A_2\partial\alpha_2}-k_1T_{11}-k_2T_{22}+\Omega^2\rho_{11}w+q_3=0; \quad (2)$$

$$\frac{\partial M_{11}}{A_1\partial\alpha_1}+\psi(M_{11}-M_{22})+\frac{\partial H}{A_2\partial\alpha_2}-\frac{1}{\varepsilon_1}Q_{11}+\Omega^2\rho_{12}\varepsilon_1\varphi_1=0;$$

$$\frac{\partial H_{11}}{A_1\partial\alpha_1}+2\psi H+\frac{\partial M_{22}}{A_2\partial\alpha_2}-\frac{1}{\varepsilon_1}Q_{22}+\Omega^2\rho_{12}\varepsilon_1\varphi_2=0;$$

Efficient physical and mechanical properties of a composite are determined by the properties of the isotropic components that make up this composite.

In the case of fiber-laminated composite, let E_f, ν_f, G_f, V_f be the Young's modulus, Poisson's ratio, shear modulus and volumetric fiber are the content of high modulus, respectively; E_m, ν_m, G_m, are the complex analogs of the polymer matrix, respectively.

For a monolayer composite with the fibers, arranged along the axis of symmetry (the winding angle $\beta = 0$), effective characteristics are present in [3]:

Longitudinal modulus : $E^k{}_1=E_fV_f+E_mV_m$;

Transverse module : $E_2^k = \dfrac{eE_1^k}{(V_f + eV_m)(V_m + eV_f) - (e\nu_f - \nu_m)^2 V_f V_m}$

Poisson's ratio : $\nu^k{}_1= \nu_fV_f+\nu_mV_m$

Shear moduli $G_{12}^k = \dfrac{g(1+V_f)+V_m}{gV_m+1+V_f} G_f$; (3)

$G_{23}^k = \dfrac{gG_m}{V_f+gV_m}$; where $e = \dfrac{E_f}{E_m}$, $g = \dfrac{G_f}{G_m}$.

If we admit a sufficiently high saturation of the composite with the reinforcing fibers, we can claim with high accuracy that each monolayer is an orthotropic body,

$$\sigma^k{}_{11}=C^k{}_{11}\varepsilon^k{}_{11}+C^k{}_{12}\varepsilon^k{}_{22},\ \sigma^k{}_{22}=C^k{}_{12}\varepsilon^k{}_{11}+C^k{}_{22}\varepsilon^k{}_{22}; \tag{4}$$

$$\sigma^k{}_{12}=2C^k{}_{66}\varepsilon^k{}_{12},\ \sigma^k{}_{23}=2C^k{}_{44}\varepsilon^k{}_{23};\ \sigma^k{}_{13}=2C^k{}_{55}\varepsilon^k{}_{13};$$

where $\sigma^k{}_{ij}$, $\varepsilon^k{}_{ij}$ are the amplitudes of the corresponding stresses and strains, respectively; $C^k{}_{ij}$ are the components of a complex matrix with the average stiffness of the reinforced layer:

$$C^k{}_{11}=\frac{E_1^k}{1-\nu_1\nu_2};\ C^k{}_{22}=\frac{E_2^k}{1-\nu_1\nu_2};$$

$$C^k{}_{12}=\frac{\nu_1 E_2}{1-\nu_1\nu_2}=\frac{\nu_2 E_1}{1-\nu_1\nu_2}; \tag{5}$$

$$C^k{}_{44}=G^k{}_{23};\ C^k{}_{55}=G^k{}_{13};\ C^k{}_{66}=G^k{}_{12}.$$

These equations are valid:

$G^k_{12}=G^k_{13}$; $\nu^k_1E^k_2=\nu^k_2E^k_1$.

Constitutive equations for stress and moments can be written as

$$T_{11}=B_{11}E_{11}+B_{12}E_{22}+A_{11}K_{11}+A_{12}K_{22};$$

$$T_{22}=B_{12}E_{11}+B_{22}E_{22}+A_{12}K_{11}+A_{22}K_{22};$$

$$M_{11}=A_{11}E_{11}+A_{12}E_{22}+D_{11}K_{11}+D_{12}K_{22};$$

$$M_{22}=A_{12}E_{11}+A_{22}E_{22}+D_{12}K_{11}+D_{22}K_{22}; \quad (6)$$

$$S=B_{33}E_{12}+2A_{33}K_{12};\ H=A_{33}E_{12}+2D_{33}K_{12};$$

$$Q_{11}=G_{13}E_{13};\ Q_{22}=G_{23}E_{23};$$

Here A_{ij}, B_{ij}, D_{ij} are the stiffness parameters defined by the formulae:

$$B_{ij}=h\sum_{k=1}^{m}C^k_{ij};\ A_{ij}=h\sum_{k=1}^{m}C^k_{ij}z_k;\ D_{ij}=h\sum_{k=1}^{m}C^k_{ij}z^2_k;\ i,j=1,2;$$

$$B_{33}=h\sum_{k=1}^{m}C^k_{66};\ A_{33}=h\sum_{k=1}^{m}C^k_{66}z_k;\ D_{33}=h\sum_{k=1}^{m}C^k_{66}z^2_k;$$

where *m* is the number of monolayers in the composite material; z_k is the coordinate of the gravity center of the monolayer section.

The stiffness matrix in the case of symmetric reinforcement for $\beta\neq0$ is defined with the formulae of the coordinate system transformation [3]:

$$C'_{11}=C_{11}c^4+2(C_{12}+2C_{66})c^2s^2+C_{22}s^4;$$

$$C'_{22}=C_{11}s^4+2(C_{12}+2C_{66})c^2s^2+C_{22}c^4;$$

$$C'_{12}=(C_{11}+C_{22}-4C_{66})c^2s^2+(c^4+s^4)C_{12};$$

$$C'_{66}=(C_{11}+C_{22}-2C_{12})c^2s^2+2C_{22}(c^2-s^2)cs+C_{66}(c^2-s^2)^2; \quad (7)$$

$C'_{44}=C_{44}c^2+C_{55}s^2$; $C'_{55}=C_{55}c^2+C_{44}s^2$;

where C'_{ij} are the components of transformed matrix; $c = \cos\beta$; $s = \sin\beta$; β is the angle of rotation around the z-axis of the new coordinate system with respect to the old one.

Further constitutive relationships for polymer binder are used [4]. Components of the complex compliance under I', I'' shear are represented by the functions of the cyclic frequency $\omega = 2\pi f$ as

$$I'(\omega,T)=\int_{r_1}^{r_2}\frac{C(r)}{G(r,\upsilon,T)}\left(1-H(r)\frac{\omega^2}{(\varphi(r,\upsilon,T))^2+\omega^2}\right)dr,$$

$$I''(\omega,T)=\int_{r_1}^{r_2}\frac{C(r)H(r)}{G(r,\upsilon,T)}\left(\frac{\varphi(r,\upsilon,T)\omega}{(\varphi(r,\upsilon,T))^2+\omega^2}\right)dr.$$

The mechanical properties of the polymer binder correspond to a typical viscoelastic material, in particular, propylene. The dependences of the complex modules and the loss angle tangent on the frequency and temperature are shown in Figures 1, 2.

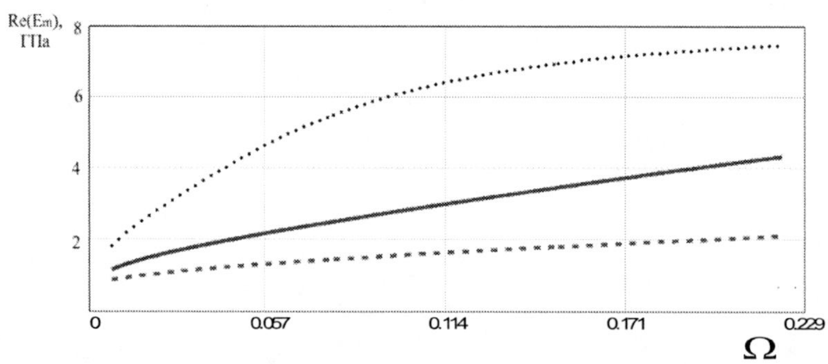

Figure 1. Dependence of the complex modulus on the frequency and temperature: .....T = 250 K;___ T = 290 K; - - - T = 330 K.

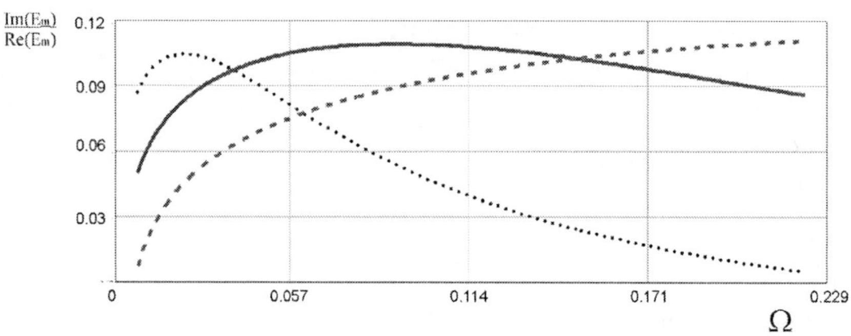

Figure 2. Dependence of the loss angle tangent on the frequency and temperature: .....T = 250 K;___ T = 290 K C; - - - T = 330 K.

For elastic fiber, we will take: E_f = 186 GPa, ν_f = 0.3, ρ_f= 1730kg/m^3, V_f= 0.8.

For the numerical study, an approach associated with the representation of required functions in the form of expansions in the Fourier series is used:

$$X(\alpha_1,\alpha_2) = \sum_{n=0}^{N}(X_{+n}(\alpha_1)\cos(n\alpha_2) + X_{-n}(\alpha_1)\sin(n\alpha_2)).$$

After the separation of coordinates and rewriting the equations into a dimensionless form, a complex-valued system of the usual differential equations of the 10th order is obtained, which is solved by the orthogonal convolution method.

3. Numerical Experiment

The main parameters that determine the structure of the composite shell are the reinforcement angles and the volume content of the fiber in the composite. In the following, we consider a cylindrical shell, which is under the influence of a harmonic surface load distributed over a cylindrical panel in the central part of the shell and has an amplitude intensity of p_3 = 0.1 MPa. The wall of the shell consists of five layers, thus we consider that there is symmetrical reinforcement with angle β. The geometric parameters of the shell: L = 5R, R = 0.2m, the total thickness of the package h = 0.002m.

Below, we consider the forced oscillations of the composite cylindrical shell for three variants of reinforcement: unidirectional along the generatrix β = 0°, β = ±45° and orthogonally reinforced β = 0° – 90°. For certainty, the boundary conditions corresponding to the rigid clamping are taken into account. Figures 3-8 present the results of calculating the frequency response for three cases of the mechanical properties of the polymer matrix.

In the first case, the polymer is considered an elastic material with characteristics corresponding to the instantaneous elastic state of the polymer E_0, ν_0, $tg\delta_0$. In the second case, we consider long elastic characteristics E_∞, ν_∞, $tg\delta_\infty$.. The case of $tg\delta$ = 0, i.e., the absence of mechanical losses, is also considered. The values of limiting characteristics, obtained from thermoviscous models, respectively, with $\omega\to\infty$ and $\omega\to 0$ at T = 290 K, are considered, too.

Figures 3, 4 present the results of calculations for β = 0. Figure3 corresponds to the instantaneous elastic scheme. Figure 4 corresponds to the long-term characteristics, in comparison with calculations for the viscoelastic model. Similar results are shown in Figures 5, 6 for β = ±45° and in Figures 7, 8 for β = 0° – 90°. Tables 1 – 3 present the values of the first three natural frequencies Ω_i^∞, Ω_i^0 and the values of the corresponding amplitudes $|W_i|^\infty$, $|W_i|^0$.

The calculations allow us to draw the following conclusions. Calculations of the instantaneous elastic characteristics E_0, ν_0, $tg\delta_0$ give satisfactory results in the calculation of the first eigenfrequencies.

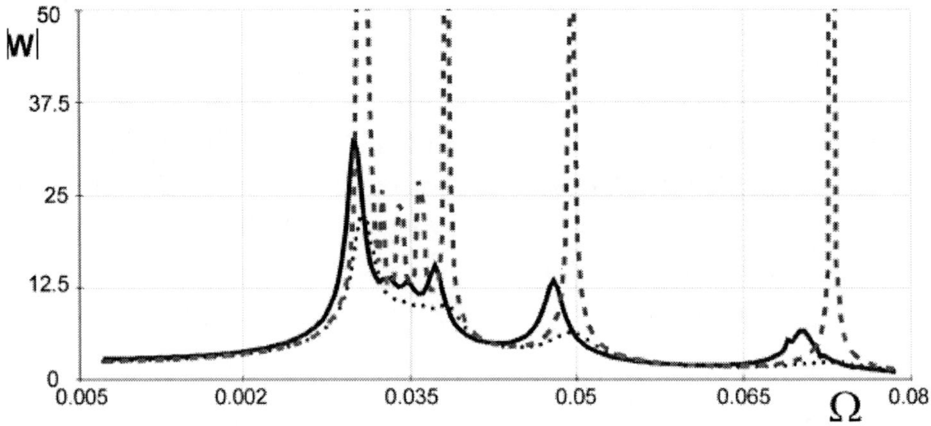

Figue 3. Results of calculations for $\beta = 0$ and instantaneous elastic scheme.

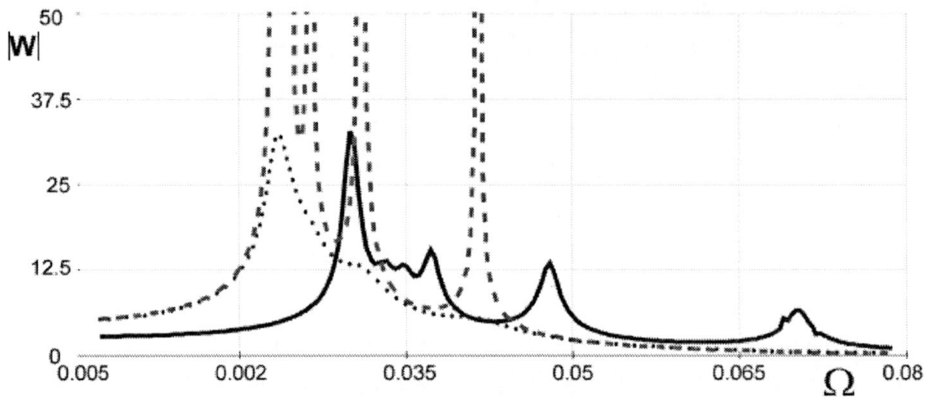

Figure 4. Results of the calculations of long-term characteristics for $\beta = 0$.

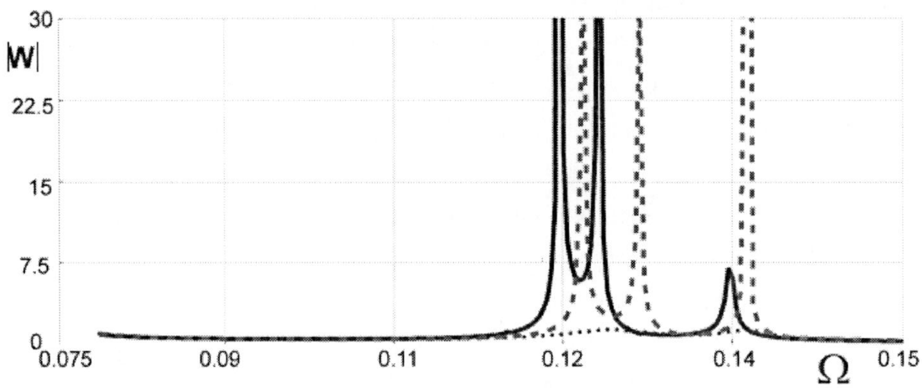

Figure 5. Results of calculations for $\beta = 45°$ and instantaneous elastic scheme.

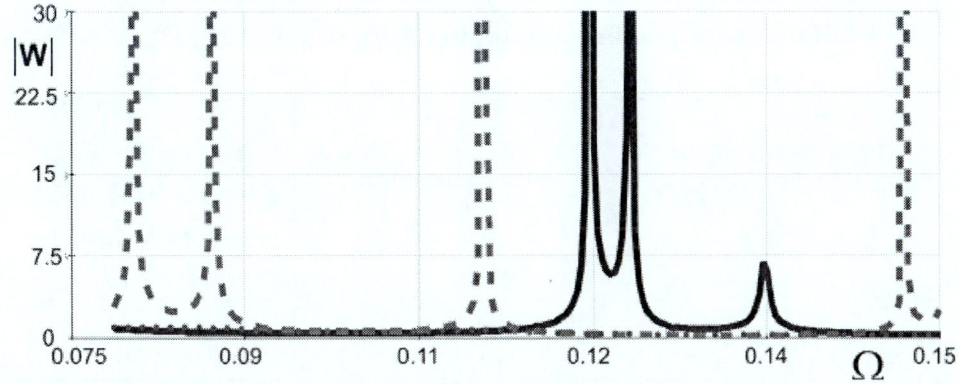

Figure 6. Results of the calculations of long-term characteristics for $\beta = \pm 45°$.

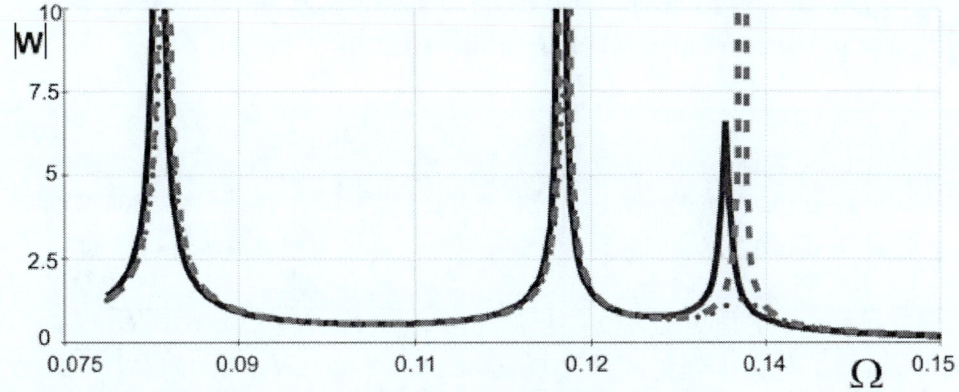

Figure 7. Results of calculations for $\beta = 0° - 90°$ and instantaneous elastic scheme.

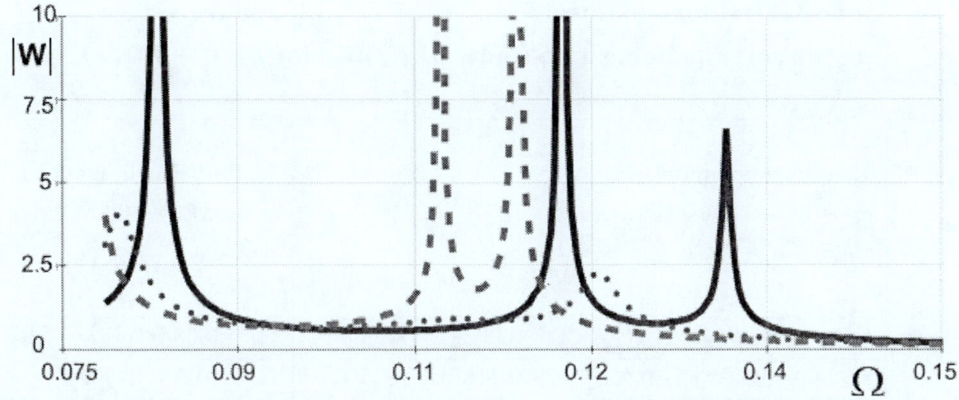

Figure 8. Results of the calculations of long-term characteristics for $\beta = 0° - 90°$.

In this case, the discrepancy with the results of calculations for a viscoelastic scheme decreases with the change β from 0° to 90°, which is associated with the increase of the coefficients B_{22} and D_{22}. The influence of viscoelastic properties of the polymer weakens due to the strengthening of the reinforcing factor of elastic fibers.

Table 1. First three natural frequencies Ω_i^∞, Ω_i^0 and the values of the corresponding amplitudes $|W_i|^\infty$, $|W_i|^0$ for $\beta = 0°$

	Long-elastic model $\omega \to 0$: E_∞=1000 MPa; $tg\delta_\infty$=0.2; ν_∞=0.45	**Viscoelastic model**	**Instant-elastic model** $\omega \to \infty$: E_0=3300 MPa; $tg\delta_0$=0.1; ν_0=0.34
$\beta=0°$	Ω_1^∞ =0.023; $\|W_1\|^\infty$ =32.5	Ω_1=0.03; $\|W_1\|$=32.7	Ω_1^0 =0.031; $\|W_1\|^0$ =22.7
	Ω_2^∞ =0.03; $\|W_2\|^\infty$ =13.1	Ω_2=0.03; $\|W_2\|$=15.3	Ω_2^0 =0.038; $\|W_2\|^0$ =10.1
	Ω_3^∞ =0.042; $\|W_3\|^\infty$ =5.3	Ω_3=0.048; $\|W_3\|$=13.4	Ω_3^0 =0.049; $\|W_3\|^0$ =6.4

Table 2. First three natural frequencies Ω_i^∞, Ω_i^0 and the values of the corresponding amplitudes $|W_i|^\infty$, $|W_i|^0$ for $\beta = \pm 45°$

	Long-elastic model $\omega \to 0$: E_∞=1000 MPa; $tg\delta_\infty$=0.2; ν_∞=0.45	**Viscoelastic model**	**Instant-elastic model** $\omega \to \infty$: E_0=3300 MPa; $tg\delta_0$=0.1; ν_0=0.34
$\beta=\pm 45°$	Ω_1^∞ =0.054; $\|W_1\|^\infty$ =3.74	Ω_1=0.07; $\|W_1\|$=15.6	Ω_1^0 =0.071; $\|W_1\|^0$ =4.2
	Ω_2^∞ =0.083; $\|W_2\|^\infty$ =1	Ω_2=0.12; $\|W_2\|$=204	Ω_2^0 =0.124; $\|W_2\|^0$ =1.3
	Ω_3^∞ =0.11; $\|W_3\|^\infty$ =0.7	Ω_3=0.123; $\|W_3\|$=53.1	Ω_3^0 =0.136; $\|W_3\|^0$ =1.3

Table 3. First three natural frequencies Ω_i^∞, Ω_i^0 and the values of the corresponding amplitudes $|W_i|^\infty$, $|W_i|^0$ for $\beta = 0° - 90°$

	Long-elastic model $\omega \to 0$: E_∞=1000 MPa; $tg\delta_\infty$=0.2; ν_∞=0.45	**Viscoelastic model**	**Instant-elastic model** $\omega \to \infty$: E_0=3300 MPa; $tg\delta_0$=0.1; ν_0=0.34
$\beta=0°$–90°	Ω_1^∞ =0.08; $\|W_1\|^\infty$ =4.1	Ω_1=0.083; $\|W_1\|$=33.6	Ω_1^0 =0.084; $\|W_1\|^0$ =10.8
	Ω_2^∞ =0.109; $\|W_2\|^\infty$ =0.9	Ω_2=0.118; $\|W_2\|$=250	Ω_2^0 =0.118; $\|W_2\|^0$ =8
	Ω_3^∞ =0.12; $\|W_3\|^\infty$ =2.3	Ω_3=0.131; $\|W_3\|$=6.6	Ω_3^0 =0.133; $\|W_3\|^0$ =1.3

Calculations of long-term elastic characteristics give a significant discrepancy with calculations on the viscoelastic model for both eigenfrequencies and vibration levels. It cannot be recommended for their determination. However, the values Ω_i^∞ and Ω_i^0 define the boundaries of the intervals in which Ω_i are located and can be used to estimate them.

CONCLUSION

Mathematical and computer models for multilayered composite polymeric shells of revolution under the influence of vibration loadings have been developed. In particular, numerical analytical approaches based on the Fourier-series method, as well as the convolution method, were used. Forced vibrations of cylindrical shell, made of a polymer fiber composite material, are considered. The influence of viscoelastic, instant-elastic and long-elastic properties of a polymeric matrix on the amplitude-frequency characteristics of a shell are studied numerically.

REFERENCES

[1] Grigolyuk, E. I.; Kulikov, G. M. *Multilayer Reinforced Shells. Mashinostroenie,* Moscow. 1988, pp. 1 - 288 (in Russian).

[2] Yudin, A. S.; Safronenko, V. G. *Vibroacoutics of Structurally-Nongomogeneous Shells*, Southern Federal University Press, Rostov-on-Don. 2013, pp. 1 - 424 (in Russian).

[3] Cristensen, R. M. Mechanics of Composite Materials. *Mir,* Moscow. 1982, pp. 1 - 336 (in Russian).

[4] Stepanenko, Y. P.; Isaev, K. V.; Azarov, A. D. In: *Proc. II Int. Conf. Modern Problems of Continuum Mechanics*, Rostov-on-Don, 1996, vol.1, 118.

In: Proceedings of the 2018 International Conference … ISBN: 978-1-53615-862-5
Editors: Ivan A. Parinov et al.

Chapter 24

REDUCTION OF STRESSES IN THE SHELL BY THE PLASTIC CHANGE OF ITS FORM

***A. S. Yudin*[1,*] *and D. V. Shchitov*[2]**
[1]I. I. Vorovich Institute of Mathematics, Mechanics and Computer Sciences, Southern Federal University, Rostov-on-Don, Russia
[2]North-Caucasus Federal University Branch, Pyatigorsk, Russia

ABSTRACT

The problem of changing the shape of the shell and modeling the bottom of the axisymmetric tank for transportation of liquid cargo is considered. The original shell version of the composite geometry is converted to a spherical segment. The purpose of changing the form is to reduce the level of stress intensity. Mathematical modeling is performed on the base of geometrically and physically nonlinear equations using the semi-inverse method.

Keywords: large deformations, shape change, mathematical model, semi-inverse method, functional control

1. INTRODUCTION

There is a wide class of structures, which are containers for transporting liquid cargo. During transportation, the body of the thin-walled tank experiences both static and dynamic loads caused by vehicle vibrations. Multiple variable loads can cause fatigue

* Corresponding Author's E-mail: 2915132@aaanet.ru.

cracks and integrity. Taking into account the specific nature of the transported substances, this can cause environmental and economic damage. Therefore, it is necessary to analyze the causes of destruction of tanks that are widely used in practice, estimate the shortcomings of some structures, search for technology to eliminate them, and perform comparison with alternative constructions.

The conducted studies show that static stresses caused by the weight of the load have a significant impact on the resource of structures. Two-hundred-liter barrels belong to widespread and typical capacities (see Figure 1, on the left hand). The most loaded and weak element of their design is the bottom. There is a common variant of the bottom of the composite geometry, in which there is a section of the annular plate (edge), a sloping spherical dome (center) and a torus insert between them (see Figure 1, on the right hand). The Gaussian curvature $K = k_1k_2$ of such a shell changes the sign on the joint line between the sphere and torus; k_1, k_2 are the principal curvatures.

The shell is subjected to radial axisymmetric deformation, so that all functions depend only on the meridional coordinate α_1. The study of the shell of this form shows that its stress state is not optimal. In the area of the torus insert, there is a peak stress intensity, which leads to the appearance and development of fatigue cracks during operation, therefore leading to destruction. At the same time, the bottom in the form of a sloping spherical shell, without sections of another geometry, leads to a smoothed stress state, which increases the resource [1 – 4].

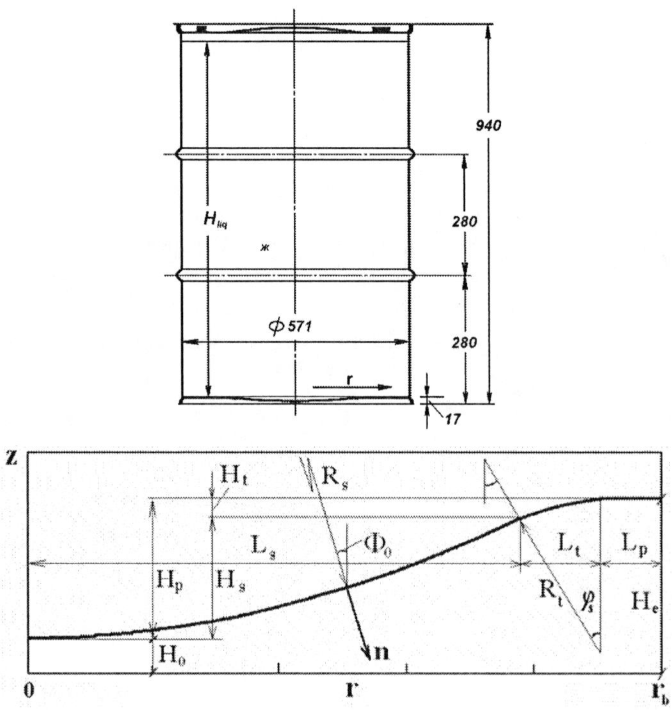

Figure 1. Sketch of the container and the initial shape of the bottoms.

Hundreds of thousands of tanks with composite geometry bottoms are manufactured in mass production. Due to commercial reasons, the manufacturer does not benefit from a very long service life of the products. The customer (carrier) is interested in the opposite, both in durability and reliability. He is responsible for the environmental consequences of leakage of harmful liquids in the environment.

If the carrier has a large number of such containers, is it possible to improve their properties by changing the shape of the bottoms? Let us set the following problem: to obtain a spherical (or close to it) shell ($K > 0$) from the shell of composite geometry having alternating Gaussian curvature K, which gives the best characteristics of the stress state.

2. Equations and Method

For the mathematical modeling of the deformation of the shell, both geometrically and physically, nonlinear equations are used, which have large displacement and rotation angles, a change in the metric of the middle surface, and compression of the normal material. When you build defining equations, use exponential approximation diagrams of the hardening material and physical deformation ratio with a logarithmic relative elongation [5]. A semi-inverse method of type [6] adapted to the specifics of the problem is used.

The kinematic relations have the form:

$$\varepsilon_1^\zeta = (\varepsilon_1 + \zeta\kappa_1),\ \kappa_1 = (\Phi_o' - \delta_3\Phi')/\alpha_o,$$
$$\varepsilon_1 = (w'\sin\Phi + u'\cos\Phi)/a_o + \cos(\Phi - \Phi_o) - 1, \tag{1}$$

$$\varepsilon_2^\zeta = (\varepsilon_2 + z\kappa_2),\ \varepsilon_2 = u/r_o,\ \kappa_2 = (\sin\Phi_o - \delta_3\sin\Phi)/r_o.$$
$$\gamma = \delta_1^{-1}(\gamma_o + \zeta\gamma_1),\ \gamma_1 = \varepsilon_3'/\alpha_o,$$
$$\gamma_o = (w'\cos\Phi - u'\sin\Phi)/\alpha_o - \sin(\Phi - \Phi_o). \tag{2}$$

$$K_1 = \Phi'/\alpha_o,\ K_2 = (\sin\Phi)/r_o \tag{3}$$

Here all functions depend on the curvilinear coordinate $\alpha_1 = x$. The prime denotes the derivative with respect α_1; α_o is the Lame coefficient of the original shell; Φ_o is the angle between the normal to the reference shell and the z-axis; Φ is the angle between the normal to the deformed membrane and the z-axis; $\delta_j = 1 + \varepsilon_j,\ j = 1, 2, 3$. It is assumed that the condition of incompressibility is satisfied: $\delta_1\delta_2\delta_3 = 1$, $\delta_3 = \delta_1^{-1}\delta_2^{-1}$.

Equilibrium equations have the form:

$$(r_o\overline{V}^o)' + \alpha_o r_o p_w^o = 0, \tag{4}$$

$$(r_o\overline{H}^o)' - \alpha_o\overline{N}_2^o + \alpha_o r_o p_u^o = 0, \tag{5}$$

$$(r_o M_1^o)' - \alpha_o M_2^o cos\Phi - \alpha_o r_o \delta_1 Q^o = 0, \tag{6}$$

$$p_w^o = \delta_1\delta_2 p_w,\ p_u^o = \delta_1\delta_2 p_u. \tag{7}$$

The internal forces $\overline{V}^o$ and $\overline{H}^o$ are oriented, respectively, vertically (along the axis of symmetry) and horizontally (along the radius of the cylindrical coordinate system), Q^o is the cutting force:

$$\overline{V}^o = \overline{N}_1^o sin\Phi + Q^0 cos\Phi,\ \overline{H}^o = \overline{N}_1^o cos\Phi - Q^0 sin\Phi. \tag{8}$$

Values $\overline{N}_1^o$, $\overline{N}_2^o$ are the generalized forces associated with tangential internal forces N_1^o, N_2^o and moments M_1^o, M_2^o :

$$\overline{N}_1^o = N_1^o + \delta_1^{-1}(K_1M_1^o + K_2M_2^o),\ \overline{N}_2^o = N_2^o + \delta_2^{-1}(K_1M_1^o + K_2M_2^o),$$
$$N_1^o = \delta_2 N_1,\ N_2^o = \delta_1 N_2,\ Q^0 = \delta_2 Q,\ M_1^o = \delta_2 M_1,\ M_2^o = \delta_1 M_2. \tag{9}$$

Internal pressure is considered as a load. Taking into account that the positive direction of the normal is inside the shell, we have:

$$p_w = -pcos\Phi,\ p_u = psin\Phi. \tag{10}$$

The constitutive equations of the Davis-Nadai type are taken as [5]

$$N_1^o = \overline{B}_1(\overline{\varepsilon}_1 + 0.5\overline{\varepsilon}_2),\ N_2^o = \overline{B}_2(\overline{\varepsilon}_2 + 0.5\overline{\varepsilon}_1),$$
$$M_1^o = \overline{D}_1(\overline{\kappa}_1 + 0.5\overline{\kappa}_2),\ M_2^o = \overline{D}_2(\overline{\kappa}_2 + 0.5\overline{\kappa}_1); \tag{11}$$

$$\overline{\varepsilon}_1 = \ln(\overline{\delta}_1),\ \overline{\varepsilon}_2 = \ln(\overline{\delta}_2),\ \overline{\kappa}_1 = \kappa_1/\delta_1,\ \overline{\kappa}_2 = \kappa_2/\delta_2; \tag{12}$$

$$\overline{B}_1 = \overline{B}/\delta_1,\ \overline{B}_2 = \overline{B}/\delta_2,\ \overline{D}_1 = \overline{D}\,\delta_3^2\delta_2,\ \overline{D}_2 = \overline{D}\,\delta_3^2\delta_1,$$
$$\overline{B} = (4/3)\Lambda(\overline{\varepsilon})h_o,\ \overline{D} = (1/9)\Lambda(\overline{\varepsilon})h_o^3. \tag{13}$$

Here $\Lambda(\bar{\varepsilon}) = C\bar{\varepsilon}^{\eta-1}$ is the secant modulus of plasticity diagram of the material, where $\bar{\varepsilon} = (2/\sqrt{3})\sqrt{\bar{\varepsilon}_1^2 + \bar{\varepsilon}_1\bar{\varepsilon}_2 + \bar{\varepsilon}_2^2}$ is the intensity of logarithmic deformations.

The diagram of the material can be identified by characteristic values, which are usually given in the reference books. This is the modulus of elasticity E, the elastic limit σ_{02}, the limit of elastic deformation $\varepsilon_{02} = \sigma_{02}/E$, the temporary stress intensity σ_t, and the temporal intensity logarithmic deformations $\bar{\varepsilon}_t$. These data are used to determine the parameters of the power approximation of the diagram's nonlinear properties of the material: $\sigma(\bar{\varepsilon}) = C\bar{\varepsilon}^{\eta}$, where $\eta = [\ln(\sigma_t) - \ln(\sigma_{02})]/[ln(\bar{\varepsilon}_t) - \ln(\varepsilon_{02})$, $C = \sigma_t\bar{\varepsilon}_t^{\eta}$. Thus, for the material of the container (steel), the characteristics were given as follows: $E = 0.21 \times 10^6$ *MPa*, $\sigma_{02} = 210$ *MPa*, $\sigma_t = 380$ *MPa*, $\varepsilon_{02} = 0.001$, $\bar{\varepsilon}_t = 0.25$, $C = 44.1$ *MPa*, $\eta = 0.107$. The secant module was stated equal to E when $\bar{\varepsilon} < \varepsilon_{02}$ and $\Lambda(\bar{\varepsilon})$ when $\bar{\varepsilon} \geq \varepsilon_{02}$.

We integrate equations (4) and (5). Taking into account (8), we obtain:

$$r_o(\bar{N}_1^o \sin\Phi + Q^o \cos\Phi) = pI_{11} + C_V, \tag{14}$$

$$r_o(\bar{N}_1^o \cos\Phi - Q^o \sin\Phi) = I_{21} - pI_{22} + C_H, \tag{15}$$

where C_V and C_H are integration constants associated with vertical and horizontal forces at one end of the integration range. In this case, we will integrate from the edge of the shell to the center to state the forces on the outer contour r_b. Then I_{11}, I_{21}, I_{22} are the integrals depending on the current coordinate written as

$$I_{11} = -\int_{\xi}^{r_b} \alpha_o r_o \delta_1 \delta_2 \cos\Phi d\xi,\ I_{21} = -\int_{\xi}^{r_b} \alpha_o r_o \overline{N}_2^o d\xi,$$
$$I_{22} = -\int_{\xi}^{r_b} \alpha_o r_o \delta_1 \delta_2 \sin\Phi d\xi. \tag{16}$$

From relations (14) and (15), we express Q^o. To do this, we multiply the equation (14) by $\cos\Phi$ and subtract from it the equation (15) multiplied by $\sin\Phi$. We have:

$$Q^o = r_o^{-1}[(pI_{11} + C_V)\cos\Phi - (I_{21} - pI_{22} + C_H)\sin\Phi. \tag{17}$$

Substitute the expression (2.17) for Q^o into equation (6) and integrate it. We obtain:

$$r_o M_1^o = I_{31} + p(J_{11} + J_{22}) + C_V J_{31} - J_{32} - C_H J_{33} + C_M, \tag{18}$$

where

$$J_{11} = -\int_{\xi}^{r_b} \alpha_o\, \delta_1 l_{11} cos\Phi d\xi,\ J_{22} = -\int_{\xi}^{r_b} \alpha_o\, \delta_1 l_{22} sin\Phi d\xi,\ J_{31} = -\int_{\xi}^{r_b} \alpha_o\, \delta_1 \delta_2 cos\Phi d\xi,$$

$$J_{32} = -\int_{\xi}^{r_b} \alpha_o\, \delta_1 l_{21} sin\Phi d\xi,\ J_{33} = -\int_{\xi}^{r_b} \alpha_o\, \delta_1 sin\Phi d\xi. \tag{19}$$

C_M is a integration constant associated with the moment reaction on the circuit.

The pressure p does not depend on ξ and is taken out of the integral sign. From (18), it follows:

$$p = \frac{r_o M_1^o + J_{32} - J_{31} - C_M - C_V J_{31} + C_H J_{33}}{J_{11} + J_{22}}. \tag{20}$$

The expression (20) output to the constant is controlled by the parameters C_V, C_H, C_M and the geometry of the new shell shape.

3. Numerical Results

The initial shape of the shell is composed of segments of the sphere (central part), torus and plate (boundary zone), Figure 1. The smooth torus insertion, without breaking the angle Φ_0 of the normal **n** to the axis of symmetry, matches the central and boundary regions. The shell has a thickness $h = 1\ mm$, the outer radius $r_b = 282.75\ mm$. Other parameters were given as follows (in mm):

$L_s = 212,\ L_t = 53,\ L_p = 17.7,\ H_t = 1.8,\ H_e = 17,\ H_0 = 8,\ H_p = H_e - H_0 = 9,\ H_s = 7.2,$
$\varphi_s = arcsin(L_s/R_s) = 0.067878,\ H_s = R_s(1 - cos\varphi_s),\ R_s = (H_s^2 + L_s^2)/(2H_s) = 3126.$

The curvature of the meridian $k_1 = d\Phi_0/dr$ is a piecewise constant alternating function equal to $1/R_s$ on the spherical section, $-1/R_t$ on the torus and zero on the flat area. The second main curvature $k_2 = sin\Phi_0/r_0$ is continuous, equal to $1/R_s$ on the area of the sphere, and decreases linearly from this value to zero on the torus insert when approaching the plate and remains zero on the plate.

Take the arc length of the original shell as an independent coordinate. Lengths of the sections are:

$s_1 = R_s \cdot \varphi_s,\ s_2 = s_1 + R_t \cdot \varphi_s,\ s_3 = s_2 + L_p.$

Equations of the middle surface at different regions (vertical coordinate) are:

$Z_{os}(s) = H_0 + R_s(1 - cos(s/R_s))$ at the sphere interval $0 < s \le s_1$;
$Z_{ot}(s) = H_0 + R_s + R_t \times [cos(\varphi_s - (s - s_1)/R_t) - cos(\varphi_s)]$ at the torus interval $s_1 < s \le s_2$;

$Z_{op}(s) = H_0 + H_p$ at the plate interval $s_2 < s \leq s_3$.

Inclination angles of normal, respectively, are:

$\Phi_{os}(s) = s/R_s$, $\Phi_{ot}(s) = \Phi_{os}(s_1) - (s - s_1)/R_t$, $\Phi_{op}(s) = 0$.

Polar radiuses, respectively, are:

$r_{0s}(s) = R_s \cdot sin(\Phi_{os}(s))$, $r_{0t}(s) = s_1 + L_t - R_s \cdot sin(\Phi_{ot}(s))$, $r_{0p}(s) = s$.

Superpositions of these functions on a whole area are denoted, respectively, $Z_0(s)$, $\Phi_o(s)$, $r_0(s)$. By calculating in an arc $\alpha_0(s) = \alpha_0 = 1$.

The expected shape of the shell after deformation is given as a segment of the sphere, the meridian of which passes through the points with coordinates $(0, H_0)$ and (r_b, H_e):

$$Z_d(r) = R_d + H_0 - \sqrt{R_d^2 - r^2}\,, \tag{21}$$

where R_d is the radius of the segment, H_0 is the clearance height. In this case $R_d = 4.45$ *м*.

The elongations of the meridian are small ($\varepsilon_1 << 1$). Therefore, the equation $Z_s(s) = Z_d(r_0(s))$ will be written in Lagrangian coordinates. The total length S_d of the shell meridian in the deformed state is selected approximately equal to the original length S_0. The residual $\varepsilon_1 = (S_d - S_0)/S_0$ can be taken as the mean elongation.

The condition of compatibility of deformations follows from relations (1), (2):

$$\alpha_o(\varepsilon_1 cos\Phi - \gamma_o sin\Phi + cos\Phi - cos\Phi_o) - (r_o\varepsilon_2)'=0, \tag{22}$$

and formula:

$$w'=\alpha_o(\varepsilon_1 sin\Phi + \gamma_o sin\Phi + sin\Phi - sin\Phi_o). \tag{23}$$

Since $r_0\varepsilon_2 = u$, $\gamma_0 \approx 0$, then from (22) and (23) we have:

$$u'=\alpha_o(1+\varepsilon_1)cos\Phi - cos\Phi_o),\ w'=\alpha_o(1+\varepsilon_1)sin\Phi - sin\Phi_o). \tag{24}$$

Since the initial and deformed shell shapes are specified, relationships (24) are integrated:

$$u(s) = \int_{r_b}^{s}[\alpha_o(1+\varepsilon_1)cos\Phi(s) - cos\Phi_o(s)]\,ds \tag{25}$$

$$w(s) = \int_{r_b}^{s} [\alpha_o(1+\varepsilon_1)sin\Phi(s) - sin\Phi_o(s)]\, ds \tag{26}$$

The edges of the shell do not shift, so the integration constants are equal to zero. The direction of integration is not important.

The distribution of vertical displacements can also be obtained from the difference between the equations of finite and initial forms $w(s) = Z_s(s) - Z_0(s)$ that are an additional verification. The dependence $w(s)$ is shown in Figure 2.

Knowing the radial displacement (25), we obtain the elongation of the annular areas $\varepsilon_2(s) = u(s)/r_0(s)$, and from the condition of incompressibility $(1 + \varepsilon_1)(1 + \varepsilon_2(s))(1 + \varepsilon_3(s)) = 1$, we find compression $e_3(s)$. In this case, the stress intensity and the secant module are calculated as a function of s: $\Lambda_s(s) = \Lambda(\varepsilon(s))$.

The change of the initial shape in the spherical segment corresponds to the pressure $p = 250\ kPa$. The most active plastic deformations locate at the torus insert and its surroundings. The new form has a much lower stress intensity under elastic loading by the weight of the liquid (Figure 3, curve 2) compared with the prototype (Figure 3, curve 1). The bottom, formed by one spherical segment, is subjected to minimal tension with good uniformity of state.

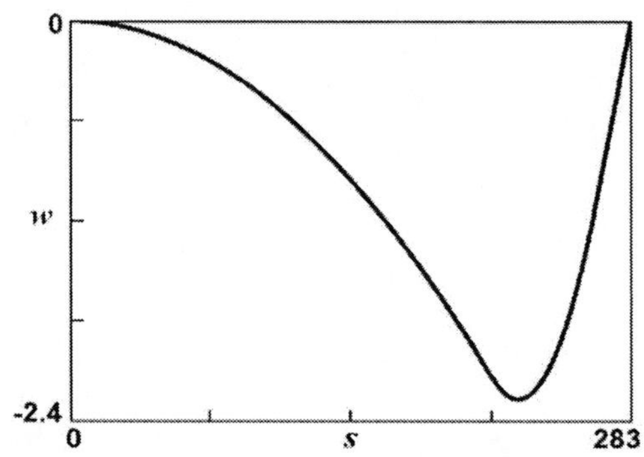

Figure 2. Dependence $w(s)$.

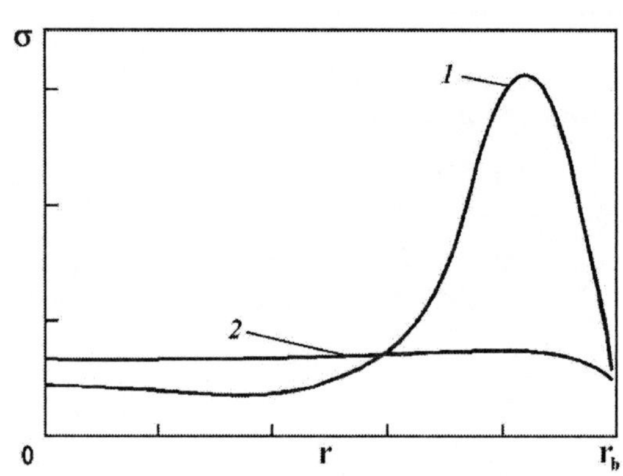

Figure 3. Dependence $\sigma(r)$.

Then we compare the forming pressure with the test pressures. Tanks are tested for air impermeability at a pressure of *20 .. 30 kPa*. Hydraulic strength tests are performed by pumping water under a pressure of *100 kPa* (up to *250 kPa* at large metal thicknesses). Experiments on the ultimate load show that destruction in the injected air occurs at a level of ≈ *300 kPa*. In this case, the cylindrical part can withstand this load being in the elastic zone. The fracture occurs due to knocking out one of the bottoms. Therefore, the technology of changing the form should be worked out with regard to security. To maintain clearance, the container must be clamped between the rigid planes on the bottoms. If the connection of the bottoms and the shell is made by the technology of rolling, it is advisable to strengthen the connection using clamps. However, this task is for specialists, if there is a need to implement such a system.

ACKNOWLEDGMENT

The work was carried out within the project No. 11.6767.2017/BCh of the basic part of the state task of Russian Ministry of Science and Education.

REFERENCES

[1] Shitov, DV. *Proc. Graduate Students and Competitors*. Rostov State University Press, Rostov-on-Don. 2001, *vol.* VII, 17-19 (In Russian).

[2] Shitov, DV. *Proc. Ist School Seminar. Mathematical Modeling, Computational Mechanics and Geophysics*. Rostov-on-Don, 14-18 October 2002. New Book, Rostov-on-Don., 2002, 170-172 (In Russian).

[3] Yudin, AS; Shitov, DV. *Izv. VUZov. North-Caucasus Region. Natural Sciences. Appendix*., 2004, No. 3, 22-29 (In Russian).

[4] Yudin, AS. *Proc. XVII International Conference*. Southern Federal University Press, Rostov-on-Don, 2014, vol. 2, 229-233 (In Russian).

[5] Yudin, AS. *Statics and Oscillations of Shells of Revolution with Liquid*. Southern Federal University Press, Rostov-on-Don., 2014, pp. 1 – 204, (In Russian).

[6] Yudin, AS; Yudin, SA; Sigaeva, TV. In: *Advanced Materials - Studies and Applications*, I. A. Parinov, S. H. Chang, S. Theerakulpisut (Eds.). Nova Science Publishers, New York., 2015, 341-351.

In: Proceedings of the 2018 International Conference … ISBN: 978-1-53615-862-5
Editors: Ivan A. Parinov et al.

Chapter 25

MODELING OF DYNAMIC PROCESSES IN WALL CONSTRUCTION WITH DEFECT IN THE FORM OF CRACK

A. V. Cherpakov*[1,2,*]*, O. V. Shilyaeva*[2]*, I. A. Parinov*[1]*,
A. V. Nesterenko*[2]*, Yu. A. Chaika*[2]*, A. N. Soloviev*[1,2]
and E. V. Rozhkov*[1]
[1]Southern Federal University, Rostov-on-Don, Russia
[2]Don State Technical University, Rostov-on-Don, Russia

ABSTRACT

In the finite-element software ANSYS, the oscillations of wall construction under pulsed loading were simulated. The non-stationary problem is considered. The process of identification of defects in the construction in the form of a wall element is investigated. The investigated model of the wall construction has a defect in the form of a through crack of certain length.

The edges of the crack do not interact with each other. The wave field displacements at control points are analyzed. We propose signs of the identification of crack-like defects.

Keywords: numerical modeling, FE software ANSYS, wall element, defect, crack, identification, finite element modeling, dynamic analysis, oscillations, wave field of displacements

---

* Corresponding Author's E-mail: alex837@yandex.ru.

1. INTRODUCTION

In modern construction, various types of structures are used, consisting of both individual block elements and large compartments. One type of integral building elements is wall construction with a variety of shapes and a huge range of materials used. To study these constructions, it is necessary to apply the requirements of both the quality of execution and the input control of material properties. This control can be applied both at the stage of building construction and during its operation.

Of scientific interest is the study of the possibilities of diagnosing and controlling the properties of a construction by using non-destructive methods. These are based on the analysis of vibration parameters at a certain vibrational excitation. The solution of this problem will ensure their quality and durability of operation.

One of the approaches to the identification of properties is the vibrodiagnostic method of research. The description of the method is given in [1-5]. The approach may be based on the use of analytical or finite element modeling [5-8]. In this case, defect identification algorithms can be applied [9-12]. The approach can be carried out based on the analysis of the natural frequencies of oscillations and modal parameters' oscillations of the construction [12-14]. Calculations, using the finite element (FE) method and the application of the FE software ANSYS, can be effectively used to analyze various types of constructions, which is confirmed by the certain examples [15-20].

A schematic of the wall construction model is shown in Figure 1. The dimensions of the design and material properties are present in Table 1.

As an example, a similar full-scale construction was considered with having some simplifications. Dimensions of the design in the simulation were 1000 × 3000mm^2. A crack-like through defect, having a rectilinear configuration of execution in the model in question, with non-interacting edges, was located symmetrically relative to the center of the construction.

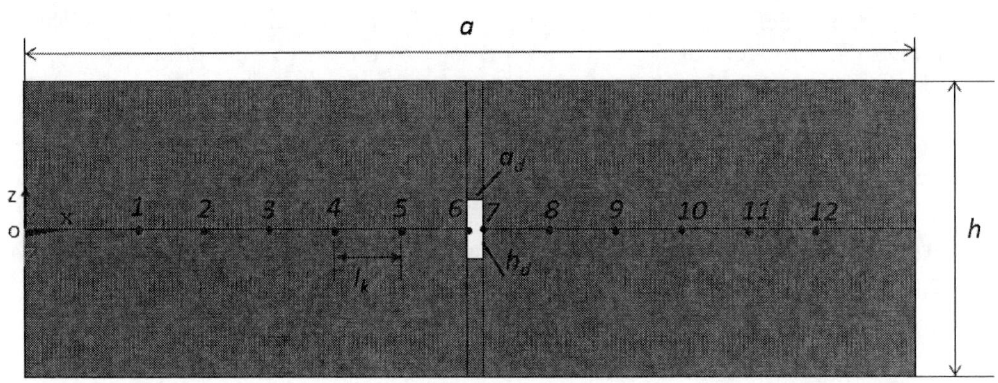

Figure 1. Scheme of wall construction: points 1 – 12 present places of control measurements; at point 1 a load pulse is applied.

Table 1. Construction parameters

Dimensions of wall construction	
Length (a), mm	3000
Height (h), mm	1000
Thickness (t), mm	80
Defect size	
a_d, mm	1
h_d, mm	200
Material properties of the structure	
Model material	Gypsum
Elastic modulus (E), GPa	15
Density (ρ), kg/m3	1200
Poison's ratio, ν	0.33

The main axes of the coordinates have a beginning at the center of the left face (O*xyz*). The axis Oy is directed perpendicular to the main plane of the construction.

2. Task of Research

On the base of numerical modeling in the FE software ANSYS, the wave processes under pulsed excitation, we analyze the vibration parameters to propose the identification signs of the presence of a defect in the form of the through crack in the wall construction.

3. Formulation of Model

In modeling, the finite element of the *SHELL281* type was used in the FE software ANSYS (see Figure 2). This type of element is suitable for the analysis of thin and moderately thick shells. The element has eight nodes with six degrees of freedom at each node: shifts along the *x*, *y* and *z* axes and turns around the *x*, *y* and *z* axes. *SHELL281* is well suited for non-linear applications with linear, large rotation, or large deformation. Changes in shell thickness are taken into account in nonlinear analysis. *SHELL281* can be used for multi-level applications by modeling composite shells or multilayer structures. The formulation of the element is based on a logarithmic strain and measurements of the actual stresses.

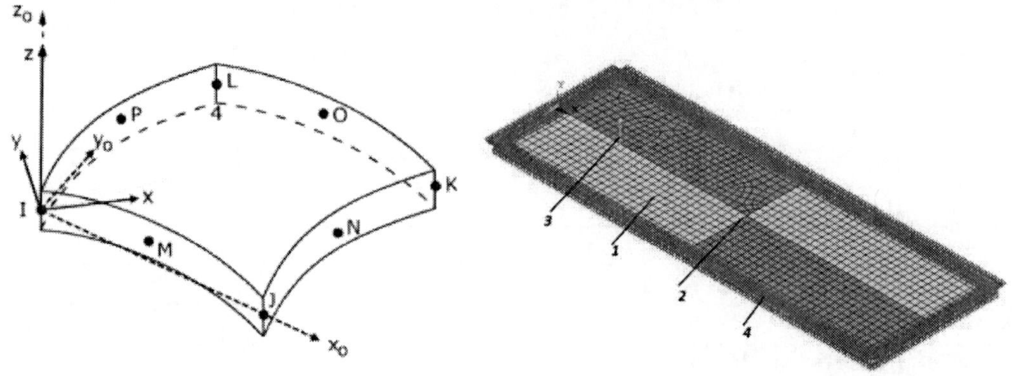

(a) (b)

Figure 2. (a) Finite element of type *SHELL281*; (b) Finite element model of wall construction: 1 – plane of wall construction, 2 – location of crack, 3 – place of application of the impact load, 4 – boundary conditions.

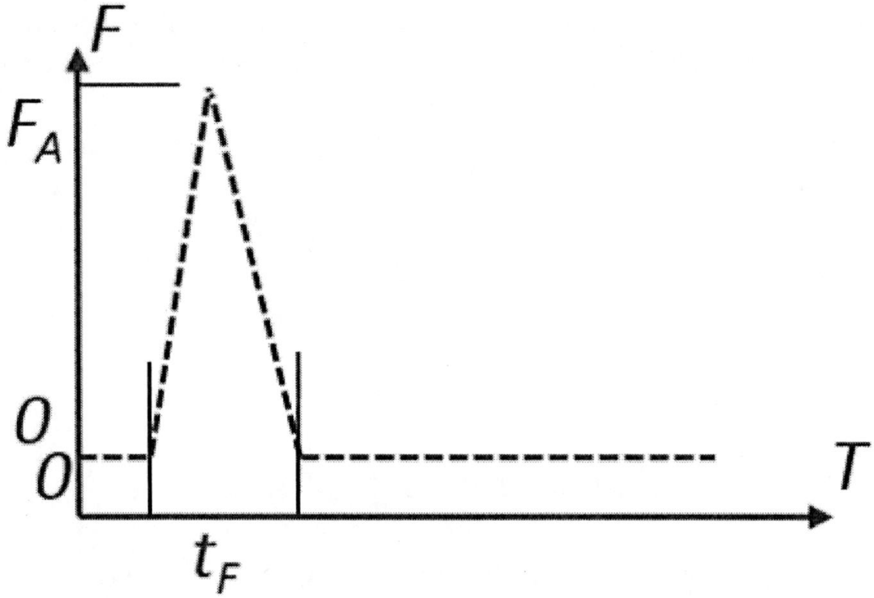

Figure 3. Pulse load in the form of force F_a applied at point 1.

A distinctive feature of the application of this element, by solving three-dimensional problems, is that the thickness deformations are taken into account in the parameters of the element itself.

Then, the model scheme and time of calculation are simplified, and the problem can be reduced to a plane one.

After formulation, the model was divided into 1216 finite elements with an average face length of 50mm. In some elements, near the nodal points of the construction, for example, in the area of the formation of a defect, the size of the FE can be reduced. The number of nodes at division was 3820.

4. Modeling

Nonstationary oscillations are modeled in the FE software ANSYS. At point 1 (Figure 1), a pulsed load is applied in the form of a force (Figure 3) with a linearly increasing load. The force F_a = 1 N is chosen as the step increment of the load.

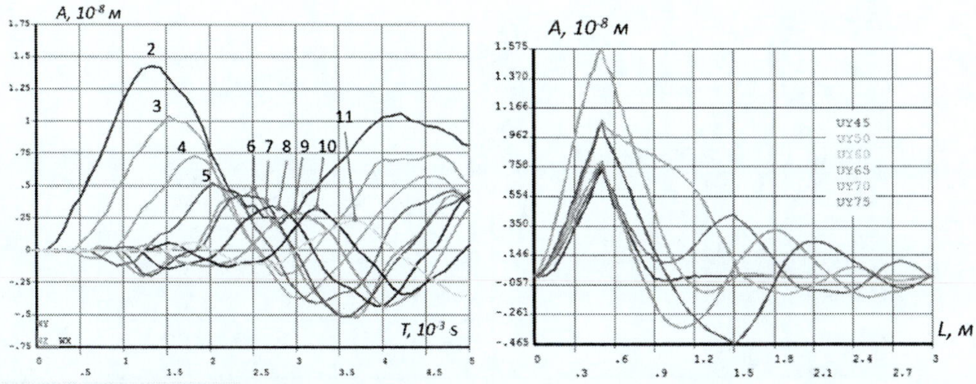

(a) (b)

Figure 4. Amplitudes of oscillations at the control points 2 – 11 (a), and the wave field of the transverse displacements of points, located on the *x*-axis at different time intervals.

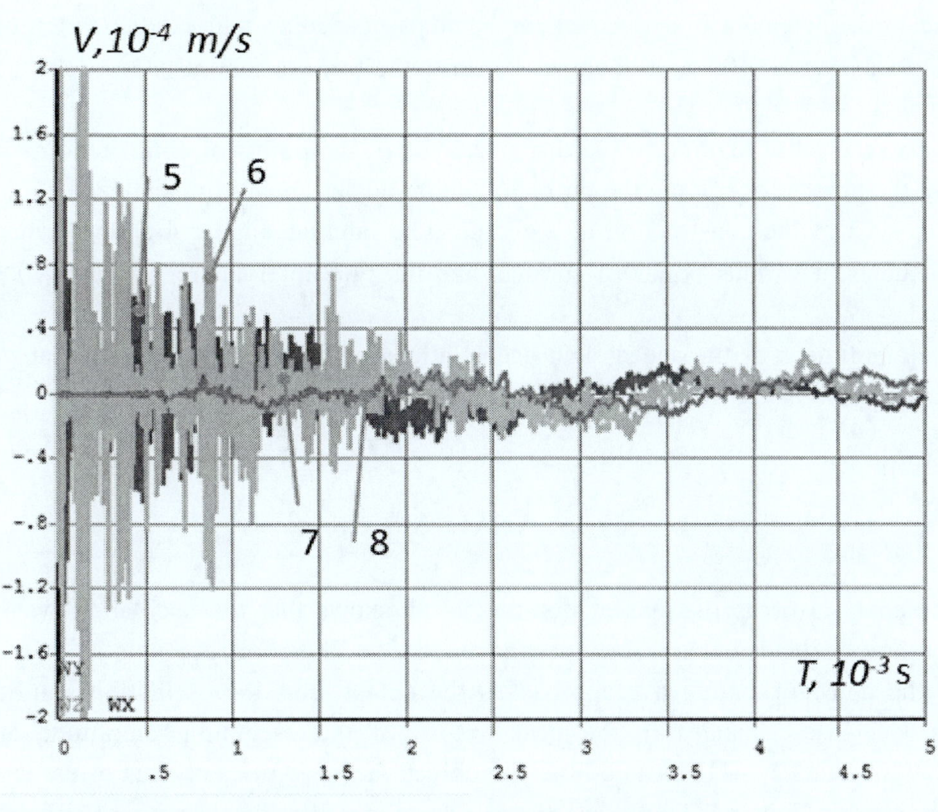

Figure 5. Amplitudes of the speeds of oscillations, measured at points 5 – 8.

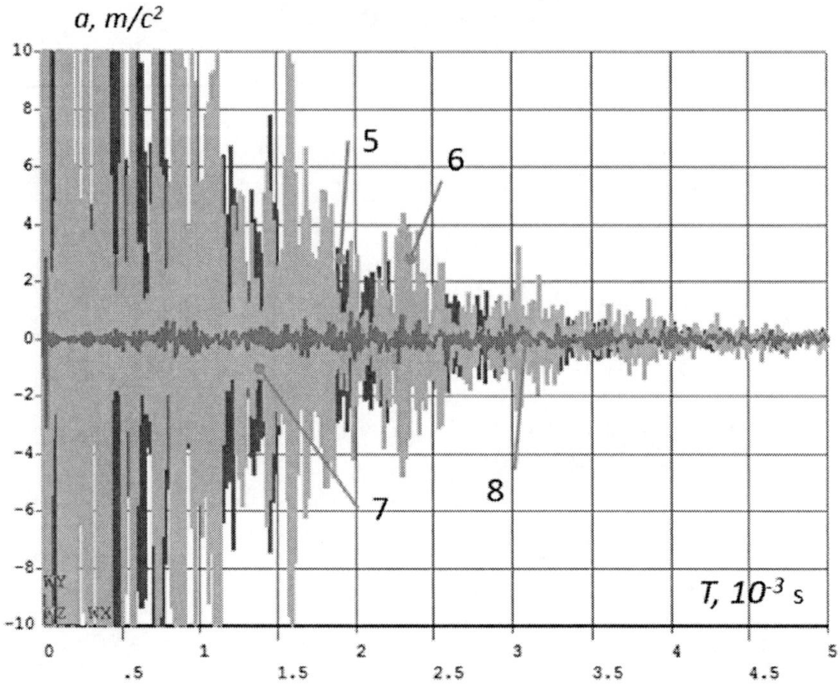

Figure 6. Amplitudes of the accelerations of oscillations, measured at points 5 – 8.

The force increment is performed for 10 steps of loading with discretization of the time interval: $dt_i = 1\times10^{-6}$ s. In the next 10 steps, the force is subtracted at each stage of the decision.

Due to the pulse loading, wave displacements of the points of construction surface arise. In the process of solving the problem, lateral displacements are analyzed at control points 1 – 12 of the construction at a certain time interval. Figure 4 shows transverse displacements at various points of surface, and the time interval from 0 to 5×10^{-3} s is analyzed.

The amplitudes of the speeds and accelerations of oscillations, measured at points 5 – 8, are present in Figures 5 and 6, respectively.

5. Analysis

The analysis of oscillations at points 2 – 11 shows that the oscillation wave, by passing through the defect area, changes the amplitude. Therefore, at points 2 – 6, located before the defect when a wave approaches the defect area, the oscillation amplitudes linearly decrease. Changes in the peak values of the oscillation amplitude at the observation points 7 – 11, located after the defect, are insignificant. This effect is quite well observed in Figure 4. This sign can serve as an identifier for the presence of a defect in the form of a crack in a wall construction. By analyzing the velocities and accelerations of oscillations, a comparison was made at points 5 and 6, located before the defect, and at points 7 and 8, after the defect. The analysis shows that the amplitudes of the wave velocities after the defect are sharply reduced. The peak values of both speed and acceleration are reduced by more than 10 times.

Conclusion

The problem of unsteady excitation of oscillations in wall construction is considered. The results of transverse displacements at the control points of the construction, during the propagation of waves from a pulsed excitation at a certain point, are given. Signs based on abrupt changes in the amplitude, speed, and acceleration of the transmitted wave can be applied after detailed analysis as identification signs for presence of defects in this type of construction.

Acknowledgments

The research was partially supported by the Russian Department of Education and Science No. 11.6767.2017/BCh, and No. 3.5378.2017/VU (organization of the implementation of researches).

References

[1] Akopyan, V. A.; Bychkov, A. A.; Rozhkov, E. V.; Soloviev, A. N.; Shevtsov, S. N. *Mechanics of Composite Materials and Constructions*, 2008, *vol. 14(1)*, 35 - 48.

[2] Cherpakov, A. V.; Shlyakhova, E. A.; Egorochkina, I. O.; Kokareva, Y. A. *Materials Science Forum*, 2018, *vol. 931*, 373 - 378.

[3] Shevtsov, Sergey N.; Arkady N. Soloviev; Ivan A. Parinov; Alexander V. Cherpakov; Valery A. Chebanenko. Piezoelectric Actuators and Generators for Energy Harvesting – Research and Development. *Innovation and Discovery in Russian Science and Engineering*, Springer Cham, Switzerland. 2018.

[4] Akopyan, V. A.; Rozhkov, E. V.; Soloviev, A. N.; Shevtsov, S. N.; Cherpakov, A. V. *Damage Identification in Elastic Constructions: Approaches, Methods, Analysis.* Southern Federal University Press, Rostov-on-Don. 2015 (In Russian).

[5] Soloviev, A. N.; Parinov, I. A.; Cherpakov, A. V.; Chaika, Yu. A.; Rozhkov, E. V. *Materials Physics and Mechanics,* 2018, *vol. 37(2)*, 192 - 197.

[6] Cherpakov, A. V.; Soloviev, A. N.; Gricenko, V. V.; Goncharov, O. U. *Defence Science Journal,* 2016, *No. 1*, 44 - 50.

[7] Soloviev, A. N.; Cherpakov, A. V.; Egorochkina, I. O.; Shlyakhova, E. A.; Parinov, I. A. In: *Proceedings of the 2016 International Conference on Physics and Mechanics of New Materials and Their Applications*, Ivan A. Parinov, Shun-Hsyung Chang, Muaffaq A. Jani (Eds.). Nova Science Publishers, New York. 2017, 439 - 446.

[8] Cherpakov, A.; Egorochkina, I.; Shlyakhova, E.; Kharitonov, A.; Zarovy, A; Dobrohodskaya, S. *MATEC Web of Conferences,* 2017, *vol. 106*, 04009.

[9] Cherpakov, A. V.; Akopyan, V. A.; Soloviev, A. N. *Technical Acoustics*, 2013, *vol. 13*, 1.

[10] Krasnoshchekov, A. A.; Sobol, B. V.; Solov'ev, A. N.; Cherpakov, A. V. *Russian Journal of Nondestructive Testing,* 2011, *vol. 47(6)*, 412 - 419.

[11] Egorochkina, I. O.; Shlyakhova, E. A.; Cherpakov, A. V.; Zarovny, A. G.; Kucherenko, D. Yu.; Lezhnev, V. N. *Engineering Journal of Don*, 2017, *No. 1(44)*, 102 (In Russian).

[12] Akopyan, V. A.; Solov'ev, A. N.; Cherpakov, A. V.; Shevtsov, S. N. *Russian Journal of Nondestructive Testing,* 2013, *vol. 49(10)*, 579 - 583.

[13] Egorochkina, I. O.; Shlyakhova, Ye. A.; Cherpakov, A. V.; Soloviev, A. N. *Engineering Journal of Don*, 2015, *vol. 38(4)*, 94 (In Russian).

[14] Soloviev, A. N.; Cherpakov, A. V.; Parinov, I. A. In: *Proceedings of the 2015 International Conference on Physics and Mechanics of New Materials and Their Applications, devoted to 100-th Anniversary of the Southern Federal University,* Ivan A. Parinov, Shun-Hsyung Chang, Vitaly Yu. Topolov (Eds.). Nova Science Publishers, New York. 2016, 515 - 520.

[15] Cherpakov, A. V.; Chebakov, M. I.; Zecheru, G. A.; Dumitrescu, A. In: *Advanced Materials – Techniques, Physics, Mechanics and Applications.* Springer Proceedings in Physics. Springer Cham, Heidelberg, New York, Dordrecht, London. 2017, 373 - 380.

[16] Verbickij, V. A.; Shilyaeva, O. V. *Exact Science*, 2017, *No. 9*, 24 - 27 (In Russian).

[17] Shilyaeva, O. V.; Muradyan, V. A.; Blyagoz, A. M.; Smorgunova, M. V. *New Technologies,* 2013, *No. 3*, 148 - 154 (In Russian).

[18] Shilyaeva, O. V.; Hunagov, R. A.; Blyagoz, A. M. *New Technologies*, 2012, *No. 3*, 149 - 153 (In Russian).

[19] Lyapin, A. A.; Seleznyov, N. M.; Shilyaeva, O. V. *Ecological Journal of Scientific Centers of Black Sea Economic Collaboration,* 2008, *No. 2*, 45 - 50 (In Russian).

[20] Seleznev, N. M.; Shilyaeva, O. V.; Vu Thi Bik Kuen. Izvestiya of High-Educational Institutes. North-Caucasus Region. *Technical Sciences,* 2009, *No. 3(151)*, 10 - 13 (In Russian).

In: Proceedings of the 2018 International Conference … ISBN: 978-1-53615-862-5
Editors: Ivan A. Parinov et al.

Chapter 26

COMPUTATIONAL AND EXPERIMENTAL APPROACH TO THE EVALUATION OF THE PROPERTIES OF CONCRETE CYLINDRICAL TUBE BASED ON THE VIBRODIAGNOSTIC METHOD

***A. V. Cherpakov*[1,2,*], *S. A. Stel'makh*[2,†], *E. M. Shcherban*[2], *M. P. Nazhuev*[2] *and I. A. Parinov*[1]**

[1]Southern Federal University, Rostov-on-Don, Russia
[2]Don State Technical University, Rostov-on-Don, Russia

ABSTRACT

The problem of identifying the mechanical properties of concrete in a cylindrical element of a tubular section is considered. The sample was manufactured using the technology of centrifugation of concrete. Numerical-experimental modeling based on the analysis of natural frequencies of oscillations is used. The own frequencies of free oscillations of a tubular sample were obtained experimentally. A numerical simulation of the eigenmodes in the finite element complex ANSYS of the sample was carried out for different variations of the elastic modulus and average density. Based on the comparison of the natural frequencies obtained experimentally and numerically, the preliminary properties of the elastic module of the object under study are calculated.

Keywords: cylindrical element, tubular section, centrifugation, concrete, experimental approach, finite element modeling, mechanical properties, evaluation of properties, vibrations, natural frequencies

---

* Corresponding Author's E-mail: alex837@yandex.ru.
† Corresponding Author's E-mail: sergej.stelmax@mail.ru.

1. Introduction

In modern building construction, one of the most demanded construction materials is heavy concrete; one of the types of the goods manufactured using this material are tubular structural elements. The use of centrifugation and special treatment in the process of manufacturing tubes allows an increase in the strength of these goods. Due to the production required controls of the output properties of goods, specialized tests and selective control are applied to them with the selection of test batches [1 – 3]. At the same time, there is a nomenclature of centrifuged thin-wall precast concrete goods of tubular configuration, which has a number of shortcomings associated with the specific distribution of properties over the cross-section of goods.

The scientific interest is the study of the possibilities of the diagnostics and control of properties in the production of centrifuged thin-walled products of ring configuration, as well as ensuring the quality and durability of such products.

For this purpose, the authors conducted a series of large-scale experimental studies in the research laboratory of the department TVVB&SK DSTU. To systematize the accumulated experimental data, tubular specimens of a circular section of concrete were molded and studied [1–7]. The scheme and example of tubular sample are shown in Figure 1. Table 1 shows the dimensions of the tubular structure: external and internal diameters, wall thickness, sample length, measurements of the sample mass, and the average density of the object.

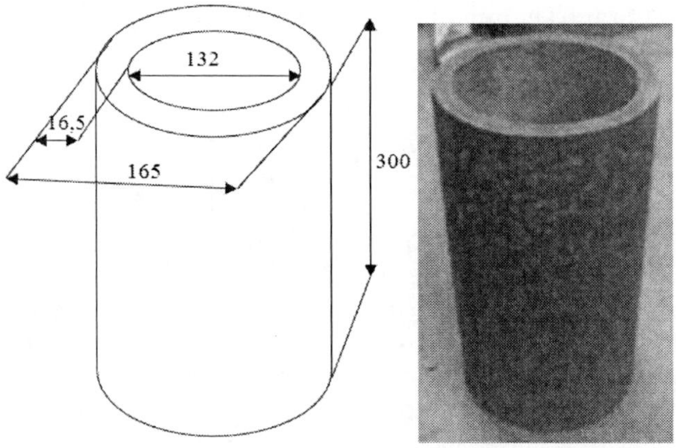

(a) (b)

Figure 1. Scheme (a) and full-scale sample (b) of centrifuged good.

Table 1. Parameters of the tubular structure

D_{out}, mm	D_{int}, mm	h_c , mm	L, mm	M, kg	ρ, kg/m^3
165	132	16.5	300	3.870	1675.6

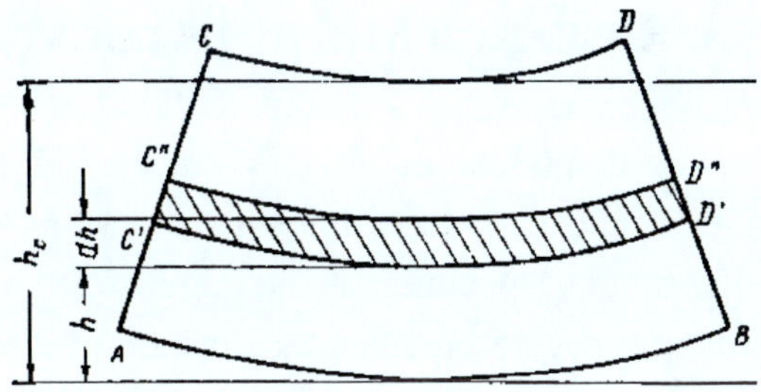

Figure 2. Diagram of single-layer and layer-by-layer molding of a centrifuged product.

The tubular sample was made using the technology of centrifuged compaction of the concrete mix. In the manufacture of the sample, it is possible to control the speed of rotation of the form over a wide range. This parameter is consistent and depends on the viscosity of the concrete mix and the pressing load (p = 0.7 kgf/cm^2; p = 1.45 kgf/cm^2; p = 2.16 kgf/cm^2) in the range from 825 to 4400 rpm. Creating a different pressing load in a laboratory setup, allowed us to obtain data on its effect on the heterogeneous structure of concrete, which can be indirectly confirmed by the spread of the average density over the cross-section of the good and other characteristics.

Figure 2 shows the conventional design scheme for single-layer and layer-by-layer moldings of a centrifuged product. By considering a certain ABCD volume with a height h_c, separated from a cylindrical element of a tubular section, the problem of the distribution of the average density of a concrete mixture, over the thickness and length of the product, was considered [1, 6, 7]. An analysis of the density of the distribution of the material over the thickness of the moldable wall shows that the density increases from the inner to outer wall of the good. The density distribution of the material can be represented by a linear relationship [1, 6, 7]. In the calculations, the characteristic of the average density of the good can be taken.

Due to the unknown density distribution over the thickness of the construction, as well as the parameters of the elastic characteristics, namely the elastic modulus, the problem arises of restoring the properties of the construction. One of the approaches for identifying properties is the vibrodiagnostic method [8 – 13]. This approach may be based on the use of analytical or finite element modeling [14 – 22].

2. Problem Formulation

Based on the comparison of the experimental approach and numerical simulation, the elastic properties, namely the average elastic modulus of the object under investigation, should be restored. The eigen-frequencies and dimensional parameters of the experimental model of the research object are used as input data for the calculation.

3. RESEARCH METHODOLOGY

Let us apply the vibration diagnostic method. We measured the eigen-frequencies of oscillations with longitudinal and transverse impact excitations of oscillations. A steel washer was attached to the tubular specimen at the edge in accordance with the scheme shown in Figure 3. A three-axis accelerometer was attached to it on a magnetic base. The accelerometer can be mounted at any point of the good. A tubular sample to imitate free vibrations was hung on an elastic fishing line. At the point 3 (Figure 3) in the direction of T1 and T2, it was performed on impact by a hammer.

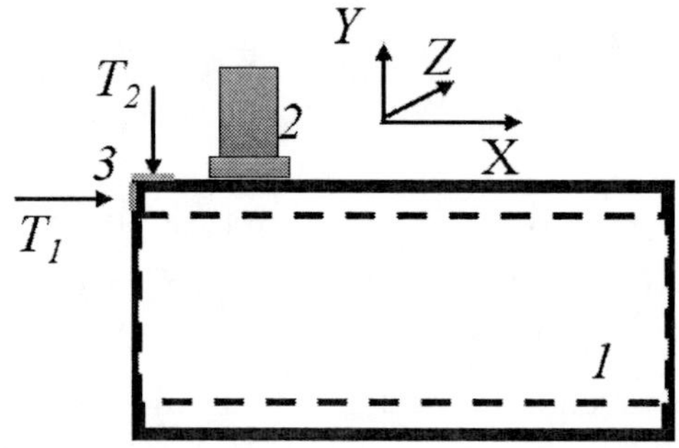

Figure 3. Scheme of tube design with attached accelerometer: 1 – tubular construction; 2 – accelerometer sensor; 3 – point of impact, T1 – face area, T2 – lateral area in the vicinity of the sensor

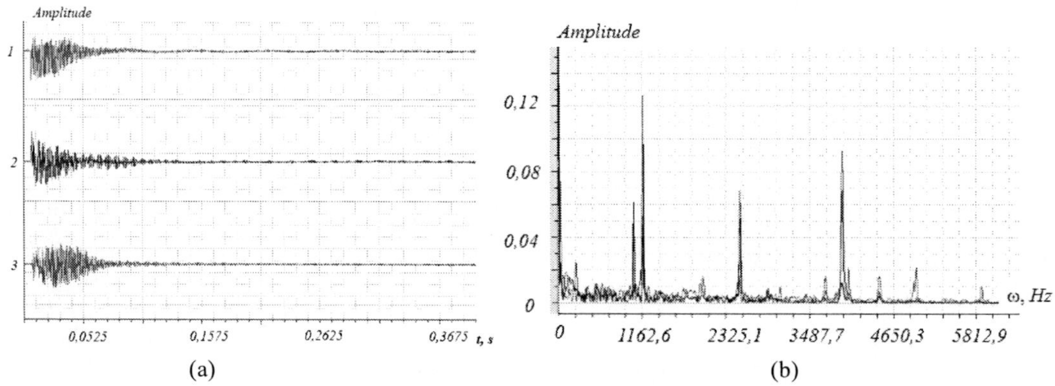

Figure 4. Acceleration amplitudes (a), recorded by the accelerometer, and eigen-frequencies (b) in the corresponding directions, at impact of a hammer on the area in the direction T1 of the tubular structure.

Table 2. Experimental eigen-frequencies of oscillations of the tubular structure

	N mode					
	1	2	3	4	5	6
ω, Hz	1034	1174	2519	3115	3699	3935

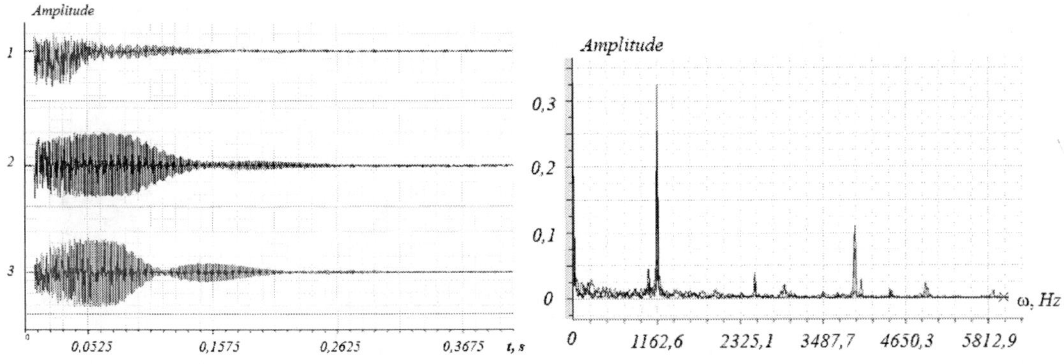

(a) (b)

Figure 5. Acceleration amplitudes (a), recorded by the accelerometer, and eigen-frequencies (b) in the corresponding directions, at impact of a hammer on the area in the direction T2 of the tubular structure.

The oscillation amplitudes of construction at a given point in three directions were recorded by using computer software [23, 24]. Figures 4 and 5 show the oscillation amplitudes at the point of attachment at the accelerometer sensor, when a hammer made impact on the face area in the T1 direction (Figure 4) and on the lateral area near the sensor in the T2 direction (Figure 5). In the results of the study, the corresponding vibration spectra were obtained. Table 2 shows the eigen-frequencies of the sample. An analysis of eigen- frequencies shows that six modes of oscillations were detected in the frequency range of 1034 – 3935 Hz.

4. Modeling

In ANSYS software, a full-bodied finite element model of a tubular structure was built. The dimensional parameters of the model were taken in correspondence to the experimental data (see Table 1). As a basic finite element, ten-node element of the SOLID92 type was used. This element has a tetrahedral structure of execution and six degrees of freedom at each of the nodes. In the computational model, it is assumed that the properties of the sample are uniformly distributed in bulk. The constructed model had 5317 nodes and 2582 elements. The average size of the side elements was assumed equal to 20 mm. Figure 6 shows the FE-model of the tube. Due to in the experimental model having absent rigid fixings, a free model was then considered without rigid fixation nodes.

The natural oscillations of the free construction without fixing were considered. The damping parameters in the sample were not taken into account. The optimization task of the elastic modulus in dependence on the eigen-frequencies of oscillations was carried out on the base of a modal calculation. A series of calculations was performed for different values of the elastic modulus and the eigen-frequencies ω_r of the first 20 modes of oscillations. The calculation results are shown in Table 3. The primary analysis shows that the eigen-frequencies of the model vary from 988 to 5039 Hz with different variations of the elastic modulus values.

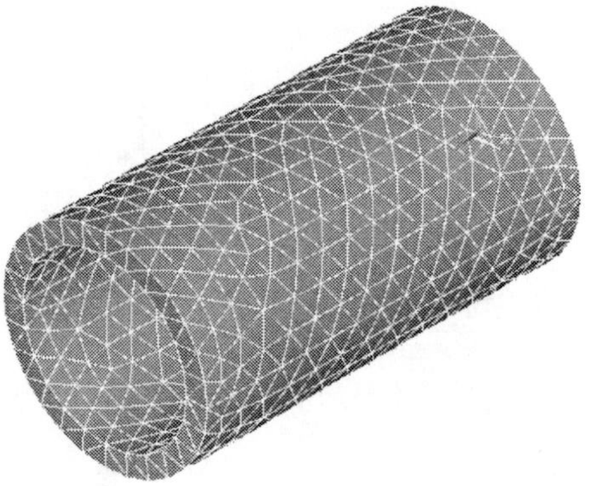

Figure 6. FE-model of tubular structure in ANSYS software.

Table 3. Dependence of the eigen-frequencies of the tubular model, calculated numerically, on the specified average elastic modulus of concrete

Mode No.	Average calculated modulus of elasticity of concrete E_r, MPa						
	12	13	13.5	13.8	14	15	16
	Natural frequencies ω_r, Hz						
1	988.3	1028.7	1048.3	1059.9	1067.6	1105.0	1141.3
2	988.8	1029.2	1048.9	1060.4	1068.1	1105.6	1141.8
3	1095.7	1140.4	1162.2	1175.0	1183.5	1225.0	1265.2
4	1095.8	1140.6	1162.3	1175.2	1183.6	1225.2	1265.4
5	2057.9	2141.9	2182.7	2206.8	2222.7	2300.8	2376.2
6	2057.9	2142.0	2182.8	2206.9	2222.8	2300.8	2376.3
7	2721.4	2832.5	2886.4	2918.3	2939.4	3042.6	3142.4
8	2721.9	2833.0	2887.0	2918.9	2940.0	3043.2	3143.0
9	2860.0	2976.8	3033.5	3067.0	3089.1	3197.5	3302.4
10	2860.9	2977.7	3034.4	3067.9	3090.1	3198.6	3303.5
11	2891.0	3009.0	3066.3	3100.2	3122.6	3232.2	3338.2
12	2891.0	3009.0	3066.3	3100.2	3122.6	3232.2	3338.2
13	2903.4	3021.9	3079.5	3113.5	3136.0	3246.1	3352.5
14	3399.2	3538.0	3605.4	3645.3	3671.6	3800.4	3925.1
15	3400.3	3539.1	3606.5	3646.4	3672.7	3801.6	3926.3
16	3565.3	3710.9	3781.6	3823.4	3851.0	3986.2	4116.9
17	3565.8	3711.4	3782.1	3823.9	3851.5	3986.7	4117.4
18	3659.5	3808.9	3881.5	3924.4	3952.7	4091.5	4225.6
19	3659.5	3808.9	3881.5	3924.4	3952.7	4091.5	4225.6
20	4364.0	4542.2	4628.7	4679.9	4713.7	4879.1	5039.1

5. Analysis

For the analysis, we chose the most suitable modes for the experimental results that have the minimum summary discrepancy ε of eigen-frequencies at one nominal value of the modulus of elasticity of concrete E_r. For a comparative analysis, 1, 3, 11, 14, 18 oscillation modes were chosen. The minimization problem was considered in the case of divergence of the eigen-frequencies of the full-scale and calculated models. The frequency discrepancy was calculated as

$$\varepsilon_i = \min_E \left| \frac{\omega_{i\,calc} - \omega_{i\,exper}}{\omega_{i\,exper}} \right| \cdot 100\% \tag{1}$$

The relative differences ε of the eigen-frequencies calculated numerically from those, and obtained on the base of an experimental approach with different variations of the average modulus of elasticity of concrete E_r,, are given in Table 4. By averaging the results, the summarized average discrepancy ε_{cp} of the natural frequencies from the experimental one is calculated as

$$\varepsilon_{cp} = \left| \frac{\sum_N \varepsilon_i}{N} \right| \tag{2}$$

The analysis of plots in Figure 7 shows that the minimum parameters ε and ε_{cp} are obtained with the modulus of elasticity E_r = 13.8 MPa.

Table 4. Relative discrepancy ε of the eigen-frequencies calculated numerically from obtained in experiment with different average elastic modulus E_r of concrete

Mode No.	Average calculated modulus of elasticity of concrete E_r, MPa						
	12	13	13.5	13.8	14	15	16
	Relative discrepancy of frequencies, ε, %						
1	4.41	0.51	1.38	2.50	3.25	6.87	10.38
3	6.67	2.86	1.01	0.09	0.81	4.34	7.77
11	7.19	3.40	1.56	0.48	0.24	3.76	7.17
14	8.10	4.35	2.53	1.45	0.74	2.74	6.11
18	7.00	3.20	1.36	0.27	0.45	3.98	7.39
ε_{cp} ,%	6.68	2.87	1.57	0.96	1.10	4.34	7.76

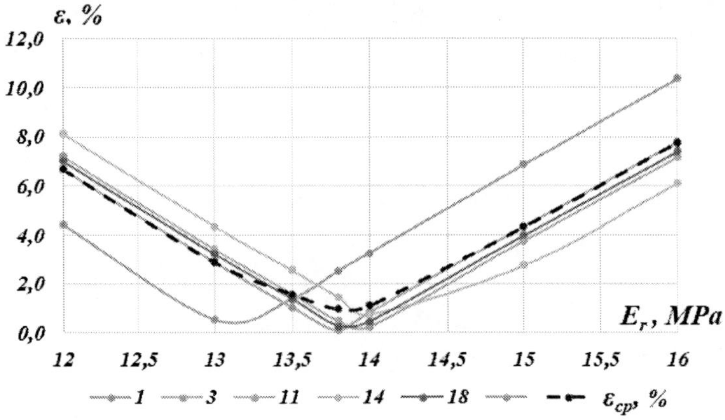

Figure 7. Relative discrepancies of eigen-frequencies, calculated numerically, from those obtained on the base of experiments with different average elastic modulus E_r of concrete.

CONCLUSION

Based on a comparison of the experimental approach and numerical simulations, the elastic modulus E_r of the tubular construction is restored. Based on solving the minimization problem, the discrepancy of the eigen-frequencies obtained experimentally, from those calculated in the FE software ANSYS at the elasticity model E_r = 13.8 MPa, was obtained.

ACKNOWLEDGMENTS

The research was partially supported by the proposal of the Russian Department of Education and Science No. 11.6767.2017/BCh and 3.5378.2017/VU (organization of the implementation of researches), and also the Russian Foundation for Basic Research (grants No. 16-08-00740-a).

REFERENCES

[1] Nazhuev, M. P.; Yanovskaya, A. V.; Kholodnyak, M. G.; Khalyushev, A. K.; Shcherban', E. M.; Stel'makh, S. A. *ESJ Engineering Journal of Don*, 2017, *No. 3* (In Russian).

[2] Mailyan, L. R.; Stel'makh, S. A.; Kholodnyak, M. G.; Shcherban', E. M. *Naukovedenie*, 2017, *No. 4* (In Russian).

[3] Mailyan, L. R.; Stel'makh, S. A.; Kholodnyak, M. G.; Shcherban', E. M. *Vestnik BGTU im. V. G. Shuhova*, 2017, *No. 10*, 52-57 (In Russian).

[4] Mailyan, L. R.; Stel'makh, S. A.; Kholodnyak, M. G.; Shcherban', E. M. *Vestnik SevKavGTI*, 2017, *vol. 3 (30)*, 134-137 (In Russian).

[5] Mailyan, L. R.; Stel'makh, S. A.; Khalyushev, A. K.; Shcherban', E. M.; Kholodnyak, M. G.; Nazhuev, M. P. *ESJ Engineering Journal of Don*, 2018, *No. 3* (In Russian).

[6] Mailyan, L. R.; Stel'makh, S. A.; Khalyushev, A. K.; Shcherban', E. M.; Kholodnyak, M. G.; Nazhuev, M. P. *Vestnik Evrazijskoj Nauki*, 2018, *No. 3* (In Russian).

[7] Mailyan, L. R.; Stel'makh, S. A.; Kholodnyak, M. G.; Khalyushev, A. K.; Shcherban', E. M.; Nazhuev, M. P. *Vestnik Evrazijskoj Nauki*, 2018, *No. 4* (In Russian).

[8] Akopyan, V. A.; Bychkov, A. A.; Rozhkov, E. V.; Soloviev, A. N.; Shevtsov, S. N. *Mechanics of Composite Materials and Structures*. 2008, *vol. 14(1)*, 35 – 48 (In Russian).

[9] Bocharova, O. V.; Vatulyan, A. O. *Acoustic Journal*. 2009. *vol. 55(3)*. 275 – 282 (In Russian).

[10] Cherpakov, A. V.; Shlyakhova, E. A.; Egorochkina, I. O.; Kokareva, Y. A. *Materials Science Forum*. 2018, *vol. 931*, 373 – 378 (In Russian).

[11] Sergey N. Shevtsov, Arkady N. Soloviev, Ivan A. Parinov, Alexander V. Cherpakov, Valery A. Chebanenko. *Piezoelectric Actuators and Generators for Energy Harvesting - Research and Development*. Innovation and Discovery in Russian Science and Engineering. Springer, Switzerland. 2018.

[12] Akopyan, V. A.; Rozhkov, E. V.; Soloviev A. N.; Shevtsov S. N.; Cherpakov A. V. *Damage Identification in Elastic Structures: Approaches, Methods, Analysis.* Southern Federal University Press, Rostov-on-Don. 2015, 1 – 74 (In Russian).

[13] Khakimov, A. G. *ESJ Oil and Gas Business*. 2015, *No. 4*, 151 – 171.

[14] Soloviev, A. N., Parinov, I. A., Cherpakov, A. V., Chaika, Yu. A., Rozhkov, E. V. *Materials Physics and Mechanics*. 2018, *vol. 37(2),* 192 – 197.

[15] Cherpakov, A. V.; Chebakov, M. I.; Zecheru, G.; Dumitrescu, A. In: *Advanced Materials - Techniques, Physics, Mechanics and Applications*, Springer Proceedings in Physics, Ivan A. Parinov, Shun-Hsyung Chang, Muaffaq A. Jani (Eds.). Springer Cham, Heidelberg, New York, Dordrecht, London. 2017, 373 – 380.

[16] Soloviev, A. N.; Cherpakov, A. V.; Egorochkina, I. O.; Shlyakhova, E. A.; Parinov, I. A. In: *Proceedings of the 2016 International Conference on Physics and Mechanics of New Materials and Their Applications*, Ivan A. Parinov, Shun-Hsyung Chang, Muaffaq A. Jani (Eds.). Nova Science Publishers, New York. 2017, 439 – 446.

[17] Cherpakov, A. V.; Akopyan, V. A.; Soloviev, A. N. *Technical Acoustics*. 2013, *vol. 13*.

[18] Krasnoshchekov, A. A.; Sobol, B. V.; Solov'ev, A. N.; Cherpakov, A. V. *Russian Journal of Nondestructive Testing*, 2011, *vol. 47(6)*, 412 – 419.

[19] Akopyan, V. A.; Soloviev, A. N.; Cherpakov, A. V.; Shevtsov, S. N. *Russian Journal of Nondestructive Testing*, 2013, *vol. 49(10)*, 579 – 583.

[20] Cherpakov, A.; Egorochkina, I.; Shlyakhova, E.; Kharitonov, A.; Zarovny, A.; Dobrohodskaya, S. *MATEC Web of Conferences*, 2017, *vol. 106*, 04009.

[21] Cherpakov, A. V., Soloviev, A. N., Gricenko, V. V., Goncharov, O. U. *Defence Science Journal*, 2016, *No. 1*, 44 – 50.

[22] Egorochkina, I. O.; Shlyakhova, Ye. A.; Cherpakov, A. V.; Soloviev, A. N. *ESJ Engineering Journal of Don*, 2015, *vol. 38(4).*

[23] Parinov, I. A.; Cherpakov A. V.; Rozhkov E. V.; Soloviev A. N.; Chebanenko V. A. *Software. Generator Signal SGenerator*. RF Certificate of State Registration of the Computer Program, No. 2018610408, 10.01.2018.

[24] Parinov, I. A.; Soloviev A. N.; Cherpakov A. V. *Software. "VibroGraf" for Registration, Visualization and Processing of Structural Vibrations.* RF Certificate of State Registration of the Computer Program, No. 2016612309, 20.03.2016.

In: Proceedings of the 2018 International Conference … ISBN: 978-1-53615-862-5
Editors: Ivan A. Parinov et al.

Chapter 27

EXPANDING ADDITIVE TO REPAIR COMPOSITIONS FOR RESTORATION OF POWER LINE SUPPORTS

***E. A. Shlyakhova*[1], *I. O. Egorochkina*[1,*] *and A. V. Cherpakov*[1,2]**
[1]Don State Technical University, Rostov-on-Don, Russia
[2]Southern Federal University, Rostov-on-Don, Russia

ABSTRACT

In this paper, we studied the possibility of creating effective expanding compositions intended for carrying out repair and restoration operations on the reinforced concrete power line supports, which are based on Portland cement, local fine sand, and components of the complex expanding additive available in regional conditions.

Keywords: effective expanding components, repair work, reinforced concrete power line supports

1. INTRODUCTION

Today, throughout the world and, in particular, Russia, considerable experience has been gained in the operation of electric transmission lines (hereinafter referred to as power lines). By the end of the 1990s, a great practical experience had been gained in

* Corresponding Author's E-mail: arin77@bk.ru.

applying various design solutions for constructing power lines, but with the collapse of the USSR, many problems had arisen. The most significant of them became that practically no funds were directed for routine inspections and repairs and a decline in the development of the country's economy led to the elimination of many power lines due to lack of demand. Unrecoverable power lines quickly fail and are subject to destruction, which in the future cannot be repaired even with the help of major repairs.

Over the past few decades, the deterioration of high-voltage power grids has increased significantly up to 40%, according to experts. This complex includes the supporting elements of power lines, namely supports and their foundations. The search for new solutions for restoration of their technical condition and the problems of modern diagnostics are the urgent tasks of the modern power industry.

The most important condition for the efficiency of repair compositions is the requirement of reliable adhesion between them and the base restored. This condition is met by compositions based on extending additives. The accumulated many-year practical experiences in using conventional solutions or sand concretes based on Portland cement for monolithing precast reinforced concrete constructions indicates that such structures do not provide sufficient required solidity. It is caused mainly due to the shrinkage of such compositions during setting and hardening. Due to the problems of creating expanding and non-shrinking cements, it has attracted great attention from numerous researchers.

Racks of power lines are the most important elements of these constructions. At present, the main part of power lines in Russia state on reinforced concrete supports.

The technical condition of many constructional elements of power lines need repair. In the result of the assessment of the actual state of the constructions, a plan of repair and restoration works has been formed and effective repair compositions are selected, to which a number of specific requirements are imposed [1-8].

2. Problem Formulation

In this paper, we studied the possibility of creating effective expanding compositions intended for carrying out repair and restoration works on the reinforced concrete power lines, based on Portland cement, local fine sand, and components of the complex expanding additive available in regional conditions.

3. Conducting Research

Treatment of the composition of expanding complex additives is based on the following discussion. In all known expanding cements, as on the base of high alumina cement and on the base of Portland cement, new formations of calcium hydrosulfophosphate type play a crucial role. Directional adjustment of the composition of Portland cement clinker is possible only under factory conditions for the manufacture of cement. Naturally, in the manufacture of mixtures for repair and restoration works in supports of power lines, it is not possible to control the kinetics of the formation of calcium hydrosulfonyl aluminate in the field conditions.

The content of gypsum in cement can be increased by adding the mixing water easily soluble sulfate-containing substances under normal conditions, for example, Glauber's salt (sodium sulfate, SS).

As an expanding component in the complex additive, it was decided to use aluminum powder. For its uniform distribution in the repair mixture, an additive of superplasticizer ST 2.1 was used in weight ratio to aluminum powder as 1:1.

Taking into account previous studies [4, 9], to obtain an expanding composition based on Portland cement binder, the content of the gas-forming additive of aluminum powder in the developed complex additive was taken equal to 0.01% of weight content of Portland cement.

In order to establish a rational amount of sodium sulphate in the composition of the expanding additive, developed for performance of repair and restoration works for racks of power line, a special calculation was made.

In the result of the analysis of previous data and the conducted theoretical calculations, and for further experimental studies of expanding compositions based on domestic raw materials, it is appropriate to use a developed expanding additive, including an additive ST.01, equal to 0.01% of weight content of Portland cement, aluminum powder and 1.5 – 2.5% of sodium sulfate.

The optimization of the composition of the developed complex additive was carried out with the performance of experiments on a two-factor simplex-summary plan of the second order's type of a regular hexagon [9]. The conditions for coding and variation of the studied factors, which correspond to the content of the components in the expanding additive developed, are present in Table 1.

Table 1. Conditions of variating and coding factors

Accepted levels and intervals of variating factors	Factor code	Natural values of factors in experiments, % of cement consumption	
		X_1 (SS)	X_2 (Al)
Main level (x_{i0})	x_i	1.9	0.01
Variation Interval (Δx_i)	–	0.6	0.0023
Upper level	+1	2.5	0.0123
Lower level	–1	1.3	0.0077

Table 2. Setting and implementation of experiments according to the plan of the experiment

No. of Test	Matrix plan in coded variables		Natural values of factors, % of cement consumption		Resultant responses (compressive strength of specimens at the age of 28 days of normal hardening) R_{com}, MPa
	x_1	x_2	SS (X_1)	Al (X_2)	
1	−1	0	1.3	0.010	32.5
2	−0.5	+0.87	1.6	0.012	29.6
3	+0.5	+0.87	2.2	0.012	32.9
4	+1	0	2.5	0.010	35.4
5	+0.5	−0.87	2.2	0.008	41.1
6	−0.5	−0.87	1.6	0.008	38.3
7	0	0	1.9	0.010	36.2

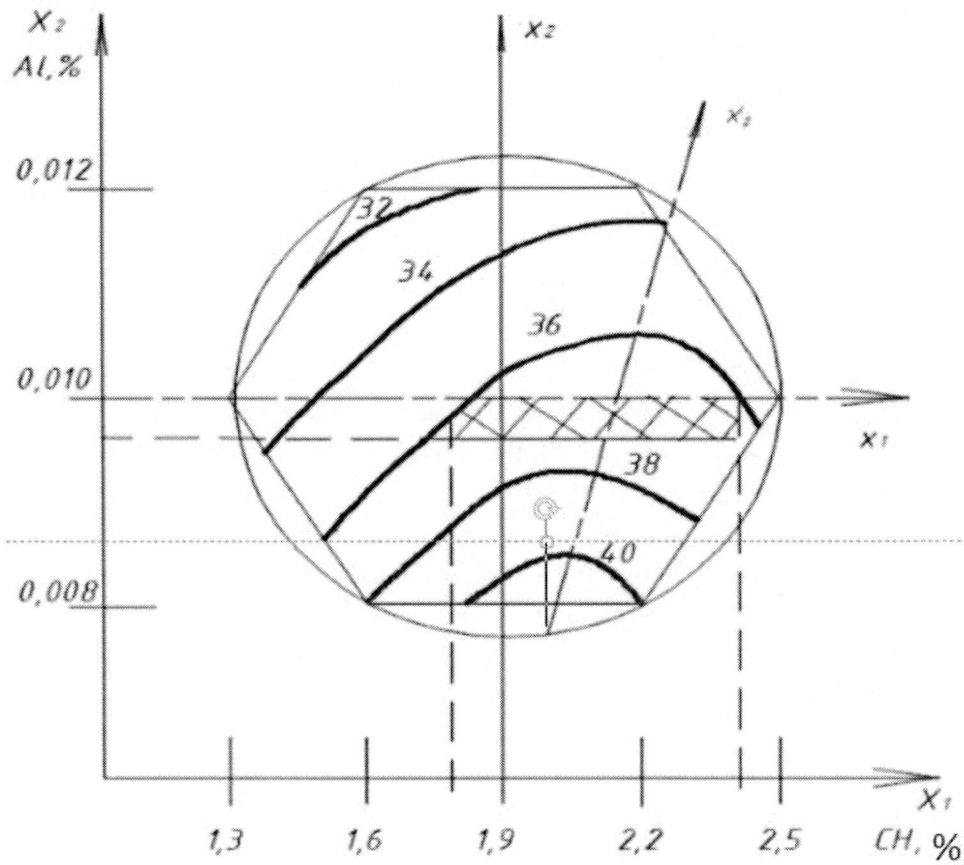

Figure 1. Isolines of the strength of the response function $R_{com} = f(x_1; x_2)$.

The planning matrix in the coded variables and the results of the implementation of the experiments with the corresponding values of the factors are present in Table 2.

In the result of the mathematical processing of the experimental data, given in Table 2, a quadratic model of the response function under adequate study at the 5% significance level was obtained:

$$R_{com} = 30.2 + 1.4\, x_1 - 4.1x_2 - 1.75x_{12} - 0.35x_{22} + 0.17x_1x_2 \quad (1)$$

By using Equation (1) and the methods of linear algebra [10], we constructed a geometric image of the studied response function $R_{com} = f(x_1; x_2)$, presented in Figure 1.

Analysis of the geometric image of the response function under the study (see, Figure 1) shows that, from the viewpoint of ensuring the required strength parameters, the shaded subregion of the factor space is optimum. It corresponds to the content of the SS-additive in the range of 1.8 – 2.4% and aluminum powder in the interval of 0.008 – 0.009% from the weight of cement.

ACKNOWLEDGMENTS

This research was carried out within the framework of financing of the Ministry of Education and Science of the Russian Federation for the project No. 11.6767.2017/BCh.

REFERENCES

[1] Egorochkina, I. O.; Shlyakhova, E. A.; Cherpakov, A. V.; Zarovny, A. G.; Kucherenko, D. Yu.; Lezhnev, V. N. *Engineering Journal of Don*, 2017, *vol. 44(1)*, 102 (In Russian).

[2] Egorochkina, I. O.; Shlyakhova, E. A.; Cherpakov, A. V.; Kucherenko, D. Yu.; Manvelyan, L. A. *Engineering Journal of Don*, 2016, *vol. 43(4)*, 172 (In Russian).

[3] Egorochkina, I. O.; Shlyakhova, E. A.; Cherpakov, A. V.; Soloviev, A. N. *Engineering Journal of Don*, 2015, *vol. 42(4)* (In Russian).

[4] Mikhailov, V. V. *Expanding and Straining Cements and Self-stressed Reinforced Concrete Constructions*. Stroyizdat, Moscow, 1974, pp. 1 - 312 (In Russian).

[5] Cherpakov, A.; Egorochkina, I.; Shlyakhova, E.; Kharitonov, A.; Zarovny, A.; Dobrohodskaya, S. In: *MATEC Web of Conferences Ser. "International Science Conference SPbWOSCE-2016" SMART City",*. 2017, 04009.

[6] Soloviev, A. N.; Cherpakov, A. V.; Egorochkina I. O.; Shlyakhova, E. A.; Parinov, I. A. In: *Proceedings of the 2016 International Conference on Physics and*

Mechanics of New Materials and Their Applications, Ivan A. Parinov, Shun-Hsyung Chang, Muaffaq A. Jani (Eds.). Nova Science Publishers. New York. 2017, 439 - 446.

[7] Egorochkina, I. O.; Shlyakhova, Ye. A.; Cherpakov, A. V. In: *Construction and Architecture – 2015, Proc. Int. Sci.-Pract. Conf.,* Rostov State University of Civil Engineering, Union of Builders of the Southern Federal District, Don Association of Builders. 2015, 410 - 412 (In Russian).

[8] Shlyakhova, E. A.; Egorochkina, I. O.; Cherpakov, A. V. *Construction and Architecture – 2017. Proc. Int. Sci.-Pract. Conf.,* Rostov State University of Civil Engineering, Union of Builders of the Southern Federal District, Don Association of Builders. 2017, 80 - 85 (In Russian).

[9] Pitersky, A. M. *Optimization of Solutions for Water Construction Problems. - NIMI,* Novocherkassk, 1995, pp. 1 - 115 (In Russian).

[10] Golovina, L. I. *Linear Algebra and Some Its Applications. Science,* Moscow, 1971, pp. 1 - 290 (In Russian).

In: Proceedings of the 2018 International Conference … ISBN: 978-1-53615-862-5
Editors: Ivan A. Parinov et al.

Chapter 28

DEVELOPMENT OF A METHOD FOR THE ANALYSIS OF STABILITY AND DIAGNOSTICS OF THE INTEGRAL CONSTRUCTION BY THE INTEGRAL INDEX

Yu. V. Esipov and E. V. Saulina*
Don State Technical University, Rostov-on-Don, Russia

ABSTRACT

The task of the theory of dynamic stability is to establish links between the load parameters and the calculated parameters, at which the dynamic stability of the working state of the structure is observed. Moreover, the technical requirements must be met. Currently, the calculations of the stability and strength of structures are conducted using the universal program complexes SELENA, ANSYS, LIRA, GEOTEK SHM Penza. Diagnostics and monitoring of the building structures are performed by using software Zetlab Seismo, Zelenograd, etc. Completeness and accuracy of the results is estimated by calculation.

Our team has developed a new method of integrated design diagnostics, which partially solved the following tasks:

- an experimental analysis of the deformation behavior of the integral structure on linear and nonlinear responses to impact and (or) periodic effects under different loading conditions;
- the control of strength (damage) of the integral structure, based on deformation responses by using two or four wide-band ferroelectric microsensors of dynamic deformation;

* Corresponding Author's E-mail: saulina86@yandex.ru.

- the studies have been carried out and the choice of integrated stability indicators for individual designations and (or) geometric shapes of structures has been performed: (1) ground, air, water, space and (2) core, frame, shell.
- information parameters have been selected for the complex diagnostic criterion of the integral structure (for example, the rod frame model).
- the method and device of diagnostics of integral structure on an integral indicator have been developed.

As a result, an information-measuring device was developed for the recording and processing of local dynamic deformations in the composition of thin-film ferroelectric sensors, such as $Pb(Zr_{0.53}Ti_{0.47})O_3$, which differ from the known analogues by unique properties in frequency and the dynamic range of deformation. The obtained technical result consisting in the establishment of information signs and parameters allow one to distinguish the stressed and (or) damaged states of rod structures at impact and harmonic deformation responses and to conduct preliminary diagnostics of constructions and structures, which will increase the safety and efficiency of their operation.

Keywords: dynamic stability, structural strength, diagnostic and monitoring systems, linear and nonlinear deformation responses, ferroelectric microsensors of dynamic deformation, integrated parameters

1. Introduction

The task of the theory of dynamic stability is to establish the ratio of the load parameters and design parameters, in which the dynamic stability of the operational state of the structure is observed. At the same time, the requirements of technical conditions (SPECIFICATIONS) must be additionally fulfilled (if they are set and checked during the design and (or) during the tests). It is known that acceptable tests for resistance to extreme influences are carried out mainly for the constructions of responsible purpose (CRP). These constructions include objects W and WT, aircraft and spacecraft (AC and SC) of dual (defense and civil) purpose, nuclear power facilities, GTN facilities, and critical infrastructure. Due to the large size of the objects, the high cost and (or) the inability to conduct field tests, the durability and (or) stability of the objects is established by computational and experimental methods.

Currently, aircrafts and spacecrafts operate on the base of the decision on the results of the tests of construction models. Moreover, during the final tests (if possible), the results of the calculation of the stability of constructions to the complex of extreme loads (operating factors of different nature) are checked [1, 2].

For another object, acceptable checks and (or) tests are carried out based on the agreement between the customer, the contractor and the supervisory authorities. It is performed with the implementation of, as a rule, the following requirements: (i) calculation and assessment of the stability of the construction at a complex of extreme loads (acting factors); (ii) carrying out analysis of the test layout and (or) structural elements.

Currently, the calculations of the stability and strength of construction is carried out using universal program complexes SELENA [2003 Selena BOS Ltd.], ANSYS [2018 ANSYS, Inc.], LIRA [2013 – 2018 LIRA soft], "GEOTEK SHM" Penza, Russian system of diagnostics, monitoring of the state of building structures, and with the software ZETLab Seismo, Zelenograd, etc. Completeness and accuracy of these results are estimated by calculation methods. What is the accuracy of the decisions?

2. Problem Formulation and Analysis of Solutions

(i) Construction and application of a set of partial (differential) criteria for the stability of the structure is the first method or stage.

(ii) Combining particular stability criteria into an integral indicator of the MAC and COMAC type is the second method or step.

(iii) Search and construction of the integral index of stability of the integral design is the third (modern) stage of development of NK and TD (SHM).

2.1. The Content of the First Method of Solving the Problem

To solve the first task, it is stated, if possible, a complete list of events, namely defects in the system that affect the loss of stability of the structure, as a whole.

The hypothetical interaction of defects in the design is described by the function of connectivity of defects with respect to the vertex outcome: stability or instability; damage or non-damage of the entire structure.

A Boolean function of the connectivity on the set (array) $\{i, n\}$: $(i = 0, 1, \dots I)/n$ (events are the defects with respect to the selected result: loss of stability) is written in the form:

$$y_n = f_n(x_{i=1,n}, x_{2,n}, \dots x_{I,n}), \tag{1}$$

where

$$y_n = 0 \vee 1,\ x_{i,n} = 0 \vee 1;\ \{i\}: i = (0,1, \dots I)/n\ ;\ \{n\}: n = 0,1, \dots N. \tag{2}$$

On the base of expressions (1), (2), we enter matrices or Boolean sets of assumptions (3) and forms of loss of stability (4):

$$X = \{x_{i,n}\} = \{x_{1,1}, x_{2,1}, \dots x_{I,1}; x_{1,2}, x_{2,2}, \dots x_{I,2}; \dots x_{1,n}, x_{2,n}, \dots x_{I,n}\}; \quad (3)$$

$$Y = \{y_n\} = \{y_1, y_2, \dots y_n\}. \quad (4)$$

When representing (3) and (4), respectively, in the form of a matrix and a vector-row, matrix calculus operations are used (for example, matrix-row multiplication operation). When representing (3) and (4) in the form of two-dimensional and one-dimensional Boolean sets, respectively, the operations of positional logical multiplication ($\wedge$) and addition ($\vee$) of elements are introduced and used.

The probability of having no damage or stability (as an integral criterion) is established on the base of the transformation of the Boolean form of the connectivity function into its probabilistic form.

In the simple case with one type of damage (loss of stability), $n = 1$, a set of defect events $\{i, n = 1\}$: $(i = 0,1, \dots I)/(n = 1)$, the simplified presentation of which has the form $\{i\}$: $(i = 0,1, \dots I)$, the expression (1) takes the form:

$$y = f(x_1, x_2, \dots x_I). \quad (5)$$

Let us suppose that the expression (5) describes a complete group of incompatible outcomes, there are no loop operations, and the logical variables $\{x_i\}$ are non-repeating. Then according to the rules of transformation [3] the probabilistic form (5) is:

$$\text{Pro}\ (y = 1) = F\ (p_1, p_2, \dots p_I), \quad (6)$$

where Pro (·) is a probability operator, F (·) is an algebraic function of real variables as a result of transformation.

For example, for the given function:

$$y_1 = x_1 \wedge x_2 \vee x_3 \quad (7)$$

in order to perform the transformation, we will consistently bring it to the form:

$$y_1 = (x_1 \wedge x_2) \vee x_3 = y_{1,2} \vee x_3\ . \quad (8)$$

The last expression represents the logical addition of two variables. While the variable $y_{1,2}$ is the conjunction of variables x_1 and x_2, let us perform the transformation of the conjunction of variables x_1 and x_2 and obtain:

$$x_1 \rightarrow p_1\,; x_2 \rightarrow p_2\,; \text{Pro}\ (y_{1,2} = 1) = p_1 \cdot p_2. \quad (9)$$

Then, let us convert the disjunction of variables $y_{1,2}$ and x_3 and get an example of function F (·) as:

$$\text{Pro}\ (y_1 = 1) = (1 - p_1 \cdot p_2) \cdot (1 - p_3) = (1 - p_1 \cdot p_2) \cdot q_3 = q_3 - p_1 \cdot p_2 \cdot q_3\,, \qquad (10)$$

Where Pro ($y_1 = 1$) is the operator of probability of condition realization ($y_1 = 1$) and represents the probability of stability loss;

Here q_3 is a probability not occurring in the background event 3; p_1 and p_2 define the probabilities of the occurrence of prerequisite events 1 and 2.

From the expression (10), it follows that, if the probability of occurrence of the defects 1, 2 and 3 is small $p_1 << 1$, $p_2 << 1$, $p_3 << 1$, then the probability of implementing the loss of stability of the structure can be estimated by the probability value of nonoccurrence of the defect 3: Pro ($y_1 = 1$) $\approx q_3$.

For comparison, here we perform the transformation of the logical function (7) into its possible form. We use the following transformations: $x_1 \rightarrow \pi_1$; $x_2 \rightarrow \pi_2$; $x_3 \rightarrow \pi_3$; $\wedge \rightarrow$ min; $\vee \rightarrow$ max.

As a result, the possible form of the logical function of the incident is found as

$$\text{Pos}\ (y_1 = 1) = \max\ (\pi_3, \min\ (\pi_1\, , \pi_2)). \qquad (11)$$

The dependence (11) shows that the possibility (uncertainty) measure of the vertex outcome 1 is determined by choosing the maximum value from the possibility measure of the defect 3 and the minimum value from the possibility measures of the defects 1 and 2.

As we can see, not knowing the integral index of stability requires us to receive and process large amounts of information and build logical schemes of relations of mutual influence of possible defects. It is a prerequisite for the loss of stability of the structure as a whole, which requires significant costs and can lead to erroneous results [4, 5].

2.2. Description of the Second Method of Analysis and Solution of the Problem

The second method of analysis of structures that has been developed within the SHM technology since Rizos et al. (1990) was associated with the introduction, justification and application of MAC and COMAC indicators. The Modal Assurance Criterion (MAC) expresses the variation of the experimental natural frequencies in comparison with the forms of the experimental modes of a damaged or stressed structure.

The value of the MAC index of two modes φ_A and φ_B (the form of modes in the intact and damaged state) is determined by the left formula:

$$MAC_{A,B} = \frac{\left|\sum_{i=1}^{n} \varphi_i^A \varphi_i^B\right|^2}{\sum_{i=1}^{n} \left(\varphi_i^A\right)^2 \cdot \sum_{i=1}^{n} \left(\varphi_i^B\right)^2} \qquad COMAC_{i,A,B} = \frac{\left|\sum_{j=1}^{m} \varphi_{i,j}^A \varphi_{i,j}^B\right|^2}{\sum_{j=1}^{m} \left(\varphi_{i,j}^A\right)^2 \cdot \sum_{j=1}^{m} \left(\varphi_{i,j}^B\right)^2} \tag{12}$$

The value of the MAC indicator expresses the degree of correlation between both modes and varies from 0 to 1. Deviation from 1 can be represented as an indicator of damage in the structure.

The modal guaranteed coordinated criterion (Co-ordinate Modal Assurance Criterion, COMAC) differs from the MAC, providing local information and combined information on different modes. The value of COMAS for modal coordinates *i* and *j* when taking into account *m* modes is defined as the right formula (12). It should be noted that this method of evaluating structures is still used [6, 7].

3. Development and Justification of Diagnostic Method Based on the Integral Indicator of Intact Structure

The idea of this method is based on the assumption that the design, as a target component of the system "design – protection – factors" itself will be able to tell researchers what dependences between the prerequisites – defects are most significant. This requires only finding and building an integral indicator of the intact structure [8 – 10].

The following tasks were set and partially solved in the development of a new method of diagnostics of constructions:

(i) it was performed an experimental analysis of the deformation behavior of an integral structure by linear and nonlinear responses to impact and (or) periodic influences under different loading conditions [11];
(ii) it was justified the control of the tension (damage) of the integral structure by deformation responses with the use of two or four wide-band ferroelectric micro-sensors of dynamic deformation [12];
(iii) it was performed research and selection of integrated indicators of sustainability for certain considered media and (or) geometric shapes designed in particular: (1) land, air, water, space, and (2) rod, frame, shell;
(iv) informative parameters were selected for diagnostics of complex criterion for an integrated construction (on the example of frame rod models) [13];

(v) the method and device for diagnostics of an integrated construction by using an integral indicator were developed.

4. Results of Experimental Determination of Stability Indicators (for Example, the Design Models)

On the base of development and approbation of a method of impact testing, the loaded constructions, and taking into account nonlinear deformations, the following results were obtained.

(1) Signs of "rapid distinction" and instability of "design as a whole":
   (i) establishment of at least one non-linear region;
   (ii) obtaining and interval estimation of the value of the informative parameter as the ratio of the nonlinear region to the linear region of phase portrait.
(2) Informative parameters for distinguishing the stress states of the layout and (or) elements of the core structure, which include:
   (i) the ratio of the nonlinear region of the phase portrait in damaged (stressed) area to the nonlinear region of the phase portrait at the initial state of construction;
   (ii) the ratio of the nonlinear region to the linear region of the phase portrait at deformation responses in damaged (stressed) construction;
   (iii) a frequency shift of the peaks (maxima) of the linear regions of the Fourier image at the deformation response of construction (its element).
(3) The following criteria for distinguishing the stress (damaged) states of construction were obtained [14]:
   (i) the presence of a nonlinear Fourier region of the image in the deformation responses of construction (its element) to serial impact testing as a qualitative criterion;
   (ii) the condition of finding the indicator $\boldsymbol{\eta}$ in the range of values: $\boldsymbol{\eta} \geq 0.1$;
   (iii) the condition of finding the shift Δf at a frequency from 0.5 to 2 Hz (or more) of maximum for linear Fourier regions due to the deformation responses of construction (its element), when the criteria (ii) and (iii) are quantitative criteria of distinction.

Conclusion

Due to the development of method and differential methodic for the stress (damaged) states of construction, elements and the layout were first obtained and checked by the following diagnostic signs: (i) phase portraits of the nonlinear regions; (ii) the frequency shifts of the linear regions of the Fourier image of the deformation response due to impact excitation and also informative parameters allowing us to distinguish stress (damaged) states [15, 16].

It was a developed information-measurement device for the registration and processing of local dynamic deformation in thin-film ferroelectric sensors, based on $Pb(Zr_{0.53}Ti_{0.47})O_3$, which differs from the known analogues by its unique properties in the frequency and dynamic ranges of deformation.

We establish information features and parameters that allow us to distinguish stress and (or) damaged states of rod structures at impact and harmonic deformation responses. It will help to register and analyze the processes and states of tension of models from a viewpoint of the integral property “damage – no damage” of construction, as well as increase the safety and efficiency of the operation of objects.

References

[1] Balitsky, F. Ya.; Barkov, A. V.; Sokolova, A. G.; et al. In: *Non-Destructive Testing and Diagnostics*. vol. 7. Vibrodiagnostics, V. V. Klyuev (Ed.), *Mashinostroenie,* Moscow, 2005, pp. 1 – 829 (In Russian).

[2] Makhutov, N. A. *Structural Strength, Resource and Technogenic Safety: For 2 Hours.* Science, Novosibirsk, 2005 (In Russian).

[3] Ryabinin, I. A.; Cherkesov, G. N. Logical-probabilistic Methods of Studying the Reliability of Structurally Complex Systems. *Radio and Communication*, 1981, pp. 1 – 266 (In Russian).

[4] Voronkov, V. N. *Mechanics of Solid*, 2005, No. 4, 178 – 187.

[5] Grudev, I. D. *Stability of Core Elements in Steel Structures*. MIK, Moscow, 2005, pp. 1 – 320 (In Russian).

[6] Charles R. Farrar, *Nonlinear System Identification for Damage Detection*. Los Alamos National Laboratory, Contract DE-AC52-06NA25396. November, 2007.

[7] Adams, D.; Nataraju, M. *International Journal of Engineering Science*, 2002, vol. 40, 1919 – 1941.

[8] Esipov, Yu. V.; Mukhortov, V. M. *Journal of Technical Physics*, 2009, vol. 51(1), 78 – 81.

[9] Esipov, Yu. V. *Measuring Technique*. 2009, No. 12, 44 – 48.

[10] Poyda, I. I., Esipov, Yu. V.; Mukhortov, Vas. M. *Russian Journal Non-destructive Testing,* 2010, No. 7, 76 – 81.

[11] Esipov, Yu. V.; Saulina, E. V.; Cheremisin, A. I. In: *2015 International Conference on "Physics and Mechanics of New Materials and Their Applications"* (PHENMA 2015). Abstracts & Schedule. I. A. Parinov, S. H. Chang (Eds.), Southern Federal University Press, Rostov-on-Don, 2015, 84.

[12] Saulina, E. V.; Esipov, Yu. V.; Dzhilyadzhi, M. S. *2016 International Conference on "Physics and Mechanics of New Materials and Their Applications"* (PHENMA 2015). Abstracts & Schedule. Muaffaq A. Jani, Ivan A. Parinov, Shun-Hsyung Chang (Eds.), UNTAG Press, Surabaya, 2016, 231 – 232.

[13] Yesipov, Yu. V.; Saulina, E. V.; Cheremisin, A. I. In: *II All-Russian Congress on Fundamental Problems of Theoretical and Applied Mechanics.* D. A. Gubaidullin, A. I. Elizarov, E. K. Lipachev (Eds.), 2015, 1345 – 1346 (In Russian).

[14] Esipov, Yu. V. *Applied Mechanics and Technical Physics*. 2013, vol. 54(2).

[15] Saulina, E. V.; Esipov, Yu. V. In: *Physics and Mechanics of New Materials and Underwater Applications Abstracts & Schedule*, I. A. Parinov, T. Somnuk, S. H. Chang (Eds.), NKMU Press: Kaohsiung, 2013, 83.

[16] Saulina, E. V.; Yesipov, Yu. V. *Control. Diagnostics.* 2014, No. 1(187), 33 – 39 (In Russian).

In: Proceedings of the 2018 International Conference … ISBN: 978-1-53615-862-5
Editors: Ivan A. Parinov et al.

Chapter 29

CALCULATION OF STRESSES AND DEFORMATIONS IN MASONRY WITH VARIOUS TYPES OF BRICKS

A. A. Matrosov, D. A. Nizhnik, N. A. Poryadina, I. A. Serebryanaya and A. N. Soloviev*
Don State Technical University, Rostov-on-Don, Russia

ABSTRACT

The paper analyzes the process of deformation of bricks with different cavities. We investigated the stress-strain state of the goods tested under axial compression. The numerical implementation of this problem in the software package of the finite element analysis ANSYS is given.

Keywords: brick, cavity, stress, strain, axial compression, ANSYS

1. INTRODUCTION

The bricks are used for construction due to a number of reasons, including the universality of the building materials compared with concrete and wood. Another reason is the ease of use when constructing a building object. The safety and environmental friendliness of this material is also one of the important indicators for its use in many

* Corresponding Author's E-mail: solovievarc@gmail.com.

areas of life, from agriculture to the construction of high-rise housing. At present, it has produced a huge number of different types of bricks. These goods have different compositions, shape, thermal conductivity, frost resistance, and other properties.

2. Problem Formulation

An important task of calculating structures is, on one hand, the determination of the stress-strain state in the entire brickwork of construction [1-3]. However, on the other hand, a brick is one of the constituent components of brickwork. The most important indicator of this good's quality is the brand, namely the strength characteristic. In accordance with the Russian State Standard [4], the experimental determination of the strength characteristic is carried out by the destructive method on a single brick. Therefore, it is of interest to determine the distribution of stresses and strains by the volume of one single brick.

In this regard, there is a need to determine the stress-strain state, as well as to calculate the strength, frost resistance, thermal conductivity, and water permeability of one brick.

This paper discusses the problem of determining stresses and strains that occur in bricks with different cavities under testing by axial compression [5, 6].

3. Modeling

There are options when the same load acts on bricks with different cavities (Figure 1). In the figure, dots show the vertices of right angles.

Figure 1 presents the following types of bricks:

(i) a solid brick without cavities, that is, there are no voids in the volume of the brick;
(ii) a brick has three blind cylindrical voids with a hole diameter of 52 mm and the volume of these voids equal to 15% of the total brick volume;
(iii) a brick has eighteen slit-like voids and the volume of these voids is equal to 29% of the total brick volume.

The main characteristics and properties of the material for the calculation are taken from the Russian State Standard for Building Materials GOST 530-2012 "Brick and Stone, Ceramic". The sizes and main characteristics of the considered bricks are present in Table 1.

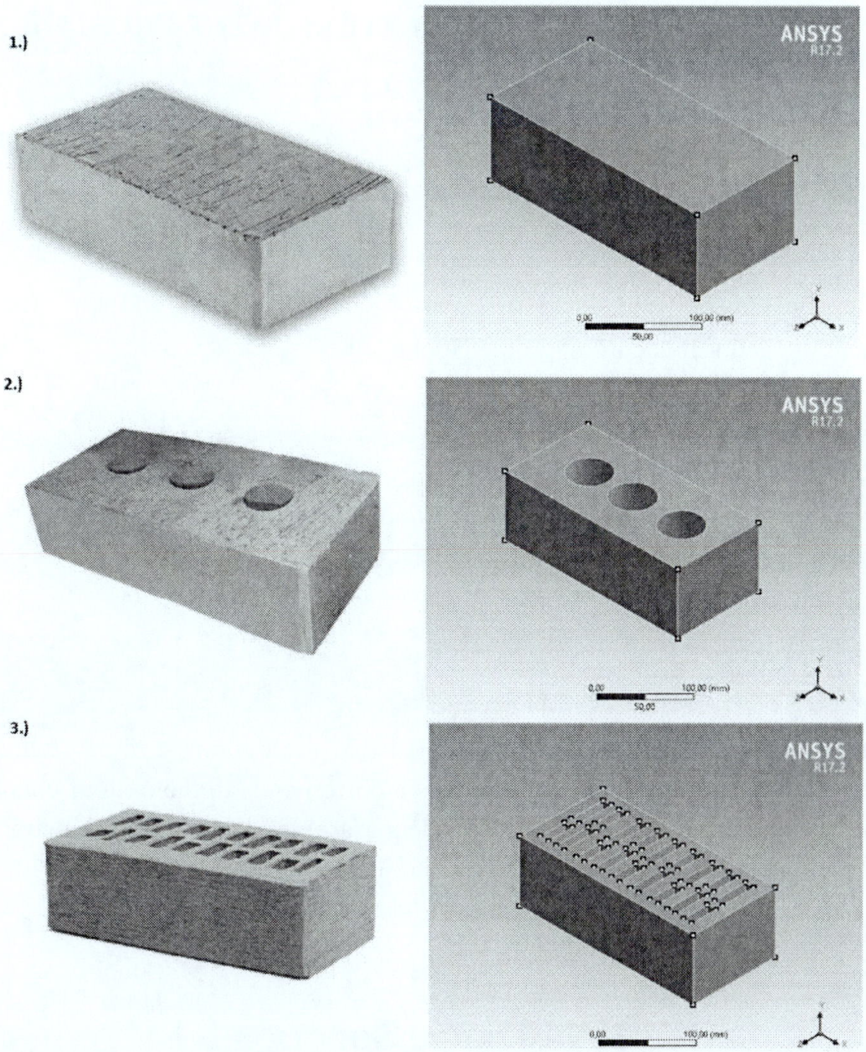

Figure 1. Types of bricks and their models: (1) solid brick, (2) brick with cavities of 15%, (3) brick with cavities of 29%.

Table 1. Sample Parameters

No.	Product Name	Dimensions, mm	Mark M	Density, kg/m^3	Young's modulus, GPa
1	Full bodied brick	250×120×65	M150	1800	1.768
2	Ceramic bricks, hollow (15%, voids)	250×120×65	M150	1700	1.768
3	Ceramic bricks, hollow (29%, voids)	250×120×65	M150	1200	1.768

4. Mathematical Formulation of the Problem

The task is considered in the static setting:

$$\frac{\partial \sigma_{ij}}{\partial x_j} + X_j = 0 \tag{1}$$

The connection between the stress tensor components σ_{ij} and the strain tensor components ε_{kl} is linear according to the Hooke law:

$$\sigma_{ij} = C_{ijkl}\varepsilon_{kl} \tag{2}$$

The components of the strain tensor are connected with the components of the displacement vector u_i by the Cauchy formulas:

$$\varepsilon_{ij} = \frac{1}{2}\left(\frac{\partial u_i}{\partial x_j} + \frac{\partial u_j}{\partial x_i}\right). \tag{3}$$

As the boundary conditions, it is assumed that the bottom surface of the brick is rigidly fixed on an absolutely solid base, which corresponds to the case of contact of the brick with a cement-sand mortar [4]. The side surfaces of the masonry are stress free. A uniformly distributed load is applied on the top surface. The magnitude of this load is chosen in such a way that it is not sufficient to destroy a brick.

5. Numerical Solution

The numerical solution of the problem is performed in the ANSYS 19.0 finite element complex in the static structural package [7]. This package allows one to determine displacements, stresses, and forces in structures or components, caused by loads that do not cause significant inertia and damping effects. When creating a finite element mesh, a three-dimensional Solid 65 element was used. This allows us to take into account the behavior of not only bricks, but also cement-sand mortar, when considering brickwork, such as crushing, cracking, plastic deformation and creep. Such an approach will allow us to further take into account both the heterogeneous structure of a brick (which is particularly important, for example, for hand-molded bricks), and its destruction with increasing load.

For solid brick and bricks with eighteen through slit holes, the finite element mesh step is made constant. For a brick with three non-through cylindrical holes, the pitch of the finite element mesh is made variable.

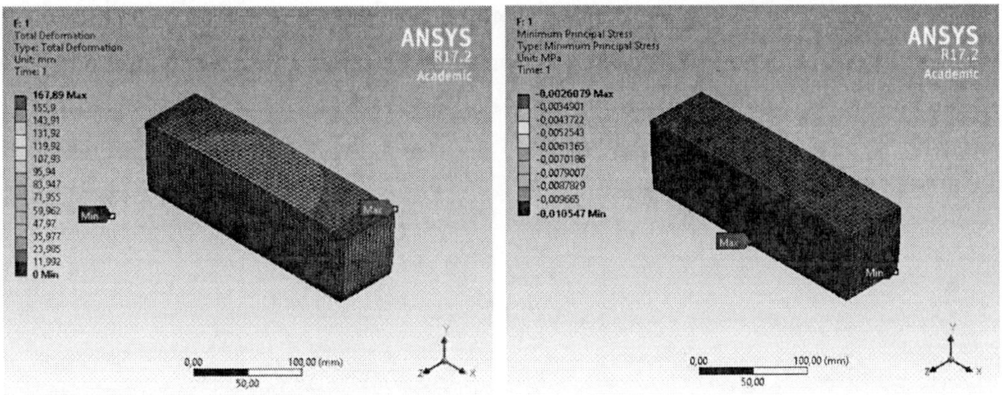

Figure 2. Stresses and strains distributions in solid brick.

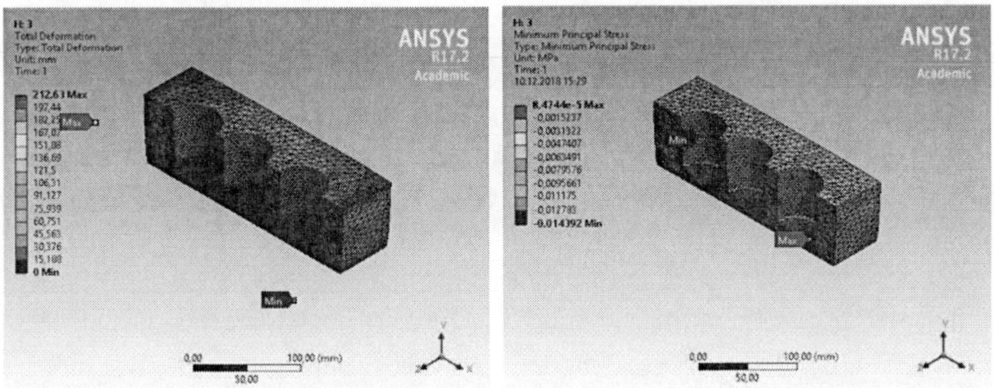

Figure 3. Distribution of stresses and strains in brick with three blind cylindrical voids.

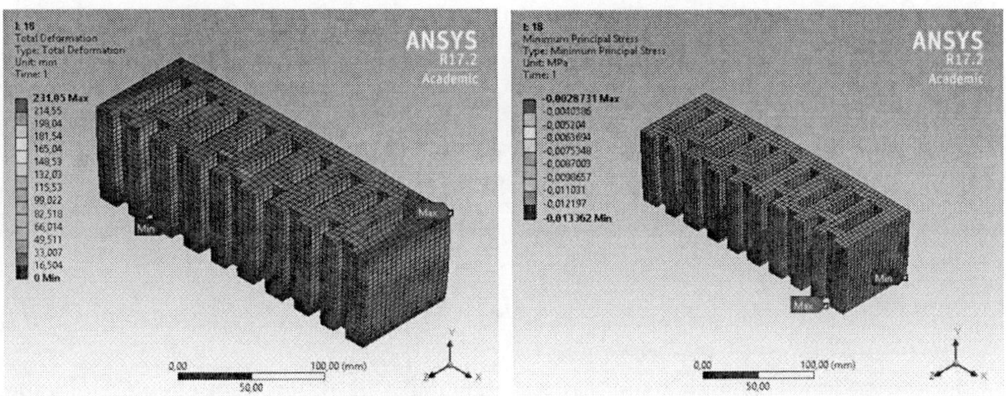

Figure 4. Distribution of stresses and strains in brick with eighteen through slit-like voids.

In the result of numerical calculations, the stress-strain state arising in a brick was determined. This solution was obtained for a distributed load and applied to the upper surface. Figures 2–4 show the generated finite element meshes. Since, according to the first theory of strength, the main stress σ_3 is of decisive importance, the figures also show the distributions of the main stress σ_3 and the displacement vector module, respectively. Extreme values σ_3 of the corresponding quantities occurring at corner points are shown by arrows in the figures.

For clarity, the figures show halves of bricks cut by a vertical plane along the greater axis of the brick.

CONCLUSION

As it was expected, the results of the calculation, when tested for axial compression, showed that the maximum values of the modulus of the displacement vector and stresses σ_3 arise along the edges of the brick, which have no restrictions from the edges.

In a solid brick, the stresses are distributed sufficiently even, which cannot be said about bricks with voids. In bricks with three cylindrical voids, the stresses increase as they approach the holes. In this case, the maximum stress values are reached at the bottom of the holes. In a brick with through slit-like holes, the stresses are distributed quite unevenly throughout the volume. Maximum stress values are achieved on transverse partitions.

Analysis of the calculation results showed that the maximum stress values in bricks with the same load do not differ by more than 10%. This allows us to conclude that the numerical values of the stresses in the calculated samples of the same order are independent of the porosity of the brick.

REFERENCES

[1] Wu, C.; Hao, H. (2006). *Int. J. Numer. Meth. Eng.*, 11: 1717-1737.

[2] Nasedkin, A. V.; Nasedkina, A. A.; Rajagopal, A. (2017). *Wave Dynamics and Composite Mechanics for Microstructured Materials and Metamaterials. Advanced Structured Materials*, 59: 205-220.

[3] Nasedkina, A. A.; Rajagopal, A. (2017). In: *Advanced Materials - Techniques, Physics, Mechanics and Applications*, Springer Proceedings in Physics, Ivan A. Parinov, Shun-Hsyung Chang, Muaffaq A. Jani (Eds.). Springer Cham, Heidelberg, New York, Dordrecht, London, 193: 347-359.

[4] Russian State Standard for Building Materials GOST 530-2012 *Brick and Stone, Ceramic*. (2013): 1-28.

[5] Lukinova, N. A.; Matrosov, A. A.; Nizhnik, D. A.; Serebryanaya, I. A. and Terekhina, Y. V. (2017). *MATEC Web of Conferences*, 132: https://doi.org/ 10.1051/ matecconf/201713203002.

[6] Lukinova, N. A.; Matrosov, A. A.; Nizhnik, D. A.; Serebryanay, I. A. and Terekhina, Yu. V. (2017). In: *Abstracts and Schedule of the 2017 International Conference on "Physics and Mechanics of New Materials and Their Applications"*, (14-16 October, 2017, Jabalpur, India), Vijay K. Gupta, Ivan A. Parinov, Shun-Hsyung Chang (Eds.). PDPM Press, *Jabalpur*: 134-135.

[7] Poryadina, N. A.; Serebryanay, I. A.; Matrosov, A. A. and Nizhnik, D. A. (2017). In: *Proc. Sci.-Pract. Conf. on Construction and Architecture,* 28 – 30 November: 124-127.

In: Proceedings of the 2018 International Conference … ISBN: 978-1-53615-862-5
Editors: Ivan A. Parinov et al.

Chapter 30

FEATURES OF THE INTERFERENCE PATTERNS, GENERATED BY LASER INTERFEROMETER, BY USING DIFFERENT TYPES OF BEAM-SPLITTERS

I. P. Miroshnichenko[1,*], I. A. Parinov[2] and V. P. Sizov[1]
[1]Don State Technical University, Rostov-on-Don, Russia
[2]I. I. Vorovich Mathematics, Mechanics and Computer Sciences Institute, Southern Federal University, Rostov-on-Don, Russia

ABSTRACT

The paper presents the results of numerical studies of an interference meter for small displacements of control object surfaces, developed on the base of a two-way laser interferometer with combined branches, using different types of beam-splitters. The analysis of the results is performed. The features of interference patterns, obtained for various types of beam-splitters, are identified and described.

Keywords: laser interferometer, beam-splitter, displacement measurement, control object, interference pattern, intensity

1. INTRODUCTION

At present, an optical interference meter for small linear and angular displacements of the surfaces of objects of control, developed on the base of a two-way laser interferometer with combined branches (see, for example, [1 – 4]), is known and used.

* Corresponding Author's E-mail: ipmir@rambler.ru.

One of the directions for further improvement of the noted meter and creation of new optical interference measuring instruments for studying defect formation processes in constructional materials, as well as diagnosing the state of materials and goods during their operation with help of acoustic active non-destructive testing methods, is the use of various types of beam splitters.

The purpose of the work, the results of which are described in the paper, is a numerical study of the prospective types of beam splitters to create new optical interference that measure instruments for linear and angular displacements.

2. Description of Studied Meter

The investigated laser interferometer (Figure 1) contains optically coupled and sequentially placed radiation source 1, optical system 2, beam splitter 3, reflector 4, associated with the test object, and photodetector 5. This device allows one to determine the linear and angular displacements of the reflector by changing the parameters of the generated interference pattern.

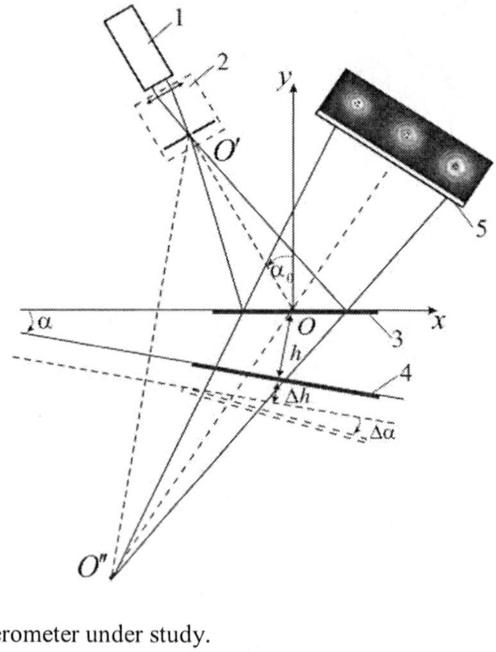

Figure 1. Scheme of interferometer under study.

For a theoretical study of the laser interferometer of displacements based on electromagnetic theory, a new mathematical model was developed [5], allowing one to take into account the characteristics of the used beam splitters, the vector character of laser radiation, complex diffraction phenomena, interference, polarization, and others that characterize the propagation of light in space.

In [5], relations were obtained that describe the interference field in the photodetector zone, the physical components of the electric field strength vector E_θ and E_ψ. These ratios include, as parameters, the quantities α and h, the determining of the position of the reflecting surface in space, and the quantities that determine the transmittance of the beam-splitter. The developed mathematical model is implemented by using a PC, which allows for numerical studies of both the interferometer as a whole and its individual elements.

The intensity distribution in the observation zone can be found by substituting the calculated values E_θ and E_ψ into the well-known expression [6]:

$$I = E_\theta E_\theta^* + E_\psi E_\psi^*, \tag{1}$$

where E_θ^*, E_ψ^* are the complex conjugate values to E_θ, E_ψ.

To reduce computational costs and greater visibility of the results obtained, examples of calculations of the intensity distribution are present for the plane E ($\psi = \pi/2$) under the assumption that the field irradiating the "beam-splitter – reflector" system is linearly polarized in the plane perpendicular to the plane of incidence (Figure 1) and does not depend on the coordinate z. The initial data during the numerical simulation were the geometrical parameters of the meter, which remained unchanged during the research. The interference field in the area of the photodetector was determined for two values of the displacement of the interferometer reflector: $h = 0.005$ m and $h = (0.005 + \lambda/4)$ m, where $\lambda \approx 0.6328$ μm is the wavelength of the laser stabilized on frequency.

The main characteristics of the beam-splitter were taken: the dependences of their transmittance factors T and reflection R on the coordinate x. The reflection factor of the beam splitter, neglecting absorption in it, can be determined from the well-known expression [6]:

$$R(x) = 1 - T(x) \tag{2}$$

As the beam-splitters, which carry out the formation of the reference and object optical beams by splitting the initial beam by amplitude, models of the following types were investigated:

1. Amplitude sinusoidal grating, the transmittance of which is determined by the expression:

$$T(x) = \frac{1 + a \cdot \cos\frac{2\pi}{l}x}{1 + a}, \tag{3}$$

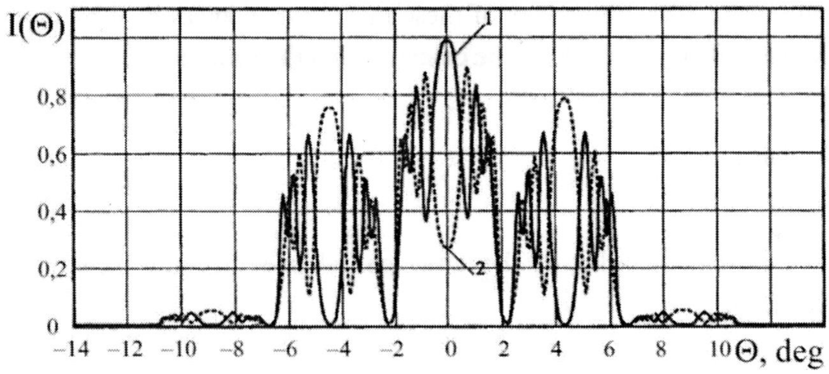

Figure 4. Interference field (beam-splitter 3, $l = 13\lambda$).

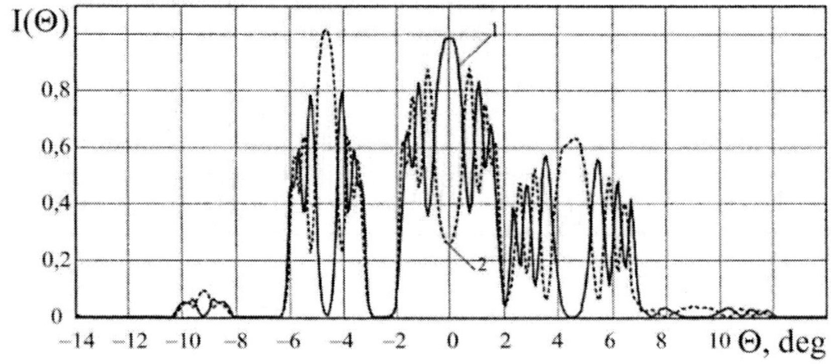

Figure 5. Interference field (beam-splitter 3, $l = \lambda$, $C = 0.004$ m).

The use of non-periodic diffraction gratings (beam-splitters 2, 4) leads to the diffracted beams of the same order, at the same diffraction angles, having different angular apertures and different amplitude values of the intensity change (Figures 3, 5). This corresponds to theoretical positions and is explained by that in the observation zone. The interference field is formed by three pairs of waves propagating under different angles, of which: (i) two convergent, modulated in phase for a maximum of the (–1) order; (ii) two divergent, coinciding with the incident wave for a maximum of zero order; (iii) two divergent, modulated in phase for the maximum of the (+1) order [7].

By using amplitude gratings (beam-splitters 1, 2), diffraction maxima of only (±1) orders are formed (Figures 2, 3), and when using phase gratings (beam-splitters 3, 4), there are also higher orders of maxima (Figures 4, 5).

The use of phase gratings in solving measurement problems is promising, since it allows an achievement of approximately the same amplitude change in the intensity in beams of 0 and (±1) orders, and the amplitude in beams of (±1) orders higher than in the corresponding beams, formed by using an amplitude grating.

In respect to the beam-splitter 5, which, in essence, is a hologram of two point sources, exposed in colliding beams (see [7]), it should be noted that its diffraction properties depend on the exposure parameters of the hologram, which are determined by values of α_Z, h_Z, y_{0Z} and C in the adopted model. In this case, depending on the values of these parameters, such a beam splitter will have either the properties of a diffraction grating (Figure 2) or the properties of a zone plate (Figure 3).

The developed mathematical model, algorithms, and programs for numerical simulation allow us to investigate how the type, characteristics of the beam-splitter, and hologram exposure parameters affect the interference field generated in the observation area. This ensures the determination of the parameters of the meter and its constituent elements that are optimal for a given measuring circuit and a specific measuring task.

From the analysis of the presented simulation results, it follows that if the radiation source is located on the y-axis ($x_0 = 0$; $\alpha_0 = 0$), then the intensity change in the beams of (±1) order on movement of the reflector and is of the same nature in amplitude, phase and frequency (Figures 2 – 5). The course of experimental modeling was also confirmed.

The result of the calculations is shown in Figure 6, where the dependences of the intensity on the linear displacement of the reflector are shown at the corresponding points of the zero and (±1) order maxima.

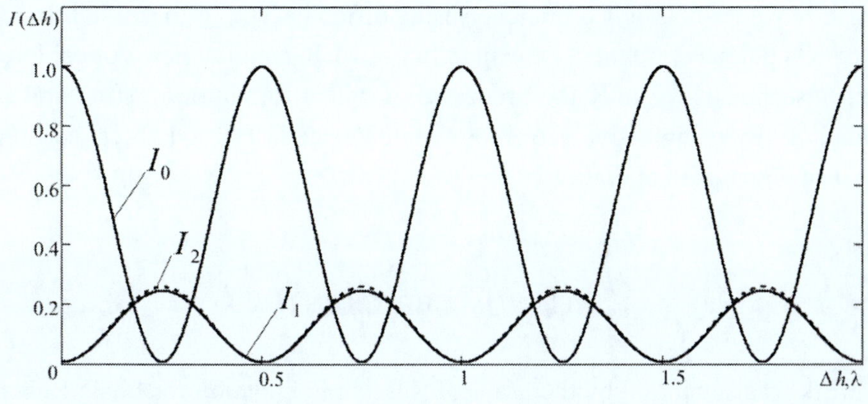

Figure 6. Change in intensity at the corresponding maximum points of zero (I_0), +1(I_1) and −1(I_2) orders at $x_0 = 0$; $\alpha_0 = 0$.

If the radiation source is shifted along the x-axis ($x_0 = 0$; $\alpha_0 = 0$), then the intensity change in the beams of zero and (±1) order has a significantly different character, which was also confirmed in the course of experimental modeling.

The result of the calculations is shown in Figure 7, where the dependences of the intensity on the linear displacement of the reflector are shown at the corresponding points of the zero and (±1) order maxima.

The analysis of the obtained results confirms the expediency of placing photoreceivers in the directions of propagation of maxima of the (±1) order.

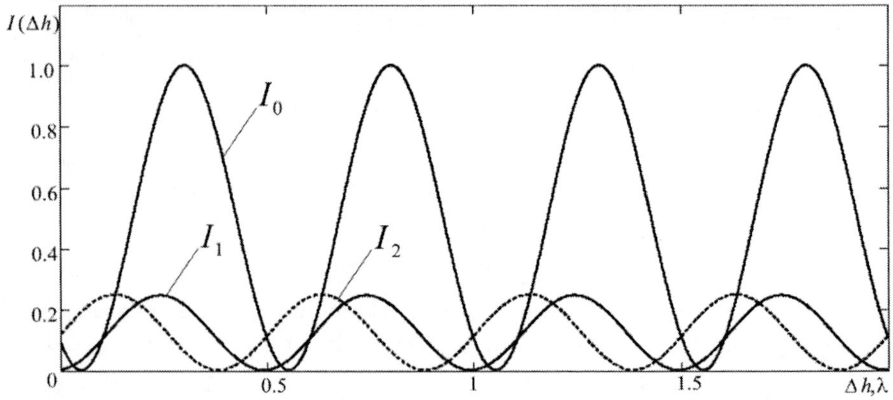

Figure 7. Change in intensity at the corresponding maximum points of zero (I_0), +1(I_1) and −1(I_2) orders at $x_0 = 0.01$; $\alpha_0 \approx 2.9$.

Conclusion

Thus, it can be noted that the conducted numerical simulation revealed the features of the interference patterns, which was formed by the interference meter of small displacements of the surfaces of the test objects and developed on the base of a two-way laser interferometer with aligned branches using different types of beam-splitters.

The results obtained are most expedient to use when creating new optical interference measuring instruments to study the processes of defect formation in structural materials and to diagnose by using active acoustic methods of non-destructive testing on the state of materials and goods in operation.

Acknowledgments

The work was supported by the grant of Russian Foundation for Basic Research, No. 19-08-00365, Ministry of Education and Science of the Russian Federation: projects Nos. 3.5378.2017/VU (organization of the implementation of researches), 9.9770.2017/BCh (the basic part of the state task).

References

[1] Miroshnichenko, I. P.; Parinov, I. A.; Rozhkov, E. V.; Serkin, A. G.; P. Sizov, V. P. In: *Piezoelectrics and Related Materials: Investigations and Applications*. Ivan A. Parinov (Ed.), Nova Science Publishers, New York, 2012, 239 – 282.

[2] Miroshnichenko, I. P.; Parinov, I. A.; Rozhkov, E. V.; V. P. Sizov; Shevtsov, V. A. In: *Physics and Mechanics of New Materials and Their Applications*. Ivan A. Parinov, Shun-Hsyung Chang (Eds.), Nova Science Publishers, New York, 2013, 145 – 154.

[3] Miroshnichenko, I. P.; Parinov, I. A.; Chang, S. H. In: *Advanced Materials. Studies and Applications*. Ivan A. Parinov, Shun-Hsyung Chang, Somnuk Theerakulpisut (Eds.), Nova Science Publishers, New York, 2015, 437 – 451.

[4] Miroshnichenko, I. P.; Parinov, I. A.; Rozhkov, E. V.; Chang, S.-H. In: *Advanced Materials Manufacturing, Physics, Mechanics and Applications*, Ivan A. Parinov, Shun-Hsyung Chang, Vitaly Yu. Topolov (Eds.), *Springer Proceedings in Physics*, Springer Cham, Heidelberg New York Dordrecht London, 2016, *vol. 175*, 341 – 356.

[5] Miroshnichenko, I. P.; Serkin, A. G.; P. Sizov, V. P. *Instrument Making*. 2007, *vol. 50(12)*, 34 – 40 (In Russian).

[6] Born, M.; Volf, E. *Fundamentals of Optics*. Nauka, Moscow, 1979, pp. 1 – 719 (In Russian).

[7] Litvinenko, O. N. Литвиненко О. Н. *Radio-optics Basics*. Technique, Kiev, 1974, pp. 1 – 206 (In Russian).

In: Proceedings of the 2018 International Conference … ISBN: 978-1-53615-862-5
Editors: Ivan A. Parinov et al.

Chapter 31

IDENTIFICATION OF THE PROPERTIES OF LAYERED INHOMOGENEOUS MEDIA BASED ON NEURAL NETWORK TECHNOLOGIES

A. A. Lyapin, Yu. I. Zhigulskaya[*] and A. I. Zhigulsky
Don State Technical University, Rostov-on-Don, Russia

ABSTRACT

The problem of determining the properties of building constructions has been considered. In the result of field experiments, initial values of the structure properties were obtained. The experiment was performed with the use of a mobile vibro-measurement complex. The experimental data formed the base for the training of a system, based on neural network technology.

Keywords: neural network, multilayer medium, vibration diagnostics, inverse problems, properties identification

1. INTRODUCTION

The dynamic processes in the "road construction – soil" system, caused by impact loads on the surface of the system, can be efficiently studied with the use of mathematical models. This is based on solving the initial-boundary problems of the theory of elasticity and viscoelasticity of layered inhomogeneous media [1, 4].

[*] Corresponding Author's E-mail: julik-zhig@mail.ru.

The specifics of a natural experiment impose sufficiently high requirements to the test equipment used on automobile roads. It should be able to write on electronic storage devices, to be as compact as possible, and allow testing in various climatic conditions. Taking into account these requirements, a mobile vibration measurement complex for recording deformation characteristics of the road constructions developed in DorTransNII (Russia) has been used for the diagnostics of roads [3, 5].

The method for studying the state of road constructions includes the following stages [8]:

(i) recording the amplitude-time characteristics of the road construction surface at varying distances from the point of loading (0.25, 0.5, 1.25, and 2.5 m);
(ii) processing the data of field tests with obtaining the amplitude-frequency characteristics of acceleration, the amplitude-frequency and amplitude-time characteristics of displacement.

Initially, for recording the response, equipment installation points are allocated that most fully reflect the state of the operated road. After that, sensors (vibration accelerometers) are installed and fastened in accordance with the adopted layout (Figure 1).

To increase tests mobility, the sensors were installed on a layer of moistened sand, since it ensures reliable connection between the roadbed and the sensor for a long time.

The testing procedure includes the following stages:

(i) sensors are installed and attached to a section of the road surface;
(ii) an impact loading is fixed at a given point, and a series of impacts is performed;
(iii) the parameters of resulting vibrations are recorded.

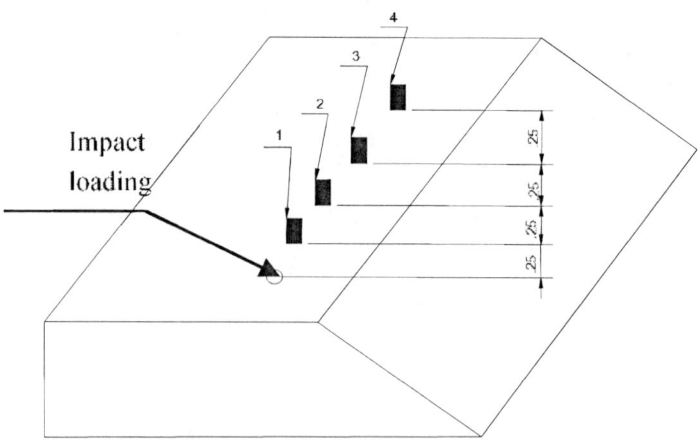

Figure 1. Sensors layout on the roadbed.

2. Problem Formulation

The paper is focused on the problem of the operative definition of the mechanical characteristics of building constructions on the base of the response to impact influence at a given point, done by using neural networks found on the example of the "road construction – soil" model.

3. Research Methodology

In accordance with the formulation of the problem, the structure and the parameters of a multilayer neural network are determined.

The general structure of the system may be shown in the scheme presented in Figure 2 [2]. The neural network itself consists of layers, which, in turn, consist of neurons. The neurons have inputs, outputs, weight coefficients, etc. [6].

The inputs of the network receive the longest vertical deflections from sensors, located at the point of impact influence and the distances of 0.25 m, 0.75 m, 1.25 m, and 2.5 m from it [7, 9]. Calculations were made using the Vibrolab software suite. The outputs of the network were the values of elasticity moduli for each layer (the remaining parameters of the structure are assumed to be specified in accordance with the design documentation for sections of the road surface).

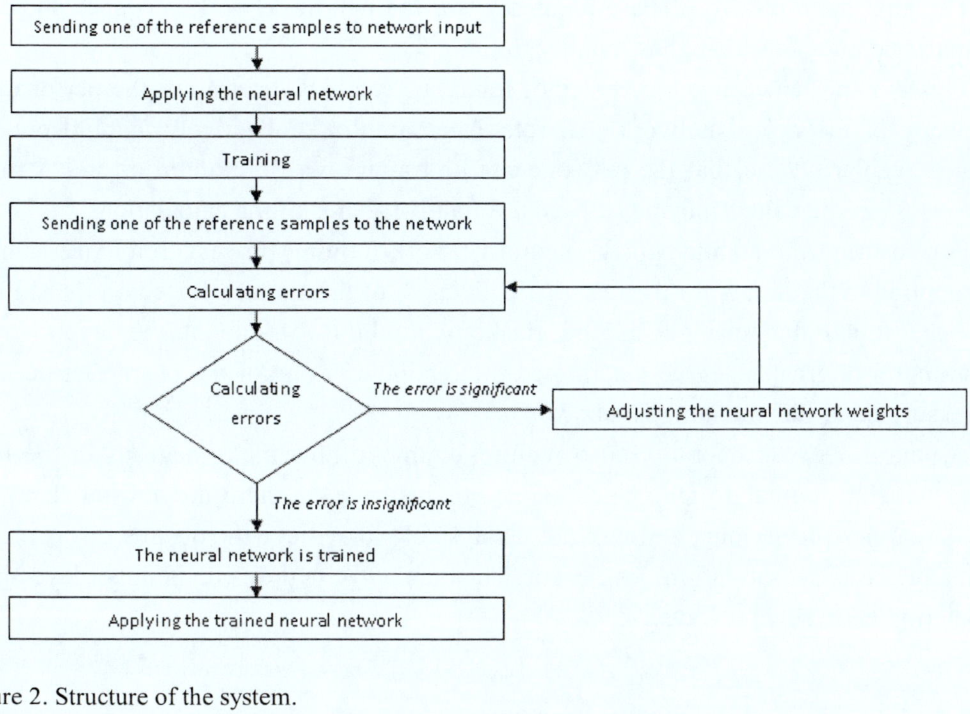

Figure 2. Structure of the system.

4. Modeling

Inputs and outputs of the network may be different values, and the results should not depend on their units of measurement. Therefore, all data are reduced to the same scale. An equating data preprocessing may also be performed before training. This may improve the speed and quality of training.

Reducing the data to the same scale ensures a linear transformation of each variable to the range of its value's variation:

$$\tilde{x}_i = \frac{x_i - x_{i,min}}{x_{i,max} - x_{i,min}}$$

to a unit segment: $\tilde{x}_i \in [0,1]$.

Linear normalization has sense if the interval is densely filled with values x_i. However, if there are "spikes" that exceed the base spread, then they namely determine the normalization scale, which results in the concentration of the main part of values near the zero.

Since, in the studied case, all values are as close to each other as possible, and at the same time are small enough, the use of linear data normalization is acceptable.

The values of elasticity moduli will be normalized by the same principle as the inputs move from 0 to 1, where 0 corresponds to the smallest value, and 1 corresponds to the largest value. This is called the standard linear interpolation.

The next main task to prepare input data for the neural network is proper choice of reference samples for the highest-quality training.

This is important, since the choice of reference samples determines the performance efficiency of the neural network. All reference samples (collectively) should create a certain "regularity", so that the network can be trained with minimum errors. A minor intersection in the values may be sufficient for a divergence of the calculations.

For ensuring the stability of the neural network training process, reference samples are supplied to the inputs for the best distinctiveness of the signals. Based on the study, it has been found that changes in the elasticity modulus of one of the layers in the construction in greater degree determined the readings in one of the characteristic areas on the surface of the medium (Figure 3).

A neural network operation also requires knowing how many neurons in the inner layers will be optimal for the calculation. For instance, if there are few of them, the calculation may be so long, and will not always able to achieve the desired error; if there are many neurons, "saturation" will occur. The number of neurons in the inner layers may be determined in several ways.

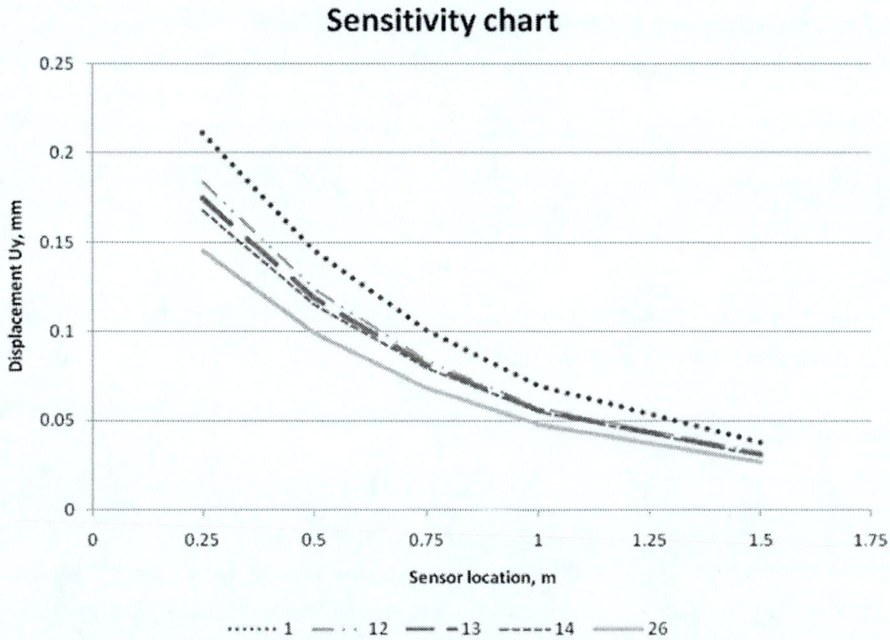

Figure 3. Sensitivity plot of values change for layered construction.

Table 1. Property selections

Set of number	1	12	13	14	26
Elasticity modulus E_1 (experiment), MPa	3500	2500	3500	4500	4500
Elasticity modulus E_2 (experiment), MPa	500	700	700	700	900
Elasticity modulus E_3 (experiment), MPa	40	70	70	70	100

To determine the number of neurons in hidden layers of the network, one can use the formula that is derived from the Arnold – Kolmogorov – Hecht-Nielsen theorem [10]:

$$\frac{N_y Q}{1+\log_2 Q} \leq N_w \leq N_y\left(\frac{Q}{N_x}+1\right)\left(N_x+N_y+1\right)+N_y$$

where N_y is the dimension of the output signal; Q is the number of elements in the set of training examples; N_w is the required number of synaptic connections; N_x is the dimension of the input signal.

However, regardless of the chosen standards and the number of neurons in the inner layers, the network did not provide reliable results for such problem formulation, since after changing all three output parameters at once (elasticity moduli of the roadbed), the values, obtained from the sensors, intersected with other values calculated previously. In this situation, training a network with three outputs was impossible.

It was therefore decided to use a different type of normalization E_i*.

Of all calculated values, the maximum set E_1 was chosen. It was considered an initial one. It is a criterion for a normalization, and all other values are normalized in respect to E_1 [9]:

$$E_i^* = \frac{E_i^1 - E_i^x}{\max\left|E_1 - E_j\right|},$$

where i is the number of the sensor, x is the number of the set, E_j is the minimum value of all sets, E_1 is the maximum value from set E_1.

5. Analysis

Figure 4 shows the diagrams for comparing the values of elasticity moduli in the layers, obtained during the experiments and the neural network calculations and Table 2 presents the property selection.

Figure 4 shows that when elasticity module E_1 is equal to 3,500 MPa, the results of the experiment and the computer data are virtually identical, since these values were chosen as standard samples for training.

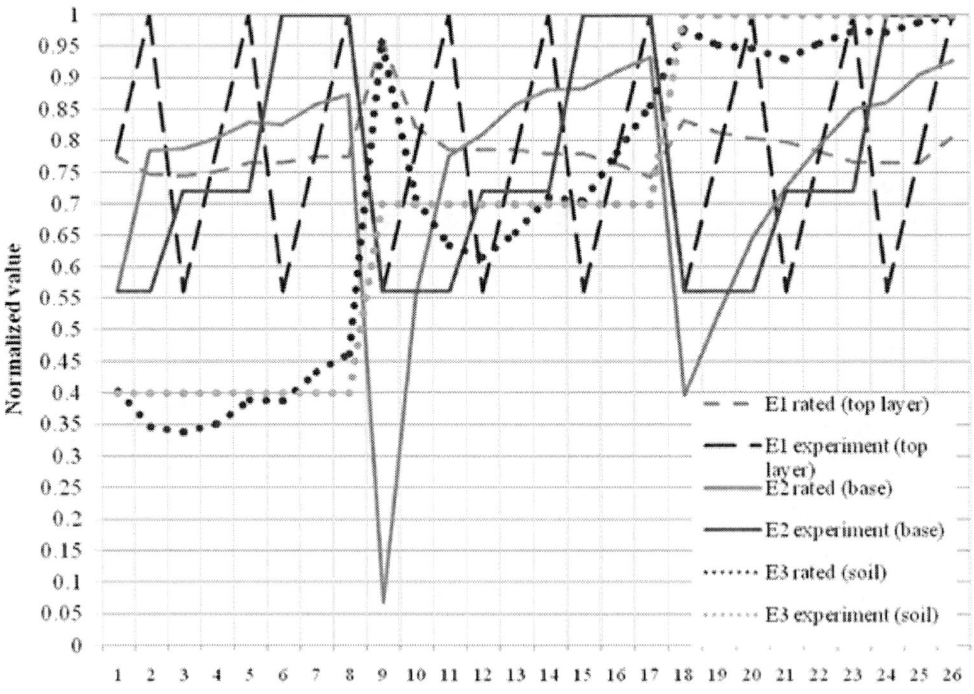

Figure 4. Error diagram.

Table 2. Property selections

Set number	1	2	3	4	5	6	7	8	9	10	11	12	13	14	15	16	17	18	19	20	21	22	23	24	25	26
Elasticity modulus E1 value (experiment), MPa	3500	4500	2500	3500	4500	2500	3500	4500	2500	3500	4500	2500	3500	4500	2500	3500	4500	2500	3500	4500	2500	3500	4500	2500	3500	4500
Elasticity modulus E2 value (experiment), MPa	500	500	700	700	700	900	900	900	500	500	500	700	700	700	900	900	900	500	500	500	700	700	700	900	900	900
Elasticity modulus E3 value (experiment), MPa	40	40	40	40	40	40	40	40	70	70	70	70	70	70	70	70	70	100	100	100	100	100	100	100	100	100

Conclusion

For the rest of the data, the following regularity is observed: the more the rigidity of the layer, the greater the error in determining the elasticity modulus for it by a computer program, due to insignificant changes of the deflections. That is, a trained network accurately calculates the elasticity modulus and determines the elasticity modulus of the crushed stone base with satisfactory precision. However, the error in determining the elasticity modulus for the asphalt roadbed is significant due to poor sensitivity to the changes in its properties.

The error in network training may be reduced by increasing in the number of observation points on the surface, however, the improvement is insignificant.

Acknowledgment

The work has been performed with financial support of the Russian Foundation for Basic Research (project 18-01-00715-a).

References

[1] Nowacki, W. *Theory of Elasticity*. Mir, Moscow., 1975, pp. 1 – 872 (In Russian).

[2] Haykin, S. *Neural Networks – a Comprehensive Foundation*. Williams, Moscow., 2006, pp. 1 – 1104 (In Russian).

[3] Kadomtsev, MI; Lyapin, AA; Shatilov, YuYu. *Engineering Journal of Don*., 2012, No. 3(21) (In Russian).

[4] Kadomtsev, MI; Lyapin, AA; Timofeev, SI. *Engineering Journal of Don*., 2012, vol. 19(1) (In Russian).

[5] Lyapin, AA; Meshcheryakov, IA. *Naukovedenie*., 2012, No. 4(13), 173 (In Russian).

[6] Baturina, NYu. *Engineering Journal of Don*., 2013, vol. 27(4) (In Russian).

[7] Maksimova, OM. *Proceedings of Irkutsk State Technical University*., 2013, No. 8 (79), 81 – 89 (In Russian).

[8] Uglova, EV; Tiraturyan, AN; Lyapin, AA. *Perm State Technical University Mechanics Bulletin*., 2016, No. 2, 111 – 130 (In Russian).

[9] Lyapin, AA; Zhigulskaya, YuI. *Science Review*., 2014, vol. 10(2), 441 – 445.

[10] Widrow, B; Lehr, MA. *Proceedings of the IEEE*., 1990, vol. 78(9), 1415 – 1442.

In: Proceedings of the 2018 International Conference … ISBN: 978-1-53615-862-5
Editors: Ivan A. Parinov et al.

Chapter 32

INDENTER SELECTION FOR THE NANOINDENTATION TESTS OF THE TWO-LAYER COATINGS

Andrey S. Vasiliev, Sergey S. Volkov, Evgeniy V. Sadyrin [*]
and Sergei M. Aizikovich
Research and Education Center "Materials", Don State Technical University,
Rostov-on-Don, Russia

ABSTRACT

Data retrieved from indentation experiments of a multilayer sample strongly depends on the indenter geometry. In the present paper, we consider an approach to select the indenter geometry, as well as the load values, to obtain the most representative data in the course of the experiment on nanoindentation of two-layer coatings on substrates. The choice of the indenter is analyzed depending on the thicknesses of the layers. The described approach is based on the constructing of the mathematical model of interaction between the indenter and the three-layer sample (two layers of the coating and the substrate layer) within contact mechanics of elastic solids. To implement the approach using the bilateral asymptotic method, the indentation stiffness functions for various axisymmetric indenters were constructed and analyzed. The mathematical model is suitable for the subsequent quantitative analysis of the parameters of the samples. As an example, calculations were made for a three-layer piecewise homogeneous sample, the lower layer of which (the substrate) is much thicker than the upper two layers.

Keywords: nanoindentation, multilayer coating, mathematical modeling, contact problem

[*] Corresponding Author's Email: ghostwoode@gmail.com.

1. Introduction

Proper increasing of the number of layers during the design of multilayer coatings can significantly improve its strength and performance characteristics. However, the difference in the values of the elastic characteristics at the interface of various materials within the coating leads to an increase in the stress concentration at these interfaces. Such increase can subsequently lead to the formation of cracks, delamination and destruction of the coating [1]. The problem of finding a compromised solution when designing a multilayer structure arises: to achieve high values of strength and performance characteristics without lowering the reliability of the product. A number of works are devoted to studyng the behaviour of the physical and mechanical characteristics of the multilayer coatings under different conditions: droplet impingement erosion [2], friction [3], corrosion [4], bacterial activity [5] etc. Today, nanoindentation as a method of material testing is gaining popularity for diagnostics of coating properties. Nanoindentation uses small precision local force application on the material and simultaneous registration of deformation responses with nanometer resolution [6]. When using nanoindentation for diagnostics of multilayer structures, the operator faces the task to choose the indenter and values of the load to obtain the most representative data suitable for the subsequent quantitative analysis of the structure parameters.

Recent investigations in contact mechanics of coated solids [7–13] can be efficiently used to model interaction of an indenter and a multilayer sample. In [14] the theory of contact mechanics for coated solids was used to model the interaction of an indenter and the sample of different temperatures. Analytical and semi-analytical methods are the most suitable choices to solve mixed boundary value problems arising in the mathematical modeling [10, 15–18]. Such methods allow us to analyze the structure of the solution and trace the dependence of the result from the multiple input parameters. In the present paper, the indentation stiffness functions for various axisymmetric indenters were constructed and analyzed in order to plan a correct experiment on nanoindention of a multilayer sample using the bilateral asymptotic method.

2. Problem Formulation and Mathematical Modeling

We consider a three-layer piecewise homogeneous sample, the lower layer of which (the substrate) is much thicker than the two upper layers: layer 1 (upper) – TiN (E = 300 GPa, ν = 0.23); layer 2 (medium) – Al (E = 70 GPa, ν = 0.34); layer 3 (substrate) – steel (E = 200 GPa, ν = 0.3), where E is the Young's modulus and ν is the Poisson's ratio. Let us introduce the parameter $E_{ef} = E/(1-\nu^2)$, which is the effective elastic modulus. The ratio of the effective elastic moduli of different layers to each other plays an important role:

$$\beta_{31} = \frac{E_{ef}^{(3)}}{E_{ef}^{(1)}}, \beta_{32} = \frac{E_{ef}^{(3)}}{E_{ef}^{(2)}}, \beta_{12} = \frac{\beta_{32}}{\beta_{31}} = \frac{E_{ef}^{(1)}}{E_{ef}^{(2)}}.$$

The indices (1), (2), and (3) hereinafter denote first, second and third layers, respectively. For the three-layer sample under consideration, the following is fulfilled: $E_{ef}^{(1)} = 316.75$, $E_{ef}^{(2)} = 79.149$, $E_{ef}^{(3)} = 219.78$, $\beta_{31} = 0.6938$, $\beta_{32} = 2.7768$, $\beta_{12} = 4.002$, $\beta_{31}{}^{-1} = 1.441$, $\beta_{32}{}^{-1} = 0.36$, $\beta_{12}{}^{-1} = 0.2498$. We denote h_i as the thickness of the i-th layer, $i = 1, 2$; $H = h_1 + h_2$ is the thickness of the coating.

We model the process of a rigid axisymmetric indenter of a spherical or conical shape indentation into an elastic sample with a two-layer coating. To do this, we use the results obtained earlier, which allow modeling of a sample with an inhomogeneous by depth or layered coating [12, 19]. The sample is modeled by an elastic half-space, and friction forces in the contact region are neglected. In the papers mentioned, it is shown, that the contact problem on indentation is reduced to the following integral equation:

$$\int_0^1 p(t)tdt \int_0^\infty L(u) \mathrm{J}_0\left(ur\lambda^{-1}\right) \mathrm{J}_0\left(ut\lambda^{-1}\right) du = \lambda E_{ef}(0)(\delta - f(r))/(2a), \quad r \le 1. \tag{1}$$

where $p(t)$ are the contact normal stresses under the indenter (punch), a is the radius of the contact area, $f(r)$ is the function describing the indenter profile, δ is the displacement of the indenter (indentation depth), $\lambda = H/a$ is the characteristic geometrical parameter of the problem, $L(u)$ is the kernel transform of the integral equation (it is also called the compliance function). For the analysis of experiments on nanoindentation, in particular in the determination of elastic moduli, it is convenient to use the concept of indentation stiffness [20]:

$$S = \frac{dP}{d\delta} = 2\varepsilon E_{ef}^{r} \sqrt{A_c} / \sqrt{\pi}, \tag{2}$$

where P is the indentation force, $\varepsilon > 1$ is the parameter characterizing the difference in the projection of the contact area from the circular region, and A_c is the area of the projection of the contact region.

For an axisymmetric contact, obviously, it is satisfied $A_c = \pi a^2$. It was shown [21] that $\varepsilon \approx 1.012$ and $\varepsilon \approx 1.034$ for a square and triangular contact region, respectively; for a circular contact area ε = 1 is satisfied. For a homogeneous sample without a coating, the parameter E^r_{ef} is a constant value and its value coincides with the value of the effective elastic modulus. For materials with coatings (homogeneous or inhomogeneous), E^r_{ef} is a function of the parameter λ. We will call it the restored effective modulus. It follows from (2):

$$E^r_{ef}(\lambda) = S/(2a)$$

On numerous numerical examples, it was shown [19] that by the behavior of the compliance function it is possible to predict the form of the indentation stiffness function. So, we separately analyze the compliance functions corresponding to a different combination of thicknesses of the two upper layers of the sample.

It is also known that $L(0) = \beta_{31}^{-1}$ and $L(u) \rightarrow 1$ as $u \rightarrow \infty$. In addition, the kernel transform does not depend on the thickness of the coating *H*, but it depends on the ratio of the thicknesses of the first and second layers. We introduce the parameter h_1/H and analyze how it affects the kernel transform (and, as consequence, the stiffness function). The stiffness function depends only on the parameter λ, but does not depend on *H*.

Figure 1 shows the kernel transforms for different values of the parameter h_1/H. The thicker the second layer, the more it affects the kernel transform and the stiffness function. The maximum values of the kernel transform increases, tending to β_{12}. The range of the values of the argument *u* expands, where the transform significantly differs from β_{31}^{-1} and 1.

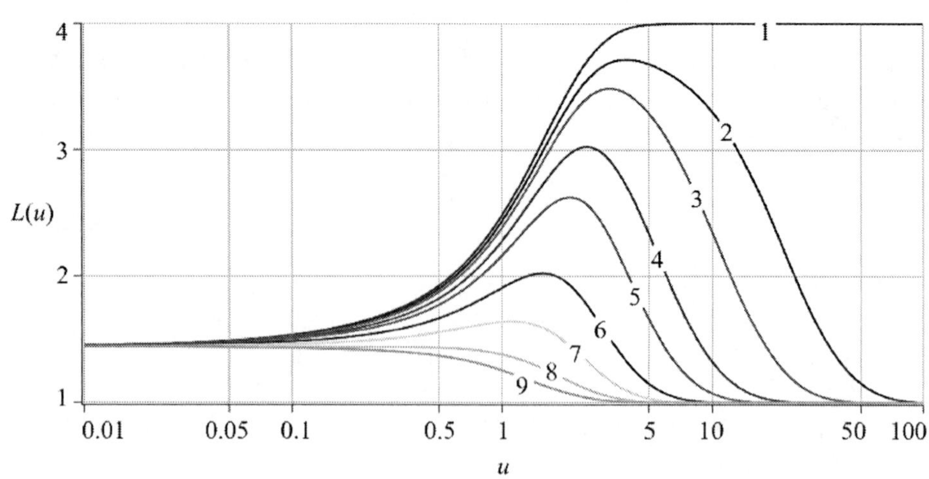

Figure 1. Kernel transforms $L(u)$ of integral equation (1) for different values of h_1/H: 0 (1), 0.05 (2), 0.1 (3), 0.2 (4), 0.3 (5), 0.5 (6), 0.7 (7), 0.9 (8), 1 (9).

For the spherical indenter, the restored effective modulus has the form [22]:

$$E_{ef}^{r}(\lambda)=E_{ef}^{(3)}\frac{1+3\sum_{i=1}^{N}C_i(\cosh(A_i\lambda^{-1})-A_i^{-1}\lambda\sinh(A_i\lambda^{-1}))}{1+2\sum_{i=1}^{N}C_i\cosh(A_i\lambda^{-1})+2\lambda^2\sum_{i=1}^{N}\left(\frac{B_i^2-A_i^2}{A_i^2B_i^2}\right)} \tag{3}$$

For a conical indenter [19]:

$$E_{ef}^{r}(\lambda)=E_{ef}^{(3)}\frac{1+2\lambda\sum_{i=1}^{N}\left(\frac{C_i}{A_i}\left(\mathrm{ch}(\frac{A_i}{\lambda})-\frac{\lambda}{A_i}\mathrm{sh}(\frac{A_i}{\lambda})\right)+\frac{D_i}{A_i}\left(\frac{\lambda}{A_i}-\frac{\lambda}{A_i}\mathrm{ch}(\frac{A_i}{\lambda})+\mathrm{sh}(\frac{A_i}{\lambda})\right)\right)}{1+\lambda\sum_{i=1}^{N}\frac{1}{A_i}\left(C_i\,\mathrm{ch}(A_i\lambda^{-1})+D_i\,\mathrm{sh}(A_i\lambda^{-1})\right)} \tag{4}$$

Here A_i, B_i are the complex-value parameters of the kernel transform approximation by the following expression:

$$L(u)\approx L_N(u)=\prod_{i=1}^{N}(u^2+A_i^2\lambda^{-2})/(u^2+B_i^2\lambda^{-2})$$

C_i, D_i are the complex constants, which are obtained from the solution of certain systems of linear algebraic equations [19, 22]. The restored effective elastic modulus, depending on the parameter λ, is shown in Figure 2 for the cases $h_1 = 0.1H$ and $h_1 = 0.3H$ for conical and spherical punches. As can be seen, the curves, corresponding to the conical and spherical indenters, differ very little. These graphs depend neither on the radius of the spherical indenter, nor on the cone angle, nor on the actual thickness of the coating (the coating thickness influence through the parameter λ).

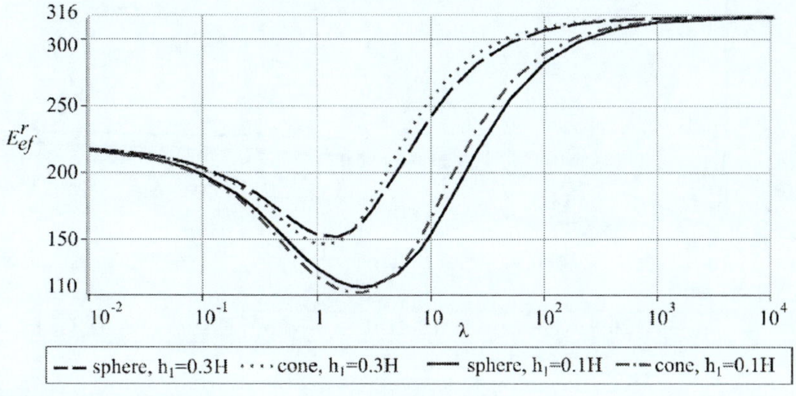

Figure 2. Restored effective modulus E_{ef}^{r} for $h_1 = 0.1H$ and $h_1 = 0.3H$.

In order to determine the ranges of indentation depth, in which to perform an experiment on nanoindentation, the dependences of the value of the restored Young's modulus on the absolute values of the indentation depth were plotted (see Figures 3–6). The coating thickness H values were considered equal to 1 μm and 2 μm. It was found that a spherical indenter with a radius of 20 μm is poorly suited for such experiments, since in order to research the properties of the upper two layers the indentation depth should be very small, which is difficult to implement in practice. For example, in the case of R = 20 μm, H = 1 μm, h_1 = 0.1 μm at indentation depths δ > 70 nm, the upper two layers have very little effect on the experiment.

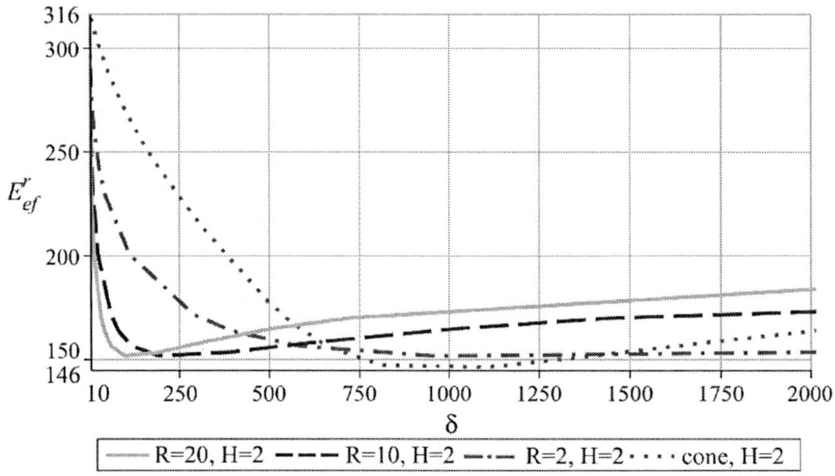

Figure 3. Restored effective modulus E^r_{ef} depending on the indentation depth δ for $h_1 = 0.3H$ and coating thickness H = 1 μm.

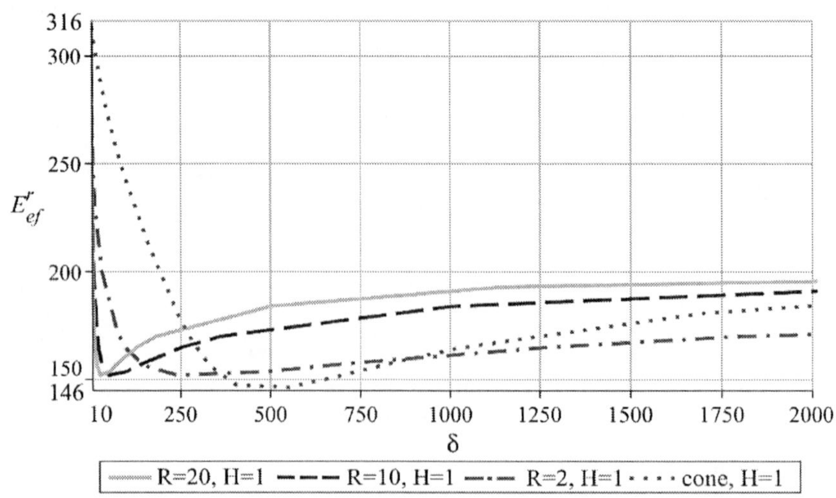

Figure 4. Restored effective modulus E^r_{ef} depending on the indentation depth δ for $h_1 = 0.3H$ and coating thickness H = 2 μm.

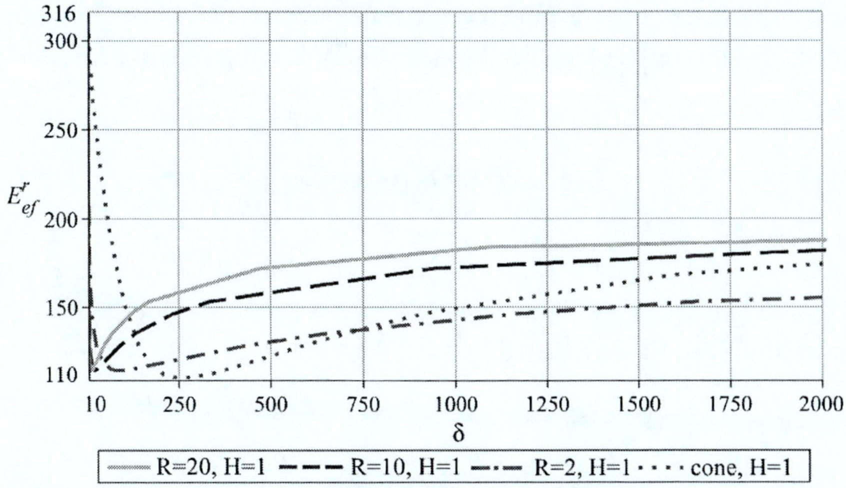

Figure 5. Restored effective modulus E^r_{ef} depending on the indentation depth δ for $h_1 = 0.1H$ and coating thickness $H = 1$ μm.

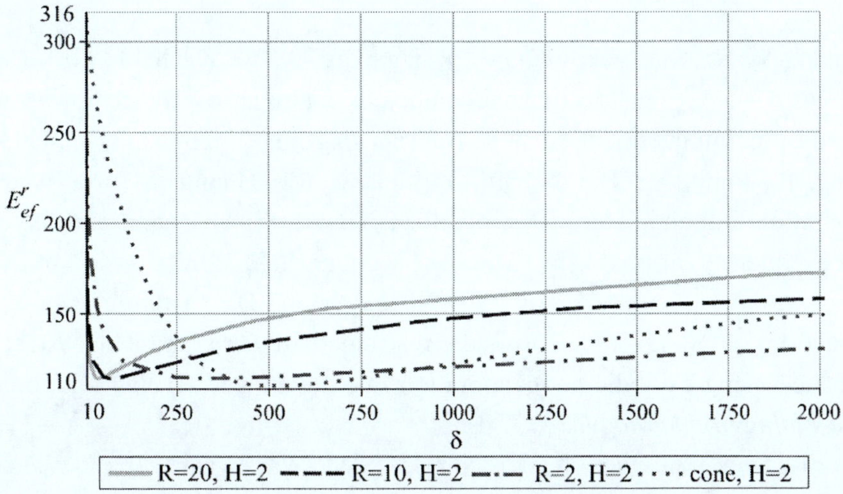

Figure 6. Restored effective modulus E^r_{ef} depending on the indentation depth δ for $h_1 = 0.1H$ and coating thickness $H = 2$ μm.

CONCLUSION

The results obtained make it possible to estimate which indenters most suit for the investigation of a three-layer sample in the case of certain thicknesses of the upper two layers. For example, for a study of $h_1 = 0.3H$, $H = 1$ μm, it is better to use a spherical indenter of small radius (for example, 2 μm) or to use a conical indenter or the Berkovich pyramid, which can be modeled with a high degree of reliability by a conical indenter [6].

For the case $h_1 = 0.3H$, $H = 2$ μm, the conical indenter fits somewhat worse, but a spherical indenter of a slightly larger radius (for example, 10 μm) will be a solid choice.

ACKNOWLEDGMENTS

The authors acknowledge the support of the grant of the Russian Science Foundation No. 15-19-10056. Sadyrin E. V. was supported by the scholarship of the President of the Russian Federation (No. SP-3672.2018.1).

REFERENCES

[1] Chi, S. H., & Chung, Y. L. (2003). Cracking in coating–substrate composites with multi-layered and FGM coatings. *Engineering Fracture Mechanics*, 70 (10), 1227-1243.

[2] Varavka, V. N., Kudryakov, O. V., Ryzhenkov, A. V., Kachalin, G. V., & Zilova, O. S. (2014). Application of nanocomposite coatings to protect power equipment from droplet impingement erosion. *Thermal Engineering*, 61 (11), 797-803.

[3] Zhang, P., Fang, Y., Dai, S., Fu, Y., Zhang, M., Huang, M., & Hu, J. (2016). Preparation and characterization of MoS2-TiL/MoS2-TiH nano multilayer coating with excellent wear properties. *Materials Letters*, 173, 35-38.

[4] Gilewicz, A., Chmielewska, P., Murzynski, D., Dobruchowska, E., & Warcholinski, B. (2016). Corrosion resistance of CrN and CrCN/CrN coatings deposited using cathodic arc evaporation in Ringer's and Hank's solutions. *Surface and Coatings Technology*, 299, 7-14.

[5] Kruk, T., Szczepanowicz, K., Kręgiel, D., Szyk-Warszyńska, L., & Warszyński, P. (2016). Nanostructured multilayer polyelectrolyte films with silver nanoparticles as antibacterial coatings. *Colloids and Surfaces B: Biointerfaces*, 137, 158-166.

[6] Golovin, Y. I. (2008). Nanoindentation and mechanical properties of solids in submicrovolumes, thin near-surface layers, and films: A Review. *Physics of the solid State*, 50 (12), 2205-2236.

[7] Ke, L. L., & Wang, Y. S. (2006). Two-dimensional contact mechanics of functionally graded materials with arbitrary spatial variations of material properties. *International Journal of Solids and Structures*, 43 (18-19), 5779-5798.

[8] Liu, T. J., Wang, Y. S., & Zhang, C. (2008). Axisymmetric frictionless contact of functionally graded materials. *Archive of Applied Mechanics*, 78 (4), 267-282.

[9] Guler, M. A., & Erdogan, F. (2004). Contact mechanics of graded coatings. *International Journal of Solids and Structures*, 41 (14), 3865-3889.

[10] Vasiliev, A. S., Volkov, S. S., & Aizikovich, S. M. (2018). Approximated analytical solution of contact problem on indentation of elastic half-space with coating reinforced with inhomogeneous interlayer. *Materials physics and mechanics*, 35 (1), 175-180.

[11] Argatov, I. I., & Sabina, F. J. (2016). Small-scale indentation of an elastic coated half-space: The effect of compliant substrate. *International Journal of Engineering Science*, 104, 87-96.

[12] Aizikovich, S. M., Vasil'ev, A. S., & Volkov, S. S. (2015). The axisymmetric contact problem of the indentation of a conical punch into a half-space with a coating inhomogeneous in depth. *Journal of Applied Mathematics and Mechanics*, 79 (5), 500-505.

[13] Krenev, L. I., Sadyrin, E. V., Aizikovich, S. M., & Zubar, T. I. (2017). Indentation of a Functionally Graded Coating on an Elastic Substrate by a Sphero-Conical Indenter. In *Advanced Materials* (pp. 397-405). Springer, Cham.

[14] Krenev, L. I., Volkov, S. S., Sadyrin, E. V., & Chizhik, S. A. (2018). Mechanical Material Tests by the Nanoindentation Method at Various Indenter and Specimen Temperatures. *Journal of Engineering Physics and Thermophysics*, 91 (3), 594-600.

[15] Volkov, S. S., Vasiliev, A. S., Aizikovich, S. M., & Sadyrin, E. V. (2018, May). Contact problem on indentation of an elastic half-plane with an inhomogeneous coating by a flat punch in the presence of tangential stresses on a surface. In *AIP Conference Proceedings* (Vol. 1959, No. 1, p. 070037). AIP Publishing.

[16] Vasiliev, A. S., Volkov, S. S., & Aizikovich, S. M. (2018, January). Approximated Analytical Solution of a Problem on Indentation of an Electro-Elastic Half-Space with Inhomogeneous Coating by a Conductive Punch. In *Doklady Physics* (Vol. 63, No. 1, pp. 18-22). Pleiades Publishing.

[17] Volkov, S. S., Litvinenko, A. N., Aizikovich, S. M., Wang, Y. C., & Vasiliev, A. S. (2016). Axisymmetric bending of a circular plate with stiff edge on a soft FGM layer. *Structural Engineering and Mechanics*, 59 (2), 227-241.

[18] Vasiliev, A. S., Swain, M. V., Aizikovich, S. M., & Sadyrin, E. V. (2016). Torsion of a circular punch attached to an elastic half-space with a coating with periodically depth-varying elastic properties. *Archive of Applied Mechanics*, 86 (7), 1247-1254.

[19] Vasiliev, A. S., Volkov, S. S., Belov, A. A., Litvinchuk, S. Y., & Aizikovich, S. M. (2017). Indentation of a hard transversely isotropic functionally graded coating by a conical indenter. *International Journal of Engineering Science*, 112, 63-75.

[20] Oliver, W. C., & Pharr, G. M. (1992). An improved technique for determining hardness and elastic modulus using load and displacement sensing indentation experiments. *Journal of Materials Research*, 7 (6), 1564-1583.

[21] King, R. B. (1987). Elastic analysis of some punch problems for a layered medium. *International Journal of Solids and Structures*, 23 (12), 1657-1664.

[22] Volkov, S. S., Vasiliev, A. S., Aizikovich, S. M., Seleznev, N. M., & Leontieva, A. V. (2016). Stress-strain state of an elastic soft functionally-graded coating subjected to indentation by a spherical punch. *PNRPU Mechanics Bulletin*, (4), 20-34.

In: Proceedings of the 2018 International Conference … ISBN: 978-1-53615-862-5
Editors: Ivan A. Parinov et al.

Chapter 33

THE EFFECT OF THE CURVATURE OF GEOMETRIC CONCENTRATOR ON REDUCING STRESS CONCENTRATION

G. A. Zhuravlev[1,*] ***and Yu. E. Drobotov***[1,2]
[1]Southern Federal University, Rostov-on-Don, Russia
[2]SPE Vibrobit LLC, Rostov-on-Don, Russia

ABSTRACT

Elastoplastic deformation of a rod with symmetric hyperbolic recesses is considered. The solution clarifies the influence of material deformation properties with different loading schemes, including transverse bending.

Keywords: deformation, stress, elastoplastic deformation, bending

1. INTRODUCTION

Approaches for increasing correlation between research methods of strengthening the machinery details and their stress state, geometric and technologic characteristics, and material properties and constructive features are valuable elements of designed systems, which can be related to the constructive strength of power systems [1]. Thus, this paper demonstrates the possibilities of applying the constructive strength techniques for

* Corresponding Author's E-mail: zhur.okp@yandex.ru.

developing a brand-new technology of deformation that strengthens the machinery details with loaded projections.

Such details have complex shapes and are loaded with transverse bending with compression of the ledges on a stressed base with its torsion, either stretching or compressing. The localization of the calculated (dangerous) cross-section depends on various factors and does not match usually the one of minimal thickness. However, the data on stress concentration correspond to generally the minimal cross-sections in the recesses zone of a rod, which approximates a real detail. The recesses zone for the projections base are limited and refer to specific cases, while solving the problems on elastoplastic and thermoplastic deformations is difficult due to non-linearity of the governing equations.

An approach for particular problems on approximately describing the stressed state of details with loaded projections is often sufficiently effective because it based on "elastic" solutions, specifically applying the widely approved kinematic hypothesis of S. P. Timoshenko [2–4]. The latter demonstrates independence of the deformations distribution law on the history of deforming and strengthening the material. Let it be written as

$$\tilde{\varepsilon}_x(y) = \varepsilon_x(y)\frac{\varepsilon_T}{\varepsilon_i(y_T)},$$

where $\varepsilon_x(y)$ and $\tilde{\varepsilon}_x(y)$ are the relative linear deformation at the current point y for some cross-section at $x = \text{const}$ for a fictitious absolutely elastic material and real material, respectively; ε_T is the relative deformation, corresponding to the material yield stress; $\varepsilon_i(y_T)$ is the criterion value to characterize the transition from an elastic deforming to non-elastic one at the boundary point $y = y_T$ of the section for ideally elastic material.

In this paper, based on [5], the problem on the stress-strain state is considered an arbitrary cross-section of a rod under transverse bending with compressing and thermal-force loading the surface layers. The problem was considered earlier in [6] for pure bending.

2. Solution of Elastic Problem

The Neuber solution for the plane problem on stress concentration is used as the elastic component. Thus, a rod with two symmetric hyperbolic recesses is loaded with pure bending, and an isometric curvilinear coordinate system is considered:

$$x = R\sinh u \cos v,\ y = R\cosh u \sin v, \tag{1}$$

where R is the focal distance parameter for $v = \text{const}$, and

$$h^2 = \left(\frac{\partial x}{\partial u}\right)^2 + \left(\frac{\partial y}{\partial u}\right)^2 = \left(\frac{\partial x}{\partial v}\right)^2 + \left(\frac{\partial y}{\partial v}\right)^2 = \frac{1}{2}R^2(\cos 2v + \cosh 2u). \quad (2)$$

In the terms of the stress function, $F = F(u, v)$; the following relationships are performed for the normal stresses σ_u, σ_v and the shear stress τ_{uv} [7]:

$$\sigma_u = \frac{1}{h}\left\{\frac{\partial}{\partial v}\left(\frac{1}{h}\frac{\partial F}{\partial v}\right) + \frac{1}{h^2}\frac{\partial h}{\partial u}\frac{\partial F}{\partial u}\right\},\ \sigma_v = \frac{1}{h}\left\{\frac{\partial}{\partial u}\left(\frac{1}{h}\frac{\partial F}{\partial u}\right) + \frac{1}{h^2}\frac{\partial h}{\partial v}\frac{\partial F}{\partial v}\right\},$$

$$\tau_{uv} = -\frac{1}{h}\left\{\frac{\partial}{\partial u}\left(\frac{1}{h}\frac{\partial F}{\partial v}\right) + \frac{1}{h^2}\frac{\partial h}{\partial v}\frac{\partial F}{\partial u}\right\}.$$

According to the Neueber approach (and, therefore, to Grodskiy and Papkovich approaches), F is represented as a combination of harmonic functions. Summarizing the results of [7], provided, separately, for simple stretching, pure bending, and pure shearing, we write the stresses for a superposition of these force factors:

$$\sigma_u = \frac{1}{h^2}(2A_1 \sin 2v + A_2 \sinh u \sin v - A_3 \cosh u \cos v)\left(-2 + \frac{\cos^2 v - \cos^2 v_0}{h^2}\right),$$

$$\sigma_v = \frac{1}{h^4}(2A_1 \sin 2v + A_2 \sinh u \sin v - A_3 \cosh u \cos v)(\cos^2 v_0 - \cos^2 v),$$

$$\tau_{uv} = \frac{1}{h^4}(2A_1 \sinh 2u - A_2 \cosh u \cos v - A_3 \sinh u \sin v)(\cos^2 v - \cos^2 v_0),$$

where A_1, A_2 and A_3 are the components of the harmonic functions:

$$A_1 = -\frac{\sigma_M \sin^2 v_0}{3(\sin 2v_0 - 2v_0 \cos 2v_0)},\ A_2 = -\frac{\tau_P \sin v_0}{v_0 \sin v_0 \cos v_0},\ A_3 = \frac{\sigma_Q \sin v_0}{v_0 + \sin v_0 \cos v_0},$$

$v = v_0$ is the equation of the free contour, and the nominal stresses are represent as

$$\sigma_M = \frac{3}{2}\frac{M}{a^2 B},\ \ \tau_P = \frac{P}{2aB},\ \ \sigma_Q = -\frac{Q}{2aB},$$

which arise, respectively, from bending moment M, shear force P and compression force Q; $2a$ is the thickness and B is the width of the rod.

Consider the free contour $v = v_0$ and the cross-section $x = \text{const}$ of the rod. Then, there is a point $y_{max} = y(u_{max}, v_0)$, at which the maximum relative deformation is achieved, and it is performed:

$$\sigma_u(y_{max})\frac{E\varepsilon_{u_{max}}}{1-\mu^2} \rightarrow \sigma_x(y_{max}) = \frac{E\varepsilon_{u_{max}}}{1-\mu^2}\cos^2(x,u),$$

where E and μ are Young's modulus and Poisson's ratio, $\varepsilon_{u_{max}}$ is the maximum relative deformation in the considered cross-section. If the deformation properties of the material are constant in the considered cross-section, then the maximal values of σ_x are at the free contour.

Let f denotes the following function:

$$f(y) = \frac{\sigma_x(y)}{\sigma_x(y_{max})}, \tag{3}$$

and the maximal stresses in the elastic core, which is bounded by points y_1 and y_2, are considered as

$$\sigma_{xy}(y_1 < y < y_2) = \frac{E\varepsilon_{u_{max}}}{1-\mu^2} f(y)\cos^2(x,u). \tag{4}$$

The stress intensity is assumed in the following form:

$$\sigma_I = \frac{1}{h^4}\{2(1-\mu+\mu^2)k_1^2(k_2^2 - 2k_2h^2 + 2h^4) + (1+2\mu-2\mu^2)k_1^2k_2(k_2 - 2h^2) + \\ +3k_2^2(2A_1\sinh 2u - A_2\cosh u\cos v - A_3\sinh u\sin v)^2\}^{1/2},$$

where

$$k_1 = 2A_1\sin 2v + A_2\sinh u\sin v - A_3\cosh u\cos v,$$

$$k_2 = \cos^2 v - \cos^2 v_0.$$

3. General Equations

Three types of values are considered: (i) those the marked with tilde denote the ones in the elastoplastic area, (ii) those with index T characterize the yield strength, and (iii) those with stroke satisfy the equation:

$$\sigma_{I_{max}} = \sigma_T. \tag{5}$$

Thus, the boundary point for the elastic kernel is found from the equation:

$$\tilde{\sigma}_I(v_T) = \sigma_T. \tag{6}$$

Let us consider an approach, introduced by Troshchenko and Getman [4], but let, for simplicity, the law for stress intensity in the elastoplastic area is assumed in the form:

$$\tilde{\sigma}_x(y) = \sigma_{x_T}\left(\frac{\tilde{\varepsilon}_x}{\varepsilon_{x_T}}\right)^{\kappa}, \sigma_{x_T} = \sigma_{x_T}(y), \kappa = \kappa(y). \tag{7}$$

We assume that the material deformation properties are defined discretely by the values of σ_T and κ as

$$\sigma_T(y) = \begin{cases} \sigma_{T_1}, y = y_{min} \div y_1; \\ \sigma_{T_2}, y = y_1 \div y_2; \\ \dots \\ \sigma_{T_n}, y = y_{n-1} \div y_{\max}, \end{cases} \quad \kappa(y) = \begin{cases} \kappa_1, y = y_{min} \div y_1; \\ \kappa_2, y = y_1 \div y_2; \\ \dots \\ \kappa_n, y = y_{n-1} \div y_{\max} \end{cases}.$$

By inserting a designation:

$$n(y) = \frac{\sigma_T(y)}{\sigma'_I(y)} = \frac{\varepsilon_T(y)}{\varepsilon'_I(y)}$$

we obtain the following relationships:

$$\sigma_{x_T} = \sigma'_x(y_{\max})n(y)f(y), \varepsilon_{x_T} = \varepsilon'_x(y_{max})n(y)\frac{\varepsilon_x}{\varepsilon_x(y_{\max})}$$

and, according to the kinematical hypothesis assumed, we find from (7) the representation:

$$\tilde{\sigma}_x(y) = \left(\frac{\tilde{\varepsilon}_{x_{max}}}{\varepsilon'_{x_{\max}}}\right)^{\kappa_n} \sigma'_x(y_{max})n^{1-\kappa}(y)f(y), \tag{8}$$

which characterizes the dependence of the normal stress on the maximal deformation in the elastoplastic area.

If we assume that

$$\bar{\bar{\varepsilon}}_{x_{max}} = \frac{\tilde{\varepsilon}_{x\max}}{\varepsilon'_{x\max}} = \frac{\sigma_T(y_2)}{\sigma'_i(y_2)},$$

then, considering specific values of $\bar{\bar{\varepsilon}}_{x_{max}}$, one can calculate y_1 and y_2 from the equation:

$$\bar{\bar{\varepsilon}}_{x_{max}} = \sigma_T(y_1, y_2) h_T{}^4 \{(1 - \mu + \mu^2) k'^2_1 2 (k'^2_2 - 2k'_2 h^2_T + 2h_T{}^4) + (1 + 2\mu - 2\mu^2) k'^2_1 k'_2 (k'_2 - 2h^2_T) + 3k'^2_2 (2A'_1 \sinh 2u - A'_2 \cosh u \cos v_T - A'_3 \sinh u \sin v_T)^2\}^{-1/2},$$

where

$$k'_1 = 2A'_1 \sin 2v_T + A'_2 \sinh u \sin v_T - A'_3 \cosh u \cos v_T,\ k'_2 = \cos^2 v_T - \cos^2 v_0,$$

$$h_T{}^2 = \sinh^2 u + \cos^2 v_T,$$

$$v = \operatorname{sgn} y \arcsin \sqrt{\frac{\sqrt{a^2+b^2} + y^2 + x^2 - \sqrt{\left(\sqrt{a^2+b^2} + y^2 + x^2\right)^2 - 4y^2\sqrt{a^2+b^2}}}{2\sqrt{a^2+b^2}}},$$

$$u = \operatorname{arcsinh} \frac{x}{\cos v \sqrt{a^2+b^2}}.$$

The calculation cycle is limited in σ_T by range $[0, y_{max}]$ for calculating y_2, and by range $[y_{min}, 0]$ for calculating y_1, by assuming the boundary point in the stretching area as y_2, and the boundary point in the compressing area as y_1. Stresses are calculated in every region separately:

1. in the elastoplastic area as $\tilde{\sigma}_x(y) = \left(\bar{\bar{\varepsilon}}_{x_{max}}\right)^{\kappa_n} \sigma'_x(y_{max}) n^{1-\kappa}(y) f(y)$;
2. in the elastic area as $\sigma_{xy}(y_1 < y < y_2) = \sigma_x(y) \frac{\tilde{\sigma}_x(y_1, y_2)}{\sigma_x(y_1, y_2)}$;
3. in the areas $y > 0$ and $y < 0$ by using different formulas with calculating $\sigma_x(y)$ and $\sigma_x(y_1, y_2)$ for fictitious (absolutely elastic) material. In general, the center of the elastic core is, displaced in relation to the rod's symmetry axis because of compressing and changing the deformation properties of material in cross-section.

The residual normal stresses $\sigma_{x0}(y)$ after the elastoplastic deforming are calculated by using the Ilyushin – Genki theorem:

$$\sigma_{x0}(y) = \bar{\tilde{\sigma}}_x(y) - \sigma_x(y),$$

with

$$\sigma_{x0}(y_1 < y < y_2) = \sigma_{xy}(y_1 < y < y_2) - \sigma_x(y).$$

To establish the compliance of $\bar{\tilde{\varepsilon}}_{u_{max}}$ with external load, checking equilibrium conditions is performed. The latter can be considered in dependence on the loading scheme (pure bending, shear or compressing), which is dominant in the specific case for the stress – strain state of the treated cross-section.

Let us consider, as an example of a short projection, which length is comparable with the width of treated cross-section at sufficiently large compression angles: $\frac{\pi}{10} \leq \alpha \leq \frac{\pi}{2}$, i.e., for the large values of Q:

$$\int_{y_{min}}^{y_{max}} \left(\tilde{\sigma}_x(y) + \sigma_{xy}(y)\right) dy = -Q.$$

Then, the iterative process for calculating $\bar{\tilde{\varepsilon}}_{x_{max}}$ is continued until the conditions are satisfied:

$$\left|\bar{\tilde{\varepsilon}}_{x_{max}}^{\kappa_n} \sigma'_{x_{max}} \left(\int_{y_{max}}^{y_{n-1}} n^{1-\kappa_n}(y)|f(y)|dy + \int_{y_{n-1}}^{y_{n-2}} n^{1-\kappa_{n-1}}(y)|f(y)|dy + \cdots + \right.\right.$$
$$+ \int_{y_k}^{y_\xi} n^{1-\kappa_k}(y)|f(y)|dy + \left(\bar{\tilde{\varepsilon}}_{x_{max}}^{\kappa_n} \sigma'_{x_{max}}\right)^{-1} \int_{y_\xi}^{y_\eta} |\sigma_x(y)| \frac{\tilde{\sigma}_x(y_\xi)}{\sigma_x(y_\xi)} dy +$$
$$\left.\left. \int_{y_\eta}^{y_m} n^{1-\kappa_{m+1}}(y)|f(y)|dy + + \cdots + \int_{y_1}^{y_{min}} n^{1-\kappa_1}(y)|f(y)|dy\right) + Q\right| \leq \lambda$$

where λ is the permissible error of calculating the compressing force, and the limits of integration are chosen taking into account the sign of the negative stresses, caused by Q.

For real units, all the components of the loading scheme are functionally related, thus the analysis with considering only one component M, P and Q characterizes the complete solution.

4. Example of Pure Bending

As an example, we consider pure bending of a rod with material properties $\kappa = \text{const}$ and $\sigma_T = \text{const}$, at the chosen cross-section $u = \text{const}$.

Let the approximate value of the bending moment $\bar{\tilde{M}}$ is assumed as

$$\overline{\widetilde{M}} = \frac{\widetilde{M}}{M'} =$$

$$= \left(\bar{\bar{\varepsilon}}_{u_{max}} \int_0^{v_T} \sigma_u(v) \sin 2v \, dv + \bar{\bar{\varepsilon}}^{\kappa}_{u_{max}} \int_{v_T}^{v_0} n^{1-\kappa}(v)\sigma_u(v) \sin 2v \, dv \right) \Big/ \int_0^{v_0} \sigma_u(v) \sin 2v \, dv$$

which characterizes changing of mechanical properties during deformation for specific values of $\sigma_{u_{max}}$, and

$$\overline{\widetilde{M}} = \frac{\bar{\bar{\varepsilon}}_{u_{max}} I_1 + \bar{\bar{\varepsilon}}'_{u_{max}} I_2}{I_3},$$

where

$$I_1 = -\int_0^{v_T} \frac{\sin^2 2v}{(\sinh^2 u + \cos^2 v)^2} (\cos 2v - \cos 2v_0 - 4\sinh^2 u - 4\cos^2 v) dv,$$

$$I_2 = \left(-\frac{\sigma_T}{A'_1}\right)^{1-\kappa} \int_{v_T}^{v_0} \frac{-\sin^2 2v(\cos 2v - \cos 2v_0 - 4\sinh^2 u - 4\cos^2 v)\sinh^2 u + \cos^2 v^{-2\kappa}}{\sqrt{\{(1-\mu+\mu^2)\sin^2 2v(k_{10}^2 + k_{20}^2) - (1+2\mu-2\mu^2)\sin^2 2v k_{10} k_{20} + 3\sinh^2 2u k_{20}^2\}^{1-\kappa}}} dv,$$

$$I_3 = -\int_0^{v_0} \frac{\sin^2 2v}{(\sinh^2 u + \cos^2 v)^2} (\cos 2v - \cos 2v_0 - 4\sinh^2 u - 4\cos^2 v) dv,$$

and zero indices demonstrate the correspondence of values for pure bending.

It is stated in [P], that the improved calculations of the moment with the lever:

$$H_t = (b^2 \sinh^2 u \cos^2 v + 0.25a^2 \sinh 2u \sin 2v) \frac{\cosh u \sin v \sqrt{a^2+b^2}(b^2-a^2) + ab^2}{a^4 \cosh^2 u \sin^2 v + b^4 \sinh^2 u \cos^2 v},$$

are matched satisfactorily with approximate values. The iterative process is continued until the calculated value of $M'\overline{\widetilde{M}}$ attains the initial value of M.

As an example, the elastoplastic deforming of a rod with geometric parameters: $a = 0.967$ mm, $b = 0.520$ mm and $B = 1$ mm, made of steel with $\sigma_T = 58$ kgf/mm^2, $E = 2.15 \times 10^4$ and $\mu = 0.3$ was considered in [P]. The calculations with the precision of 0.001 were carried out for the cross-section $u = 30^\circ$ and $\kappa = 0; 0.3; 0.6; 0.9$. The result plots are shown in Figure 1.

The decreasing of elastic core sizes leads to formation of a plastic hinge ($v_T = 0$) with the limit value of the relative maximal deformation:

$$\bar{\bar{\varepsilon}}_{u_{lim}} = \frac{(1+\sinh^2 u)^2 \sigma_T}{A'_1 \sqrt{3} \sinh 2u(1-\cos v)}.$$

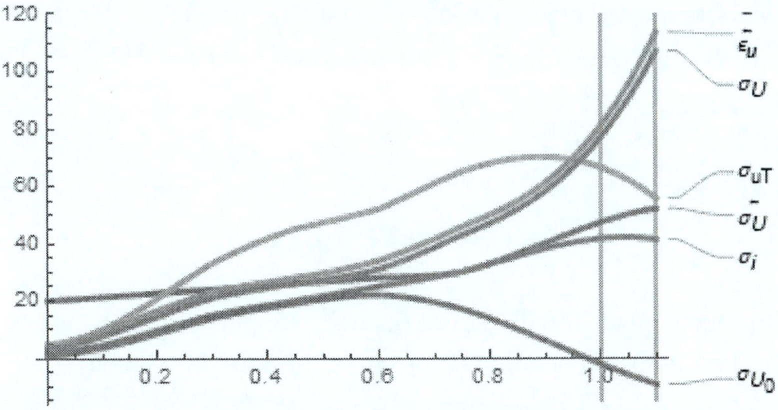

Figure 1. Stress state of a rod in the section $u = 30°$ under pure bending with moment M for $\bar{\bar{\varepsilon}}_{u_{max}} = 1.37$ and $\kappa = 0$.

It can be observed from Figure 1, that the value of $\bar{\bar{\varepsilon}}_{u_{max}}$ does not depend on choice of κ, but is determined by the geometric parameters of the rod, the treated cross-section and the material properties. Thus, $\bar{\bar{\varepsilon}}_{u_{lim}} = 2.848$.

With increasing of the bending moment, $\bar{\bar{\varepsilon}}_{u_{max}}$ increases, when the value of κ decreases. In this, the maximum of $\overline{\widetilde{M}}$ well characterizes the strengthening of detail and its sensitivity to breaking the deformation strengthening mode, while $M' = \text{const}$. The plots demonstrate that increasing in κ reduces the strengthening of detail and its sensitivity to the processing mode. Based on such an analysis and similar studies, the material for the detail under strengthening could be chosen.

The stress state of the cross-section $u = 30°$ with $\bar{\bar{\varepsilon}}_{u_{max}} = 1.37$ is considered by using the Cartesian coordinates at the calculation points and different methods for calculating $\overline{\widetilde{M}}$. The simplified computation provided the error within the range of $2 - 4$ %.

The case $\kappa = 0$ describes the rigid-plastic deformation of material, which cannot be strengthened. For illustrative purposes, there is a plot of $\sigma_u(v)$ with $\tilde{\varepsilon}_{u_{max}} = 1.0$, shown in Figure 1.

An interesting result is obtained here: on a free contour, there is $\tilde{\sigma}_u = \sigma_{u_T} = \sigma_u$ ($\bar{\bar{\varepsilon}}_{u_{max}} = 1$), while, everywhere in the elastoplastic zone, there is $\bar{\bar{\varepsilon}}_u = \sigma_{u_T}$. It should be noted that the nature of changing σ_{u_T} and σ_I is related to a high degree of heterogeneity of the stress state: for the two-dimensional stress state, there is the value of σ_{u_T}, exceeding the value of σ_T, but the change of the stress state at attainment of the center of cross-section increases the material plasticity and reduces σ_{u_T} (even at $\sigma_{u_T} < \sigma_T$). It is easy to observe that when this feature manifests itself more clearly, the greater is the value of u.

The residual compressive strength $\sigma_{u_0} < 0$ (shown in plots as the area bounded by two vertical lines) in the plastic deformed zone is defined by the greater area $S = \{v \in V: \widetilde{\sigma_u} = \sigma_{u_T}\}$.

ACKNOWLEDGMENTS

The work is accomplished with partial financial support of the Ministry of Education and Science of the Russian Federation (Project No. 11.6767.2017/BCh).

REFERENCES

[1] Zhuravlev, G. A. Izvestiya Vuzov. North-Caucasus Region. *Natural Sciences*. 1977, vol. 2 (In Russian).
[2] Del,' G. D., Chebaevskij, B. P. *Strength Problems*. 1971. No. 1 (In Russian).
[3] Uzhik. G. V. *Tear Resistance and Strength of Metals*. Moscow: DAN, 1950 (In Russian).
[4] Troschenko, V. T. *Strength Problems*. 1972, No. 2.
[5] Shnejderovich, R. M. *Strength at Static and Variable Loading*. Mashinostroenie, Moscow. 1968 (In Russian).
[6] Zhuravlev, G. A. In: *Mechanics of Solid Medium*. Rostov State University Press, Rostov-on-Don, 1981, 73 – 89 (In Russian).
[7] Neuber, G. *Concentration of Stresses*. OGIS, Moscow. 1947 (In Russian).

In: Proceedings of the 2018 International Conference … ISBN: 978-1-53615-862-5
Editors: Ivan A. Parinov et al.

Chapter 34

CONTACT APPROACH OF ELASTIC BODIES AND EFFECTS OF MATERIAL PROPERTIES ON ITS FEATURES

***G. A. Zhuravlev*[1,*], *T. M. Andreeva*[1] *and Yu. E. Drobotov*[1,2]**
[1]Southern Federal University, Rostov-on-Don, Russia
[2]SPE Vibrobit LLC, Rostov-on-Don, Russia

ABSTRACT

The impact of material properties of the elastic circular cylinders in their initial-linear touch is analyzed. The incidents of electoral effect (effect of independence of the radius is only smaller for contacting cylinders) are reviewed.

Keywords: contact, elastic cylinders, initial-linear touch, the effect of selective independence

1. INTRODUCTION

Let δ denote the approach of two parallel elastic circular cylinders under compression. In this paper, the problem of estimating the interdependence between δ, unit load q and the radii R_i of the cylinders in contact is considered. As there is no common classical solution of the problem (this problem has been never given by Hertz), various approaches are used, but the results of experiments are rather inconsistent [1 – 9].

---

* Corresponding Author's E-mail: zhur.okp@yandex.ru.

The experimental researches on the approach of two parallel rollers (cylinders of finite length) were more often carried out compared to compressing two plates and a roller with positioning the roller between the plates. Such experiments (for example, [1]) demonstrated linear relationships $\delta = cq$, where c is a constant. On the other hand, mutual nonlinearity of δ and q, with a noticeable influence of R_i on δ, follow from a number of theoretical studies ([5], as an example). However, at the same time, there are also theoretical works on the problem which state the linear dependence as universal (see [3], for example). As a result, the reliability of each of the statements is disputable. For instance, the handbook [2] points out the nonlinear relationship with a note: "… except for the approach of two parallel cylinders".

Thus, two different views on the problem of estimating the effects of the parallel elastic circular cylinders' radii on the approach of their axis exist in theory and experiment. Experimentally, for instance, by approaching two plates with compressing a roller [1], and theoretically [3] the point of view, proclaiming non-effecting is substantiated. In contrast, analytically [4 – 6], numerically [9] and experimentally [8], the principle of noticeable influence of the radii on the approach is established.

Moreover, in [6], incorrectness of the theoretical base of [3] is shown in detail and convincingly, and consequently, the first approach is criticized. As a result, the experimental results of [1] are called into question. However, while the papers [4 – 9] justify the second point of view, complete clarity is not achieved, as there are neither limitations for the scope of the reliability of the second principle revealed, nor the scope of the first principle, and the fact of full reliability of type [1] experiments is also questionable.

In [10], a previously unknown effect of approach ("EA") is found out. EA clarifies all the contradictions in estimating the approach of two parallel cylinders. EA is the selective effect, taking deal only with the great difference of the cylinders' radii: only one radius (of the smaller cylinder) does not influence on the contact approach.

Moreover, in [10] the special contact convergence mechanism of parallel cylinders is given and showed up as opposite in sign changes in elements of the total value of convergence when the radius of the smaller cylinder is being changed. For instance, it is proved that when the radius of the smaller cylinder increases, the compression value decreases together with the equal in absolute value increasing the compression of the bigger cylinder. We provided the evaluation of the radii influence well correlated with different (including the type [1]) experimental data. The results of calculations and experiments [8] have shown that the increasing of radius R (when $R = R_1 = R_2$) causes the increasing of the convergence, while the weak dependence of the convergence from R_i may be the consequence of the significant inequality in radii $R_2 \gg R_1$.

The main thing is EA leads to the following conclusions:

(i) for the first time, a reliability of [1] type experimental data is theoretically approved strictly;
(ii) it is proved, that each of the two widespread (widely published and pretending on exclusive generic) principles are, at the same time, completely contradictory principles. It is due to one of the principles points on significant effect of the cylinders' radii on the approach and other principle points on total absence of the effect. However, it does not exclude manifestation of the other principle for the cylinders of the same material, but with different radii;
(iii) each of the two principles is satisfactory only in specific limits of changing the curvature of the contacting bodies;
(iv) each of the two principles cannot be fundamental for universal use in theory and engineering practice of contact problems using the approach of two parallel elastic cylinders;
(v) the EA effect makes applying main ideas of the both principles possible for solving various problems of constructive strength and constructive flexibility.

2. Effect of Material Properties

The effect of material properties on selective manifestation of EA in the contact of two elastic circular cylinders is considered. For this purpose, consider the following three materials with different Poisson's ratio μ and Young's modulus E:

(i) Steel with $\mu = 0.3$ and $E = 2.15 \times 10^5$ MPa;
(ii) Brass with $\mu = 0.35$ and $E = 1 \times 10^5$ MPa;
(iii) Aluminum with $\mu = 0.34$ and $E = 0.69 \times 10^5$ MPa.

To analyze the influence of material properties of elastic bodies on the selective manifestation of EA in contact of elastic circular cylinders, the following analytical relation [6, 10], based on the solution of N. I. Muskhelishvili [11], will be used. It is expression for the plane deformation of a cylinder loaded with two equal and oppositely directed concentrated forces:

$$\delta = \frac{2q}{\pi}\sum_{i=1}^{2}\eta_i\left\{ln\frac{2R_i}{b} - \frac{1}{2} + \ln 2 + \frac{5}{48}\left(\frac{b}{R_i}\right)^2 + \ldots\ldots + \frac{(-1)^{n+1}}{2^n(n+1)!}\left[\frac{1}{2^n n!} - -\frac{(2^{2n-1}-1)(2n-1)!!B_{2n}}{n(2n)!}\right]\left(\frac{b}{R_i}\right)^{2n}\right\}, n = 2,3,4\ldots \quad (1)$$

where $\eta_i = (1-\mu_i^2)/E_i$; μ_i and E_i; $i = 1{,}2$, are Poisson's ratio and Young's modulus of the bodies at contact; b is the half-width of the contact area; B_{2n} is Bernoulli numbers.

The advantages of (1) are the possibility to obtain any given accuracy of calculations (with $n = 2, 3, 4, \ldots, \infty$), and its applicability for any size of the contact area without limitations of the Hertzian model. With accuracy up to $\gamma_i = q\eta_i b^2/R_i^2$ order of smallness, the solution (1) fits the solution of [4], and in the framework of the Hertzian assumption on the smallness of the contact area, (1) gives the results, close to [5]. The formula shows δ depending on R_i (but not on the reduced radius of curvature ρ), on q and η_i. The high accuracy of the formula is confirmed by calculations with FEM [9]. It also corresponds well to the experimental results [8].

In an analogy with [10], let the analysis of the approach with accuracy up to $\gamma_i = q\eta_i b^4/R_i^4$ order of smallness be performed, which is enough for a reliable solution of the problem. Let also the estimation of the influence of R_1 and R_2 on the three components δ_1, δ_2 and $\delta = \delta_1 + \delta_2$ be provided separately.

When the material properties are constant, parameter b is a function of both parameters R_i and q, and it specifies their complex influence on each of δ_i. Some of the calculation results [10] are shown in Figures 1 and 2 (the plots have logarithmic scales on the abscissa axis), with $q = const$ for the steel cylinders, $R_1 \neq R_2$. The plots illustrate the convergence of elastic bodies.

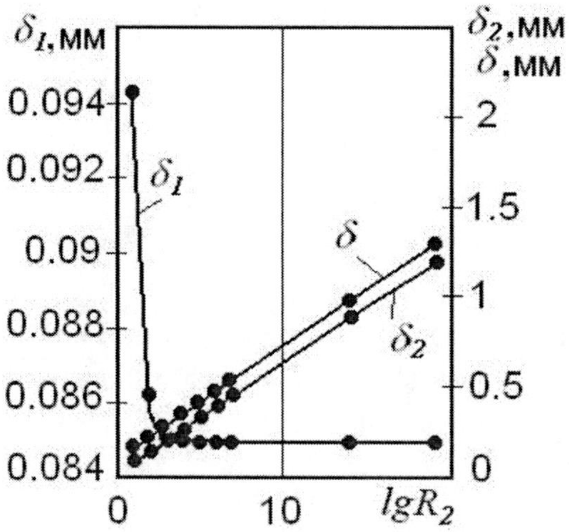

Figure 1. Influence of R_2 (the radius of the bigger cylinder) on components δ_i and δ at $R_1 = 10$ mm and $q = 10^4$ N/mm.

The increase of the strain of the single cylinder, δ_1 for Figure 2 or δ_2 for Figure 1, occurs with increasing the cylinder radius, specifically, R_1 or R_2, respectively. With increasing the radius of the bigger cylinder R_2 (Figure 1), the convergence δ increases. With increasing the radius of the smaller cylinder R_1 (Figure 2), the influence of the radius of the other cylinder on the convergence reduces, and the dependence of convergence on the unit load tends to become linear. Note that also in this case the radii of cylinders influence on the values of δ_1 and δ_2 significantly, but the function $\delta_1 = \delta_1(R)$ is increasing (with $q = const$ and $\eta_i = const$), while $\delta_2 = \delta_2(R_1)$ is decreasing [7] (see Figure 2). When $R_2 \to \infty$ and the cylinder contacts with a half-space (for instance, when $R_2 = 10^{20}$), the radius R_1 has almost no effect on $\delta = \delta_1 + \delta_2 = 1.41074$ (Figure 2) that corresponds to the experimental results [1].

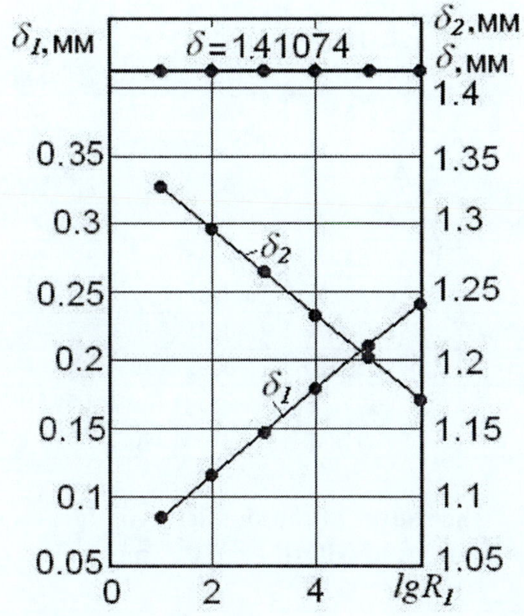

Figure 2. Influence of R_1 (the radius of the smaller cylinder) on components δ_i and δ at $R_2 = 10^{20}$ mm, $q = 10^4$ N/mm.

Decreasing of the deformation function for the larger cylinder with increasing the radius of smaller cylinder can be explained by growth of the deformation of first cylinder δ_1 with increasing its radius R_1, and decreasing of the deformation of second cylinder δ_2 because of the increasing contact area and decreasing the penetration depth of the first cylinder. The latter, actually, plays a role of an indenter with $R_1 \ll R_2$ [10]. The revealed effect of the convergence at contact manifests itself as changes of the deformation of different nature, if one of the radii changes, and as an approximate modulo equality (Figure 2), if $\eta_1 = \eta_2$ and the limiting case $R_2 \to \infty$ takes a place. It can be shown easily that manifestation of the effect in the most general case depends also on the material properties of the bodies in contact.

It is found, that each of the two common approaches, connected with: (i) significant influence of the radii of cylinders on the convergence of their axes, and (ii) absence of such an effect, is of special nature and is satisfied only in specific ranges of R_i and combinations of materials. Therefore, none of them are universal.

Tables 1 and 2 provide the calculation results for the cylinders of brass and aluminum, respectively. For a matter of convenience, the calculations use the same load.

Table 1. Influence of R_1 (the radius of the smaller cylinder) on components δ_i and δ for cylinders of brass with $R_2 = 10^{20}$ mm, $q = 10^4$ N/mm

R_1	δ_1	δ_2	δ
0.0001 R_2	1.120412864	1.634933654	2.755346518
0.001 R_2	1.184753087	1.570643680	2.755396767
0.005 R_2	1.229818805	1.525800747	2.755619552
0.01 R_2	1.249318198	1.506578593	2.755896791
0.05 R_2	1.295357385	1.462709130	2.758066515
0.1 R_2	1.316017541	1.444647739	2.760665280
0.5 R_2	1.369635018	1.408356567	2.777991585
0.6 R_2	1.376530238	1.405066687	2.781596925
0.7 R_2	1.382529276	1.402454346	2.784983622
0.8 R_2	1.387855565	1.400321119	2.788176684
0.9 R_2	1.392655634	1.398541429	2.791197063
R_2	1.397031240	1.397031240	2.794062479

Table 2. Influence of R_1 (the radius of the smaller cylinder) on components δ_i and δ for cylinders of aluminum with $R_2 = 10^{20}$ mm and $q = 10^4$ N/mm

R_1	δ_1	δ_2	δ
0.0001 R_2	1.621096403	2.372642192	3.993738595
0.001 R_2	1.715076324	2.278735666	3.993811990
0.005 R_2	1.780902528	2.213234880	3.994137408
0.01 R_2	1.809384734	2.185157628	3.994542362
0.05 R_2	1.876632856	2.121078760	3.997711616
0.1 R_2	1.906810556	2.094697003	4.001507559
0.5 R_2	1.985128070	2.041687526	4.026815596
0.6 R_2	1.995199720	2.036882093	4.032081813
0.7 R_2	2.003962342	2.033066321	4.037028663
0.8 R_2	2.011742300	2.029950378	4.041692678
0.9 R_2	2.018753623	2.027350835	4.046104458
R_2	2.025144945	2.025144945	4.050289890

The analysis of the obtained results demonstrates that influence of R_1 (the radius of the smaller cylinder) on the convergence of components is typical for cylinders of different materials. The features of the EA effect are demonstrated in Figure 2.

ACKNOWLEDGMENTS

The work was performed at partial financial support of the Ministry of Education and Science of the Russian Federation (Project №11.6767.2017/BCh).

REFERENCES

[1] Hoeprich, MR; Zantopulos, H. *Proc. ASME. Problems of Friction.* Mir, Moscow., 1981, vol. 103(1), 23 – 27 (In Russian).

[2] *Strength, Stability, Oscillations. Handbook.* I.A. Birger, Ya.G. Panovko (Eds.). Mashinostroenie, Moscow., 1968, vol. 2.

[3] Ajrapetov, EL. *Vestnik of Mashinostroeniya* [*Messenger of Mechanical Engineering*], 988, No. 6, 6 – 10 (In Russian).

[4] Loo, TT. *J. Appl. Mech.*, 1958, vol. 25, 122 – 124.

[5] Koval`skij, BS. *Calculation of Details on Local Compression*. Khar'kov. XVKIU. 1967 (In Russian).

[6] Zhuravlev, GA; Karpenko, VA. *Tekhnika Mashinostroeniya* [*Engineering Machinery*]., 2000, vol. 6(28), 51 – 55 (In Russian).

[7] Zhuravlev, GA. *Visnik of the East Ukrainian National University*. Lugansk. SNU. 2001, No. 5(39), 19 – 22 (In Ukrainian).

[8] Zhuravlev, GA. In: *Proc. V Russian Conf. with Int. Participation "Mixed Problems of Solid Mechanics", 23-25 August, 2005*. Saratov State University Press, Saratov. 2005, 148 – 151 (In Russian).

[9] Zhuravlev, GA; Tvarilidze, OV; Grebenyuk, AA. In: *Proc. VIII Int. Conf. "Modern Problems of the Mechanics of Continuous Medium"*, Rostov-on-Don., 2003, vol. 2, 77 – 80 (In Russian).

[10] Zhuravlev, GA. *Proc. Int. School on Models of the Mechanics of Continuous Medium".* N.F. Morozov (Ed.). Saratov State University Press, Saratov., 2007, 127 – 129 (In Russian).

[11] Mushelishvili, NI. *Several Main Problems of Mathematical Theory of Elasticity.* Nauka, Moscow, 1966, pp. 1 – 706 (In Russian).

Part 4. Applications

In: Proceedings of the 2018 International Conference … ISBN: 978-1-53615-862-5
Editors: Ivan A. Parinov et al.

Chapter 35

MATHEMATICAL MODELING OF OPTOACOUSTIC TRANSFORMATION IN MOVING FLUID IN THE PRESENCE OF NANOSCALE OBJECTS AS CONTRAST AGENTS

D. V. Orda-Zhigulina[1], M. V. Orda-Zhigulina[1] and I. B. Starchenko[2,*]

[1]Southern Scientific Center of the Russian Academy of Sciences, Rostov-on-Don, Russia
[2]Parametrika LLC, Taganrog, Russia

ABSTRACT

The optoacoustic effect in moving model liquid in the presence of nanoparticles was studied. The theoretical model was obtained by summing influence of three main processes: (i) laser excitation, (ii) flowing of investigated medium and (iii) sound generation by nanoparticles conglomerates. The results obtained showed that it is necessary to take into account all effects for most accurate modelling.

Keywords: optoacoustics, laser, nanoparticles, blood flow, contrast agent

* Corresponding Author's E-mail: star@sfedu.ru.

1. Introduction

The optoacoustic blood study with the use of nanoparticles as contrast agents is currently not yet widely used, but is often used in the laboratory, which suggests the feasibility of theoretical studies of this method. V. P. Zharov, D. A. Nedosekin, A. A. Karabutov, E. V. Savateeva, E. I. Galanzha, Yu. A. Minaeva, M. Seimology, J. S. Suen, V. Melerzanov, T. Oracle, etc. have conducted optoacoustic (OA) research of blood. According to the analysis of literature sources, one of the most significant shortcomings of the published works is that authors provided experiments without the development of theoretical models. In addition, the authors never take into account the effect of blood flow velocity on the level of the acoustic signal resulting from the OA transformation. Therefore, it is of interest to develop a theoretical model that would allow the influence of blood flow velocity on the level of the total optoacoustic field arising in the process of laser excitation of acoustic signals in the patient's blood to be taken into account.

2. Problem Formulation

The process of acoustic wave excitation in the blood, which is a suspension, but for theoretical calculations can be considered as a viscous liquid model, is shown in Figure 1.

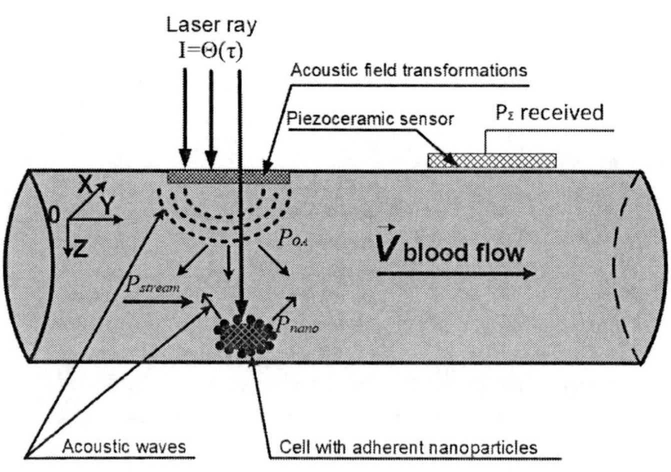

Figure 1. Scheme of the sound excitation process due to OA effect in the near-surface layer of a moving model fluid in the presence of nanoscale objects as contrast agents.

Blood vessels can be represented as a tube with infinitely thin walls, through which the flow of model fluid of speed $\vec{V}$ spreads. In the result of the optoacoustic interaction of laser radiation with the model liquid, a thermal change in blood volume occurs, which leads to the appearance of acoustic waves. The interaction of the laser optical oscillations with nanotubes, which are in the investigated model fluid, also causes an acoustic field. Another component of the total acoustic field occurs due to the flow of the fluid (blood). The total acoustic pressure, resulting from all the above effects, can be measured by using a piezoceramic receiver.

In this paper, it is proposed to use nanoscale particles as contrast agents. The treatment of nanoparticles allows them to become adhesive [1] to a certain type of human blood particles, for example, to cancer cells or blood components. Thus, the blood component surrounded by nanoparticles becomes noticeable when exposed to acoustic waves.

The relative position of the optical radiation source and piezoceramic receiver of the acoustic signal is shown in Figure 1. The diameter of the vessel is much smaller than its length, so we take into account only the transverse component of the acoustic wave, and the longitudinal component is neglected. The flow rate of the model fluid (blood flow rate) is assumed constant $|\vec{V}| = const$.

The pumping source of waves in the medium is a pulsed laser, which can excite single sound pulses with a duration of 10^{-8} to 10^{-3} seconds or a periodic sequence of such pulses, and the length of the excited sound wave depends on the duration of the laser pulse.

3. RESEARCH METHODS

Methods of modern nonlinear acoustics allow us to calculate the wave interactions that occur in the result of the OA effect in a stationary medium [2]. In addition, when analyzing the propagation of acoustic waves in a moving medium, it is accepted in the literature [3] that the effect of a smoothly inhomogeneous flow on the acoustic pressure value is reduced to the main effect, namely the scattering of acoustic waves in the flow. It depends on the parameters of the studied model liquid medium and has the same order of smallness. The first order is defined by the acoustic Mach number, and the second order is found by the hydrodynamic Mach number [3]. The contribution of each effect to the sound field is considered additive and can be calculated separately [4].

The calculation of the pressure amplitude P_{OA} of the acoustic wave resulting from the OA effect is carried out by the slowly changing profile method, proposed by R.. Khokhlov [2].

The calculation of the pressure amplitude P_{nano} of the acoustic wave, resulting from the OA effect in the presence of nanoparticles in the model liquid is, carried out by the method proposed by L. M. Lyamshev, which considers the generation of acoustic waves from a solid body in a liquid medium under the influence of optical radiation [5].

The calculation of the pressure amplitude P_{flow} of the acoustic wave, which occurs due to movement of model fluid, is calculated by the method proposed by D.I. Blokhintsev, which takes into account the "drift" of the sound wave and flow speed fluctuations [6]. It is shown in [24] that the total acoustic field in a moving fluid can be represented as $\vec{P}_{\Sigma flow} = \vec{P}_1 + \vec{P}_2$, where $\vec{P}_1$ is the pressure amplitude of acoustic waves in a stationary medium and $\vec{P}_2$ is the pressure amplitude of the acoustic waves from the flow rate of $\vec{V}$.

According to this assumption, it was shown [7] that the propagation of sound waves in inhomogeneous moving flows, and in media with nanoparticles, results in an acoustic field that can be represented as the sum of the components: P_{OA} is the acoustic pressure, generated in the absence of the flow in the result of the OA effect; P_{flow} is teh acoustic pressure addition to the total acoustic field due to the presence of the flow; P_{nano} is the acoustic pressure field arising in the result of the generation of acoustic waves in the conglomerate of nanotubes in liquid during the absorption of laser radiation.

Thus, the total acoustic field, recorded by the piezoceramic receiver (see Figure 1), is defined as

$$P_{\Sigma} = P_{OA} + P_{flow} + P_{nano} \tag{1}$$

4. Results and Discussion

Figure 2 shows the results of calculating the acoustic signal profiles for a fixed penetration depth $l = 3$ mm for different values of the flow rate of the model fluid. As can be seen from Figure 2, the effect of fluid velocity is small when the fluid velocity changes from 5 to 10 cm/s. If we increase the velocity to 80 cm/s, which corresponds to the velocity of blood in the aorta, the effect of velocity will be more noticeable, as shown in Figure 2. If we provide calculations for the flow rate in the range from 25 to 45 m/s, it can be concluded that an increase in the speed for every 10 m/s gives an increase of 7% in the acoustic pressure level. The values of such velocities lie beyond the range of blood flow velocities, but were theoretically investigated. The effect of the flow on the acoustic signal profile at different depths of the laser beam l at a constant flow rate of the model fluid (blood flow rate) $V = 2.89$ cm/s is shown in Figure 3.

The change in the amplitude of the acoustic signal profile when the blood flow rate changes from 2 to 10 cm (which corresponds to the blood flow rates in the average size of the vessels) is about 1–2% (see Figure 4). However, for blood flow rates of more than 10 cm/s, it makes sense to take it into account.

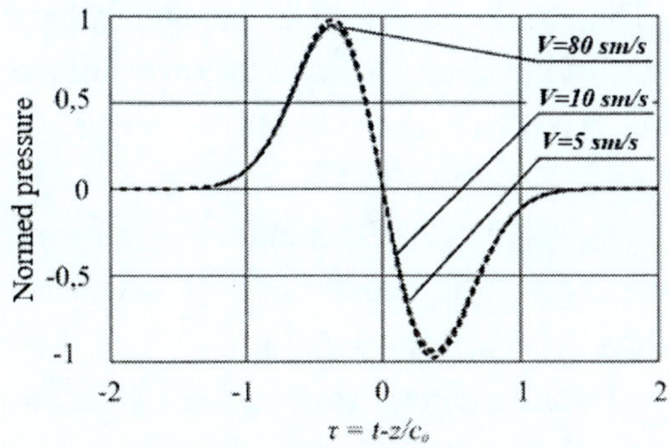

Figure 2. Profile of acoustic signal for the penetration depth $l = 3$ mm for different values of flow rate of model fluid.

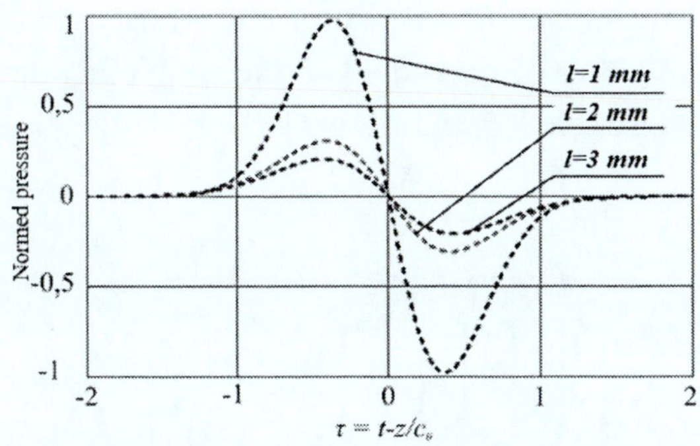

Figure 3. Profile of the acoustic signal for different depths of penetration of the laser beam at a constant rate of flow of model fluid V=2.89 см/с.

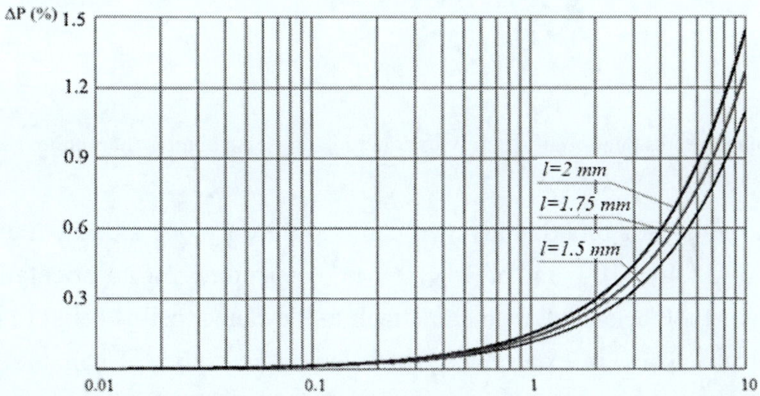

Figure 4. Dependence of the increment in sound pressure amplitude ΔP(%) from the flow rate of the model fluid (blood flow) V (cm/s) for different depths of laser beam penetration l.

In this paper, we consider the situation when laser radiation falls on a conditionally solid body in a model liquid. The multiple processes that result from OA transformation have been described above. There are conditionally solid bodies in liquid, that is, a conglomerate of nanoparticles surrounding the cell (see Figure 1), which we consider as a very thin elastic surface. The elastic properties of nanotubes have been described in [8]. Taking into account the size of nanoparticles (the average surface size of a conglomerate of nanoparticles is about 120×10^{-6} mm^2) and the vessel diameter (7 mm), this approximation is true.

The number of the conglomerates of nanotubes is determined by the number of cancer cells in the patient's blood. The number of such cells can be 2500 – 5000 pieces. The size of the cancer cell is approximately 7×2 μm^2. Using the assumption that the cell can be represented as an ellipsoid, the total surface area of the cancer cell with the above dimensions is 120.89×10^{-12} m^2. The average size of the considered carbon nanotube used in this research is 25×70 nm^2. The minimum and maximum number of nanotubes that can be adhered to the cell is from 7467 to 78500 pieces. Thus, the minimum number of nanotubes, injected into the human bloodstream for diagnosis, should be 1.19×10^{11} ps.

Let us calculate the number of nanotubes in the conglomerate that interacts with the optical radiation of the laser, in the direct projection of the laser beam, as shown in Figure 5.

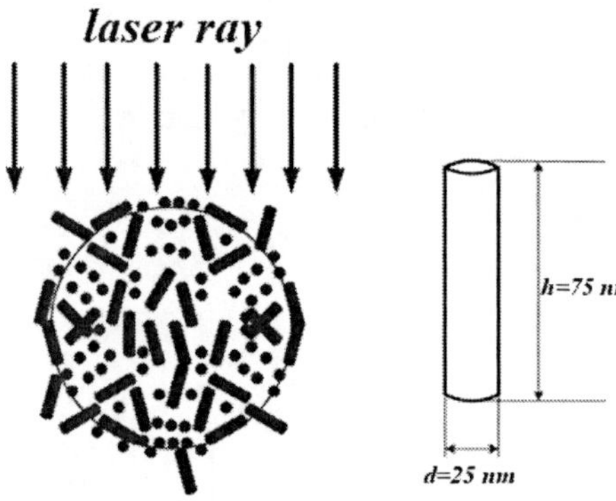

Figure 5. Schematic representation of a cell with adhered nanoscale objects (nanotubes) when exposed to laser radiation.

With OA method of blood study, we can expose only one side of the cell surface, with an area of 38.465×10^{-12} m^2 or 14×10^{-12} m^2, depending on its orientation in space. The area of the nanotube side, which can attach to the cancer cell is equal to 4.9×10^{-16} m^2 or 1.875×10^{-15} m^2. Thus, in experiment, the maximum number of nanotubes interacting with the optical radiation of the laser is 78500 PCs., and the minimum number is 7500 PCs. that 7500 – 78500 nanotubes create the acoustic pressure level, giving in expression (1) as P_{nano}.

The result of the calculation according to expression (1) with the initial data: $I_0 = 10^6$ W/m^2, $\beta = 7.2\times10^{-6}$ 1/K, $c_c = 4.174\times10^3$ J/kg·K, $\mu_1 = 0.78$, $a = 10^{-6}$ m, $m = 0.9$ gives the value of the acoustic pressure amplitude level equal to $P_{nano} = 0.1$ Pa.

The presence of a conglomerate of nanoparticles in the model liquid significantly affects the characteristics of the total acoustic field, excited by laser radiation in the liquid. In the case of the optically transparent layer and light absorption coefficient in the liquid μ, significantly greater than the wave number *k*, the presence of the layer leads to an increase in the amplitude of the acoustic field. This field arises from the OA transformation in comparison with the case, when the liquid boundary is free, and only in the direction of critical angles is the amplitude of the acoustic field significantly reduced.

Figure 6 shows the total dependence of the acoustic pressure level, obtained by summing the components of the acoustic field P_{OA}, P_{flow}, P_{nano} according to the expression (1). As a result, the time profile amplitude of acoustic signal in a model moving fluid, considering the influence of the flow and the presence of nanoscale agents, was plotted. As follows from Figure 6, the presence of nanoscale objects increases the total acoustic field by 13%, but the shape of the time profile of the acoustic signal remains unchanged.

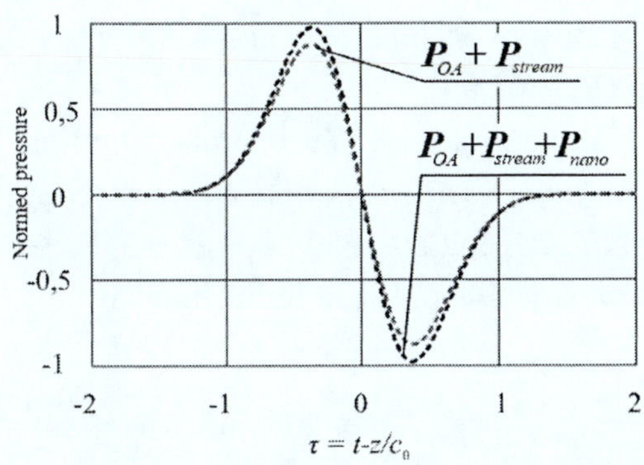

Figure 6. Time profile of the optoacoustic signal taking into account the motion of the medium and the presence of nanoscale objects in it (black curve) and without nanoscale objects (red curve) at the depth of penetration of laser beam $l = 2$ mm.

CONCLUSION

The biomechanical model of blood flow and process of optoacoustic signal generation in the result of OA transformation in a model fluid are considered. The influence of the velocity of the model fluid is taken into account. A theoretical model of optoacoustic transformation in a moving model fluid, simulating human blood flow in the presence of nanoscale particles, is developed.

It is theoretically determined that the velocity of blood flow affects the amplitude of the optoacoustic field in the presence of nanoscale objects within 1–2% for the velocities from 2 cm/s to 80 cm/s, which corresponds to the blood flow ones. It is found that the amplitude of the optoacoustic field decreases nonlinearly with increasing depth of penetration into the object.

REFERENCES

[1] Akerman, M. E., et al. *Proc. Natl. Acad. Sci. National Acad. Sciences,* 2002. vol. 99(20), 12617–12621.

[2] Novikov, B. K., Rudenko, O. V., Timoshenko, V. I. Nonlinear Underwater Acoustics. *Acoustical Society of America*, 1987, pp. 1 – 272.

[3] Voronin, V. A., Kirichenko, I. A. *Izvestiya TSURE*. 1997, No. 1(4), 118 – 120 (In Russian).

[4] Landau, L. D., Lifshitz, E. M. *Theoretical Physics. Hydrodynamics. Textbook. 5th edn*. Phyzmatlit, Moscow. 2001, Vol. 6, pp. 1 – 732.

[5] Lyamshev, L. M., Il'ichev, V. D. Laser Thermo-optical Excitation of Sound. *Nauka*, Moscow. 1989, pp. 1 – 237 (In Russian).

[6] Blokhintsev, D. I. Acoustics of Inhomogeneous Moving Medium. *Nauka*, Moscow. 1981, pp. 1 – 206 p (In Russian).

[7] Starchenko, I. B., Dzhuplina, G. Yu., Orda-Zhigulina D. V. In: *Nano- and Piezoelectric Technologies, Materials and Devices*. Ivan A. Parinov (Ed.). Nova Science Publishes, NY. 2014, 217 – 250.

[8] Chen Y. S., et al. *Nat. Commun.* Nature Publishing Group, 2017, vol. 8, 15782.

In: Proceedings of the 2018 International Conference … ISBN: 978-1-53615-862-5
Editors: Ivan A. Parinov et al.

Chapter 36

APPLICATION OF OPTOACOUSTIC METHOD FOR MODEL CALCULATION OF NON-AGGREGATED ERYTHROCYTES AND DETERMINATION OF HEMATOCRIT LEVEL

***D. A. Kravchuk*[1] *and I. B. Starchenko*[2,*]**
[1]Southern Federal University, Taganrog, Russia
[2]Parametrika LLC, Taganrog, Russia

ABSTRACT

The mathematical model of optoacoustic signals from red blood cells was calculated on the base of optoacoustic effect in the biofluid. In the developed model of the blood sample, the theoretical distribution of red blood cells was formed depending on hematocrit. The response of the optoacoustic signal from a single erythrocyte and from models of blood samples at different values of hematocrit was calculated. It was found that the amplitude of the optoacoustic signal and the spectral power density increases with the number of non-aggregated red blood cells. The obtained simulation results allow us to use an optoacoustic laser cytometer to conduct a non-invasive study of the blood composition, namely the calculation of the number of non-aggregated red blood cells and to establish the value of hematocrit in the blood. This research in the field of optoacoustics is a continuation of the earlier research aimed on the creating of optoacoustic cytometers.

Keywords: optoacoustic signal, aggregation, erythrocytes, spectral power density, laser

---

* Corresponding Author's E-mail: star@sfedu.ru.

1. Introduction

Thermal expansion is the process of absorption of light radiation in a fixed volume followed by thermal relaxation. The temperature increase creates internal stress that spreads through the tissue. For low-duration laser pulses, pressure is linearly proportional to the absorbed energy density. The type of transient optoacoustic process depends on the absorption distribution. The dominant mechanism with radiation power density less than the evaporation threshold is a thermoelastic expansion [2 – 5].

2. Problem Formulation

The optoacoustic effect is determined by the wave equation [8, 9], which can be written in terms of the velocity potential ϕ as

$$\left(\nabla^2 - \frac{1}{c^2}\frac{\partial^2}{\partial t^2}\right)\phi = \frac{\beta}{\rho_f C_p} H \tag{1}$$

where p is defined as $p = -\rho_f \partial\phi / \partial t$, c is the sound speed, ρ_f is the density of the liquid, β is a volume isobaric expansion coefficient, C_p is the heat capacity per unit mass, t is the time and H is the heating function, defined as the thermal energy over time and volume, deposited by the optical beam. Here we use the solution by using the Green's function [2] for the wave equation for the velocity potential:

$$\phi(x,t) = -\frac{\beta}{4\pi\rho_f C_P}\int_0^t\int (g(x,t|x',t')H(x',t')dx'dt' \tag{2}$$

where $g(x,t|x',t')$ is defined for infinite medium.

One-dimensional Green's function (for $z > z'$) is $g(x,t|x',t') = 2\pi cu\left[(t-t') - \frac{(z-z')}{c}\right]$, where $u(\xi)$ is the Heaviside function. Replacing the one-dimensional Green's function in Equation (2), after differentiating the resulting velocity potential, yields an optoacoustic pressure as a spatial integral:

$$p(z,t) = \frac{\beta c}{2C_P}\int H\left(z', t - \frac{(z-z')}{c}\right)dz'. \tag{3}$$

Let us consider a similar problem for spherically symmetric heat deposition. Wave equation for velocity potential, which contains only one radial spatial derivative, is reduced to one-dimensional wave equation (with replacement of *H* by *rH*) relatively of $\phi' = r\phi$. The one-dimensional above-mentioned Green's function can be used to solve the problem, but it needs to be modified to include the effect of spherical wave spreading inside, which is triggered simultaneously with the wave exiting from outside, and which is reflected at the origin, reaching a field point with inverted amplitude. The Green's function for Equation (2) should be the sum of two one-dimensional Green's functions $g(r,t|r',t') - g(r,t|-r',t')$, giving an optoacoustic pressure as

$$p(r,t) = \frac{\beta c}{2C_p r}\int r'\left[H\left(r',t-\frac{(r-r')}{c}\right) - H\left(r',t-\frac{(r+r')}{c}\right)\right]dr' \tag{4}$$

When the heating pulse is in a form of δ-function in time, we find that the optoacoustic pressure at $t > 0$ is given by the expression:

$$p(\hat{\tau}_0) = \frac{k}{2\hat{r}}\hat{\tau}_0\left(\hat{h}(-\hat{\tau}_0) + \hat{h}(\hat{\tau}_0)\right) \tag{5}$$

Let us now consider the case when the exciting optical pulse has duration much longer than the time of sound transmission along the absorption length in the irradiated body. First, we solve the problem in the frequency domain by taking the heating function in the form of $H(\hat{\xi},t) = \alpha I_0 \hat{h}(\hat{\xi})e^{-i\omega t}$, which in one dimension, according to Equation (3), gives the acoustic pressure as the space integral of the spatial heating function with a multiplier $e^{-i\omega z'/c}$. Since integration extends only by source, we approximate the exponential function as a unit giving a constant for spatial integration. Thus, the low-frequency response of the distribution and the corresponding Fourier transform are given as

$$p = \frac{\alpha\beta I_0 \xi_l c}{2C_p}e^{-i\hat{q}_l\hat{\tau}_l}, \quad p(\hat{\tau}_l) = \frac{1}{2}k\delta(\hat{\tau}_l) \tag{6}$$

respectively, where $\hat{q}_l = \omega\xi_l / c$, $\hat{\tau}_l = \frac{c}{\xi_l}\left(t - \frac{\xi}{c}\right)$, and (in *n* dimensions) $\xi_l^n = \int_0^\infty \xi^{n-1}\hat{h}(\xi)d\xi$.

In three dimensions, we find the optoacoustic pressure for spherically symmetric heat deposition by replacing the above frequency domain heating function with Equation (4). Integration is provided over an area of space small compared to the sound wavelength. Thus, we approximate the difference of two exponential functions in proportion to the radial derivative of exp(–$i\omega r$). This gives the low-frequency response of the spherical distribution and corresponding Fourier transform as

$$p = \frac{\alpha\beta I_0 \xi_l c}{C_P \hat{r}_l} \frac{d}{d\hat{\tau}_l} e^{-iq_l\hat{\tau}_l}, \quad p(\hat{\tau}_l) = \frac{k}{\hat{r}_l} \frac{d}{d\hat{\tau}_l} \delta(\hat{\tau}_l) \tag{7}$$

respectively, where $\hat{r} = r / \xi_l$.

In two dimensions, we define the high velocity potential using the Green's function for the Helmholtz equation [10], corresponding to Equation (1). The Green's function for the infinite medium is $i\pi H^{[1]}{}_0(k|\rho - \rho'|)$, which for a cylindrical-symmetric source is reduced to $J_0(k\rho')H^{[1]}{}_0(k\rho)$, where $H_0^{[1]}$ is the Hankel function of the zero order of the first kind, J_0 is the Bessel function, and k is a wave number. In the low frequency limit $J_0(k\rho') \cong 1$, that gives low-frequency optoacoustic pressure, its Fourier transform is

$$p_l = \frac{\pi\alpha\beta I_0 \xi_l c}{2C_P} \hat{q}_l H^{[1]}{}_0(\hat{q}_l \hat{\rho}_l) e^{-i\hat{q}_l\hat{t}_l}, \quad p(\hat{\tau}_l) = k \frac{d}{d\hat{\tau}_l} \frac{u(\hat{t}_l - \hat{\rho}_l)}{\sqrt{\hat{t}_l^2 - \hat{\rho}^2}_l} \tag{8}$$

respectively, where $\hat{\rho}_l = \rho / \xi_l$ and $\hat{t}_l = c\hat{t}_l / \xi_l$.

3. Research Methods

Let us consider the practical use of optoacoustic (OA) methods for the diagnosis of body tissues, using an optical cytometer that can measure hematocrit and red blood cell amount, based on the analysis of the optoacoustic signal (OAS).

If the optical radiation with intensity I_0 propagates along the x-axis and changes over time sinusoidally, the heating function can be expressed as $H(x,t) = \mu I_0 e^{-\omega t}$. Here μ is an optical absorption coefficient of the illuminated medium and ω is the modulation frequency of the optical beam. For such heating function, the steady-state pressure field is reduced to the following time-independent form [2]:

$$\nabla^2 p + k^2 p = \begin{cases} \dfrac{i\omega\mu\beta I_0}{C_p} & \text{– inside of absorber ,} \\ 0 & \text{– outside of absorber.} \end{cases} \tag{9}$$

where k is a wave number of the pressure wave. This is an inhomogeneous partial differential equation. It is possible to obtain analytical solutions of the equation for some simple geometries (e.g., layer, cylinder, sphere, etc.) using the appropriate boundary conditions, namely the continuity of pressure and normal particle velocity component. The OAS pressure for uniform illumination of the spherical absorber at distance r in the environment can be found as [2, 3]:

$$p_f^1\left(q\right)=\frac{i\omega\mu\beta I_0 v_s \alpha}{C_p\left(r/a\right)}\frac{\left[\sin q-q\cos q\right]e^{ik_f(r-\alpha)}}{q^2\left[\left(1-\rho\right)\left(\frac{\sin q}{q}\right)-\cos q+i\rho v\sin q\right]}, \tag{10}$$

where the dimensionless frequency is defined as $q = \omega\alpha/v_s$, α is the radius of the absorbing sphere, k_f is a wave number in the liquid medium for a pressure wave with a frequency ω and is determined as $k_f = \omega/v_f$. Similarly, the dimensionless values $\rho = \rho_s/\rho_f$ and $v = v_s/v_f$ are the ratios of density and sound speed, respectively. Indices s and f are used to indicate the properties of the absorber and the liquid medium, respectively, the upper index 1 in Equation (10) means that only one source of OAS is considered.

In many situations involving light scattering and ultrasonic scattering by random media, the scattering fields were expressed as superposition of spherical waves, generated by scatterers. This is essentially based on considering the particles separately and works well for a sparse medium. This approach has been widely used in many areas to explain the experimental results [2 – 4]. It was also assumed that the absorbers were illuminated by light with constant intensity regardless of their spatial position. The OAS pressure field, created by a set of absorbing spheres, can be written as a linear superposition of spherical waves, emitted by individual sources in the form:

$$p_f^{\Sigma}(q)=\frac{i\mu\beta I_0 v_s^2\alpha}{C_p}\frac{\left[\sin q-q\cos q\right]}{q^2\left[\left(1-\rho\right)\left(\frac{\sin q}{q}\right)-\cos q+i\rho v\sin q\right]}\sum_{n=1}^{N}\frac{e^{ik_f(|r-r_n|-\alpha)}}{|r-r_n|} \tag{11}$$

4. Results and Discussion

Modeling was carried out without taking into account the aggregation of red blood cells and the shape of red blood cells was considered spherical. The effects of double and multiple scattering of light waves are negligible. Optical absorption coefficients for spherical cells were considered the same. The model was built and calculated with software Matlab and the results of calculations are shown in Figure 1.

Figure 1 presents a computational model of an optoacoustic signal from a single spherical source (erythrocyte), irradiated by laser Nd: YAG [5, 6, 14, 16]. It was assumed that the absorbers were illuminated by light with constant intensity regardless of their spatial position. This means that all double and multiple scatterings of the light beam were considered insignificant. In addition, the acoustic waves, generated by the light-absorbing particle, do not interact with other particles in the medium. Under these assumptions, the OAS pressure field, created by a set of absorbing spheres, can be written as a linear superposition of spherical waves. Emitted by individual sources, the pressure field *vs* time takes the form:

$$p_f^{\Sigma}(\tau) \approx \frac{i\mu\beta F v_s^2}{2\pi C_p (r/a)} \int_{-\infty}^{\infty} \mathrm{d}q \frac{[sinq - qcosq]}{q\left[(1-\rho)\left(\frac{sinq}{q}\right) - cosq + i\rho v sinq\right]} e^{-iq\tau} \sum_{n=1}^{N} e^{-ik_f r_n}. \quad (12)$$

The sum in the above equation takes into account the interference of pressure waves, generated by sources (index Σ), distributed in space. Here r_n represents the position vector of the n-th particle and contains N OAS sources; k_f determines the direction of observation [2, 3].

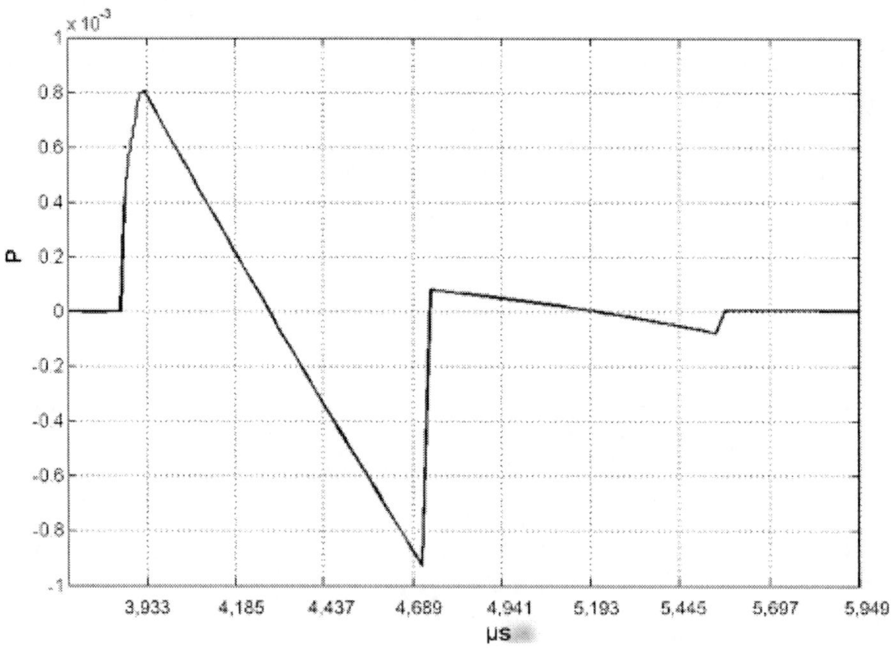

Figure 1. Optoacoustic signal from spherical absorber.

Two-dimensional simulation was provided on the example of a blood sample, consisting of 98% of red blood cells. The contribution to the absorption of optical radiation by leukocytes and platelets were not be taken into account. The average radius of erythrocytes is equal to 2.7 μm, the density in erythrocytes is 1090 kg/m^3, and sound speed is 1090 m/s. The density of the medium with erythrocytes is 1005 kg/m^3 and the sound speed within it is equal to 1498 m/s [3, 5, 6]. Notations of parameters are present in Table 1.

Table 1. Physical constants and parameters used in modeling [3]

Symbolic Notation	Value
α	2.75 μm
ρ_s	1092 kg/m^3
v_s	1639 m/s
ρ_f	1005 kg/m^3
v_f	1498 m/s
Area of investigation 100 μm × 100 μm	

Implementation of non-aggregated blood erythrocytes was performed using the Monte Carlo method. In this method, the coordinates of erythrocytes were chosen randomly with the restriction that they would not intersect with existing particles in periodic boundary conditions, although they may touch. A plasma sample of 100 × 100 μm^2 was taken into account. Erythrocytes in the blood sample can be considered as dominant absorbers of incident optical radiation, as well as sources of OAS radiation [3, 14, 15].

Figure 2 shows a two-dimensional spatial organization of red blood cells, simulating the implementation of tissue with non-aggregated red blood cells with hematocrit H = 10%, 35%, 40% and 45%. Each circle is a red blood cell, excluding aggregation.

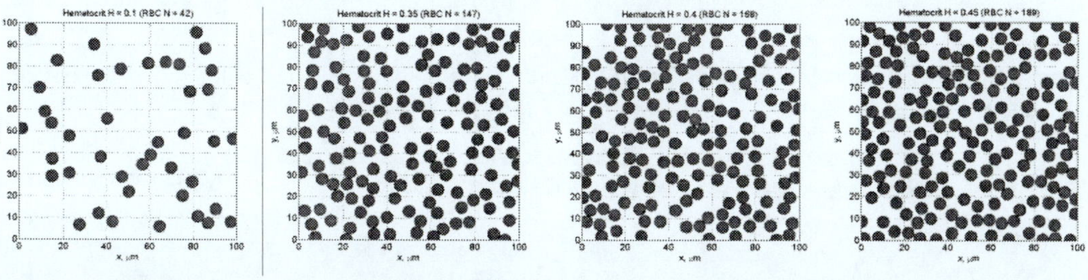

(a) (b) (c) (d)

Figure 2. Distribution of non-aggregated red blood cells in the plane of 100 × 100 μm^2 depending on hematocrit: (a) 10%; (b) 35%; (c) 40%; (d) 45%.

The optoacoustic signal in Figure 3 presents superposition of signals from erythrocytes in the result of laser exposure. Summation of signals from sources (erythrocytes) occurs depending on the distance and direction of the vector r_n, which determines the contribution to the total signal of each source [12 – 14].

Figure 4 shows the dependence of the spectral power density of hematocrit. With the growth of hematocrit, the number of red blood cells that absorb laser radiation increases, due to there being an increase in the amplitude of the signal. A large increase in the spectral power density clearly indicates that the level of hematocrit can be established in non-aggregated blood samples using the optoacoustic method. Power spectral density in

Figure 5 increased as the number of cells (hematocrit) in the area of interest increased. A monotonous increase in OAS was observed at an increased concentration of red blood cells. There was also a change in the frequency of the main maximum of the energy spectrum, depending on the number of red blood cells and their size.

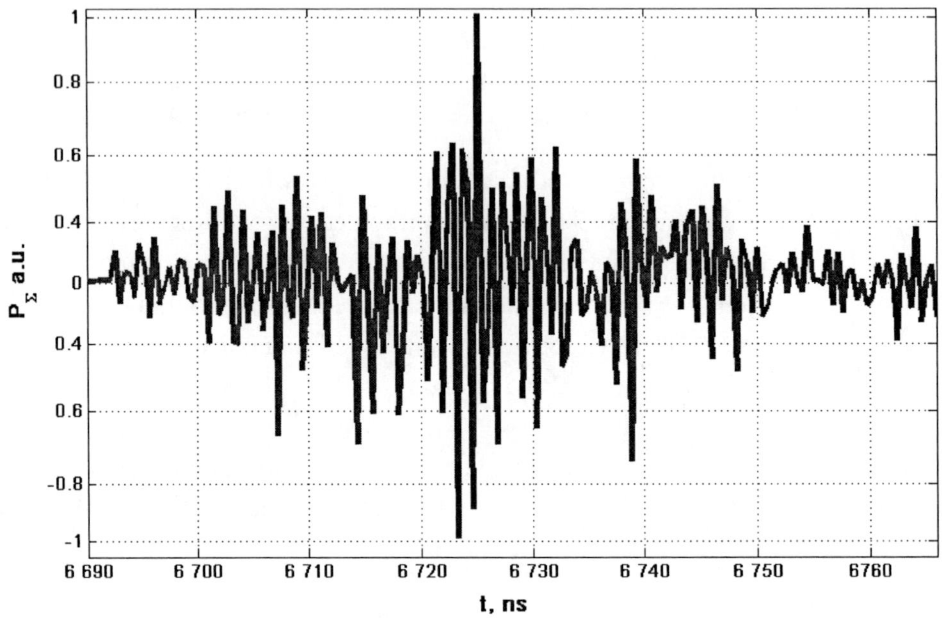

Figure 3. Opto acoustic signal in hematocrit 45% and 189 red blood cells (red color shows the envelope of the OOS).

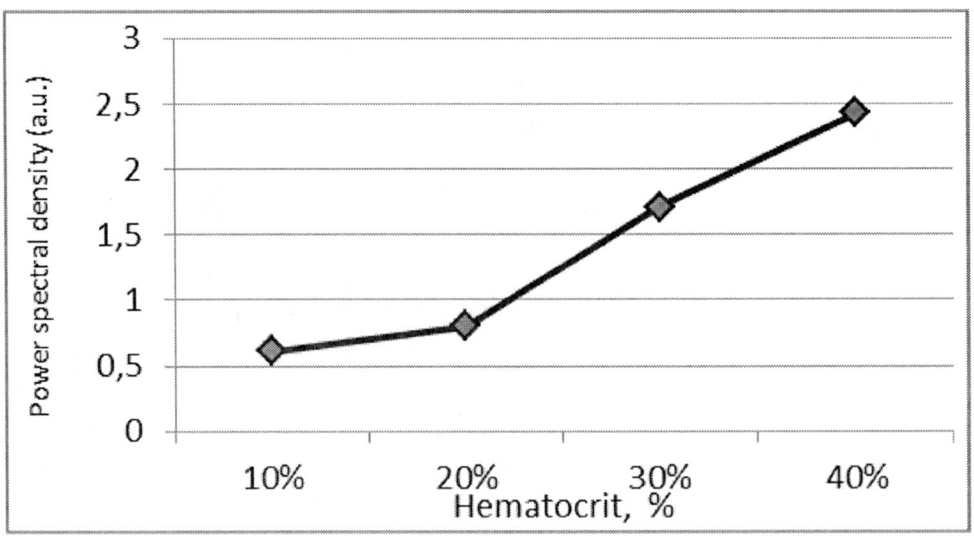

Figure 4. Dependence of power spectral density at different hematocrit values.

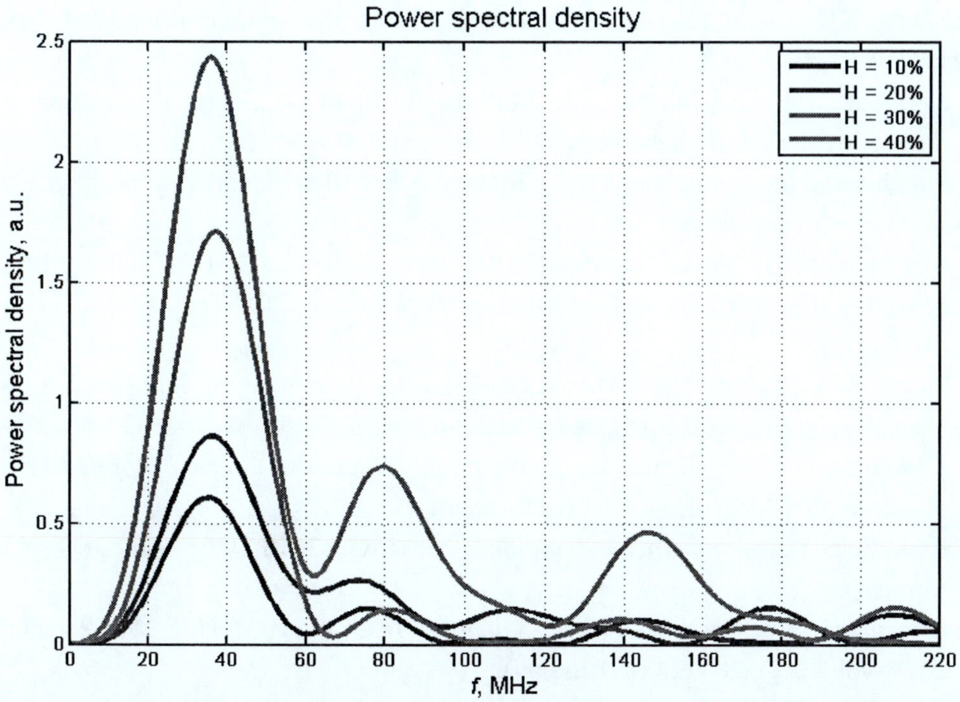

Figure 5. Spectral power density at different hematocrit values.

Conclusion

Thus, we can talk about the possibility of fixation of different levels of hematocrit at the calculation of red blood cells using an optoacoustic cytometer [7 – 10]. Consequently, the estimated mathematical model of the optoacoustic signals from red blood cells requires further development for cases with different degrees of aggregation of erythrocytes that change their form and confirmation of mathematical model with experiments on facility LIMO 100-532/1064U [11 – 17].

References

[1] Zhang, HF; Maslov, K; Sivaramakrishnan, M; Stoica, G; Wang, LV. *Appl. Phys. Lett.*, 2007, vol. 90, 1 – 3.

[2] Diebold, GJ. In: *Photoacoustic Imaging and Spectroscopy*, L. V. Wong (Ed.). Taylor & Francis Group, LLC, London., 2009, 3 – 17.

[3] Saha Ratan, K; Kolios, MC. *The Journal of the Acoustical Society of America*. 2011, vol. 129(5), 2935 – 2943.

[4] Seiyama, A; Hazeki, O; Tamura, M. *J. Biochem.*, 1988, No. 103, 419 – 424.

[5] Shung, KK; Yuan, YW; Fei, DY; Tarbell, JM. *J. Acoust. Soc. Am.*, 1984, No. 75, 1265 – 1272.

[6] Karpiouk, AB; Aglyamov, SR; Mallidi, S; Shah, J; Scott, WG; Rubin, JM; Emelianov, SY. *J. Biomed. Opt.*, 2008, vol. 5(13), 1 – 8.

[7] Starchenko, IB; Kravchuk, DA; Kirichenko, IA. *Biomedical Engineering.*, 2017, No. 5, 4 – 7 (in Russian).

[8] Kravchuk, DA; Orda-Zhigulina, DV; Orda-Zhigulina, MV; Starchenko, IB. *Modeling, Optimization and Information Technology.*, 2018, vol. 6(3), 21 – 29 (in Russian).

[9] Starchenko, IB; Kravchuk, DA; Kirichenko, IA. *Biomedical Engineering* Springer New York., 2018, vol. 51(5), 308 – 312.

[10] Kravchuk, DA; Starchenko, IB. *Scientific and Technical Journal. Applied Physics.* Moscow., 2018, No. 4, 89 – 94 (in Russian).

[11] Kravchuk, DA. *Biomedical Photonics.* Moscow., 2018, vol. 7(3), 36 – 42 (in Russian).

[12] Kravchuk, DA; Starchenko, IB. *Scientific Instrument Making.* St. Petersburg., 2018, vol. 28(2), 20 – 24 (in Russian).

[13] Kravchuk, DA. *News of South-West State University. Series: Management, Computer Engineering, Computer Science. Medical Instrument.* Kursk., 2018, vol. 8(2), 82 – 90 (in Russian).

[14] Kravchuk, DA. *Laser Medicine.* Moscow., 2018, vol. 22(1), 57 – 61 (in Russian).

[15] Kravchuk, DA; Starchenko, IB. *Bulletin of New Medical Technologies.* Tula, 2018, No. 1, 96 – 101 (in Russian).

[16] Kravchuk, DA; Starchenko, IB. *Scientific Instrument Making.* St. Petersburg., 2018, vol. 28(1), 30 – 36 (in Russian).

[17] Kravchuk, DA. *Quality and Life.* Moscow., 2018, No. 1(16), 41 – 43 (in Russian).

In: Proceedings of the 2018 International Conference … ISBN: 978-1-53615-862-5
Editors: Ivan A. Parinov et al.

Chapter 37

Condition of Absolute Stability of Control System for Deformation of Electromagnetoelastic Actuator for Nanotechnology

S. M. Afonin*
National Research University of Electronic Technology
(Moscow Institute of Electronic Technology MIET), Moscow, Russia

Abstract

The present work is concerned with stability for the control system of the deformation of the electromagnetoelastic actuator. The investigation is carried out for the Yakubovich absolute stability criterion of the control system with the hysteresis nonlinearity of the electromagnetoelastic actuator. The condition of the absolute stability on the derivative for the control system of the deformation of the electromagnetoelastic actuator is obtained. The electromagnetoelastic actuator is used for the precise matching in the nanotechnology and the adaptive optics. The piezoactuator serves for the nano- and micrometric movements in the nanotechnology.

Keywords: electromagnetoelastic actuator, control system, condition of absolute stability, deformation, piezoactuator, stationary set

* Corresponding Author's E-mail: eduems@mail.ru.

1. Introduction

The application of the electromagnetoelastic actuator, based on the electromagnetoelasticity as the piezoelectric, piezomagnetic, electrostriction, and magnetostriction effects, is promising in the nanotechnology, the nanobiology, and the adaptive optics for the precision matching. The piezoactuator solves problems of precise matching in the nanotechnology, the compensation of the temperature and gravitational deformations, and the atmospheric turbulence by the wave front correction. The piezoactuator for the nanonotechnology is used in the scanning tunneling microscopes, the scanning force microscopes, and the atomic force microscopes [1, 2].

The investigation of the static and dynamic characteristics of the piezoactuator of the nano- and micrometric movements as the control object is necessary for the calculation of the control systems of the nano- and micrometric movements for the nanotechnology [4–7]. The problems of using criteria absolute stability of the control system for the deformation of the electromagnetoelastic actuator of the nano- and microdisplacements are discussed. The stationary set of the control system of the deformation of the electromagnetoelastic actuator is determined.

The obtained condition of the absolute stability on the derivative for the control system with the electromagnetoelastic actuator allows one to estimate and calculate the characteristics of the control system of the deformation of the actuator.

2. Problem Formulation

In present work, the conditions of the absolute stability on the derivative for the control system with the electromagnetoelastic actuator are considered. The stationary set of the control system of the deformation of the electromagnetoelastic actuator is the segment of the straight line. The absolute stability conditions with the condition on the derivative of the control systems, with the deformation of the piezoactuator in the case of the longitudinal, transverse and shift piezoeffect for the hysteresis characteristic of the deformation of the piezoactuator, are obtained. The condition of the Yakubovich absolute stability on the derivative for the control system of the deformation of the electromagnetoelastic actuator is determined.

3. Stationary Set

The analytical expressions for the sufficient absolute stability conditions of the system with the hysteresis nonlinearity of the electromagnetoelastic actuators are written using the Yakubovich absolute stability criterion with the condition on the derivative, which is the development of the Popov absolute stability criterion. For the Lyapunov-stable control system, the Yakubovich absolute stability criterion for the systems with the single hysteresis nonlinearity provides the simplest, pictorial representation of results of the investigation of the stability and possibility of the synthesis of the correcting devices of the system, therefore ensuring the stability of the strain control systems with the electromagnetoelastic actuators.

We use the transfer function of the linear part of the system $W_{ij}(p)$ and the hysteresis function of the relative deformation S_j of the electromagnetoelastic actuator [4–7] for description of the system. The hysteresis nonlinearity of the electromagnetoelastic actuator is described in the general form:

$$S_j = F[\Psi_i|_0^t, t, S_j(0), \mathrm{sign}\dot{\Psi}_i], \quad (1)$$

where S_j is the relative displacement of the cross-section of the actuator along j-axis, and Ψ_i is the control parameter of the actuator along i-axis.

The hysteresis function S_j at each time instant t depends on the behavior of the function $\Psi_i = E_i$ or $\Psi_i = H_i$, where E_i and H_i are the electric field strength and the magnetic field strength on the interval $[0, t]$, the value of t, the initial value $S_j(0)$, and the sign of the rate $\dot{\Psi}_i$ of the field strength variation.

We consider the alternating-sign hysteresis characteristic of the piezoactuator in Figure 1. The set $L[E_i(0)]$ is the vertical segment $[S_j^0, -S_j^0]$ bounded by the points of intersection of the ordinate axis with the hysteresis loop at the maximum admissible field strength in the piezoactuator.

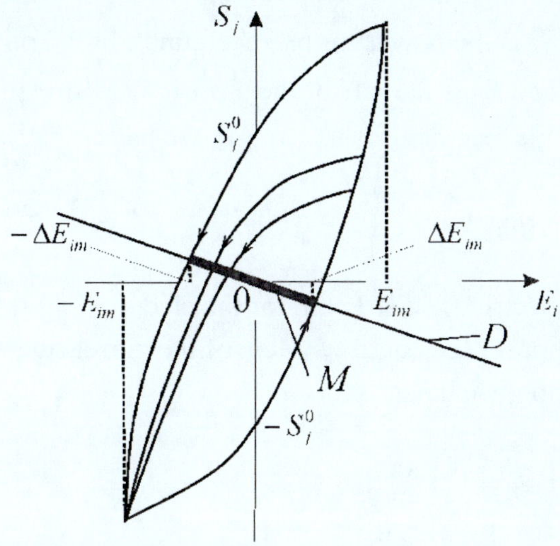

Figure 1. Generalized alternating-sign hysteresis characteristic of piezoactuator.

Let us find the stationary set for the deformation of the piezoactuator for the stable linear part of the control system. We draw the straight line D with the equation:

$$E_i + \mathrm{W}_{ij}(0)S_j = 0. \tag{2}$$

Accordingly, $\mathrm{W}_{ij}(0)$ is the value of the transfer function of the linear part of the control system for the deformation of the piezoactuator aspiring to 0. The set of points M for intersection of this straight line D with the hysteresis characteristic represents the segment of the straight line marked in Figure 1. For $E_{i0} = E_i(0)$ and $S_{j0} = S_j(0)$. We have the stationary solution for the control system.

Then, for the stationary set, we have

$$E_{i0} + \mathrm{W}_{ij}(0)S_{j0} = 0. \tag{3}$$

The stationary set M of the system is marked by the segment of straight line D in Figure 1 with the set of pairs (E_{i0}, S_{j0}). Each point of intersection of the hysteresis nonlinearity with the partial loops and the straight line D corresponds to one equilibrium position with the coordinates (E_{i0}, S_{j0}).

The analytical expression for the main generalized hysteresis loop [4–7] of the piezoactuator in Figure 1 is written as

$$S_j = d_{ij}E_i - \gamma_{ij}E_{im}\left(1 - \frac{E_i^2}{E_{im}^2}\right)^{n_{ij}} \operatorname{sign}\dot{E}_i, \tag{4}$$

where S_j is the relative displacement of the piezoactuator, E_i is the electric field strength, d_{ij} is the piezomodule, E_{im} is the electric field strength amplitude, and γ_{ij} is the hysteresis coefficient, n_{ij} is the power for piezoceramics, based on lead zirconate titanate PZT. We have $n_{ij} = 1$ and $\dot{E}_i$ as the rate of the electric field strength variation.

Therefore, for PZT piezoactuator from (3), (4), we have

$$\Delta E_{3m} + \mathrm{W}_{33}(0)E_{3m}[d_{33}\Delta - \gamma_{33}(1 - \Delta^2)], \tag{5}$$

where Δ is the relative electric field strength for the electric field strength amplitude at the extreme equilibrium point for the control system of the piezoactuator deformation.

Then, after conversion, we have

$$\Delta + \mathrm{W}_{33}(0)[d_{33}\Delta - \gamma_{33}(1 - \Delta^2)], \tag{6}$$

Thus, we obtain the quadratic equation for determination of the width of the rest of the domain 2Δ of the piezoactuator for the longitudinal piezoeffect in the form:

$$\Delta^2 + \frac{(1+W_{33}(0)d_{33})}{W_{33}(0)d_{33}}\Delta - 1 = 0, \tag{7}$$

The width of the rest domain of the piezoactuator for the longitudinal piezoeffect is:

$$2\Delta = -\frac{(1+W_{33}(0)d_{33})}{W_{33}(0)d_{33}} + \sqrt{\frac{(1+W_{33}(0)d_{33})^2}{W_{33}^2(0)d_{33}^2} + 4}\,, \tag{8}$$

The characteristics of the electromagnetoelastic actuators are the alternating-sign hysteresis type for the piezoactuators and the constant-sign butterfly type for the electrostriction actuators. In the magnetostriction and electrostriction actuators, their initial operating point is chosen on one wing of the butterfly, for example, in the first quadrant, for the deformation range symmetric at both sides of the initial point. For the butterfly characteristic, the initial working point displaces at the half deformation range. The values of the tangents of the inclination angles of the tangent line to the hysteresis nonlinearity on the butterfly wing for the electromagnetoelastic actuator are determined similar to the hysteresis characteristic.

4. Condition of Absolute Stability of System

The analytical expression for the sufficient absolute stability condition of the system with the hysteresis nonlinearity of the electromagnetoelastic actuator is written using the Yakubovich absolute stability criterion with the condition on the derivative. The Yakubovich criterion is the development of the Popov absolute stability criterion. For the Lyapunov-stable control system, the Yakubovich absolute stability criterion for the system with the single hysteresis nonlinearity provides the simplest, pictorial representation of results of the investigation of the stability. The Yakubovich criterion gives the possibility of the synthesis of the correcting devices of the system, ensuring the stability of the strain control systems with the electromagnetoelastic actuators [4–7]. The correcting devices are chosen for providing high quality of the control system of the piezoactuator deformation.

The function $S_j(E_i)$ of the hysteresis nonlinearity of the piezoactuator is continuous. We then have the quantities:

$$\nu_{1ij},\ \nu_{2ij} \in [0, \nu_{ij}],\ \nu_{ij} = \max\left[\frac{dS_j}{dE_i}\right], \tag{9}$$

where the quantities ν_{1ij} and ν_{2ij} are calculated using the hysteresis static characteristic in Figure 1, measured at the maximum admissible electric field strength in the piezoactuator.

The quantities $\nu_{1ij} = 0$ and $\nu_{2ij} = \nu_{ij}$ are the minimum and the maximum values of the tangent of the inclination angle of the tangent line to the hysteresis nonlinearity of the piezoactuator. Thus, we obtain

$$\nu_{33}:\nu_{31}:\nu_{15} = d_{33}:d_{31}:d_{15}, \tag{10}$$

where the ratios of the tangents of the inclination angle of the tangent line to the hysteresis nonlinearity of the piezoactuator for longitudinal, transverse and shift piezoeffects are proportional to the ratios of the piezomodules.

Then, the sufficient Yakubovich absolute stability condition of the control system of the deformation [4–7] of the actuator in Figure 2 have the form:

$$\operatorname{Re} \nu_{ij} W_{ij}(j\omega) \geq -1, \tag{11}$$

where in brackets j is the imaginary unit and ω is the frequency.

The amplitude-phase characteristic of the open-loop system $\nu_{ij} W_{ij}(j\omega)$ should be situated in Figure 2 at the right hand of the straight line:

$$\operatorname{Re} \nu_{ij} W_{ij}(j\omega) = -1, \tag{12}$$

for all values of $\omega \geq 0$.

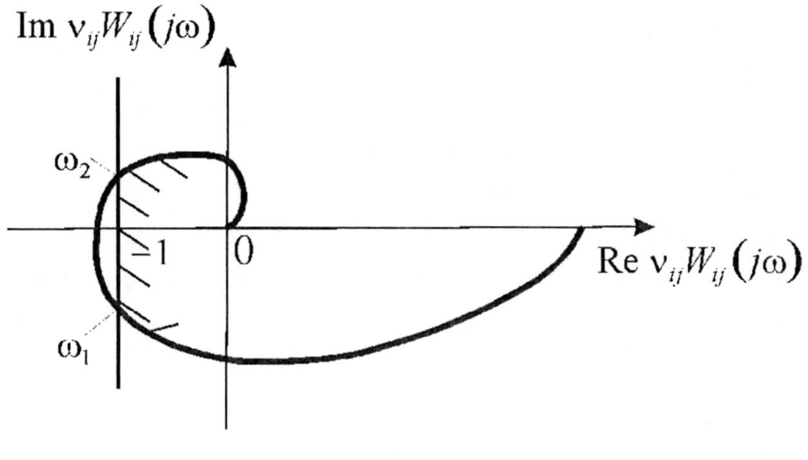

Figure 2. Absolute stability criterion for control system of deformation of electromagnetoelastic actuator in the plane, shaded corrected amplitude frequency characteristic for which absolute stability condition is satisfied.

The Yakubovich absolute stability criterion (11) in Figure 2 is simple and convenient for the synthesis of correcting devices [4, 5] for the control system. For the piezoactuator from lead zirconate titanate piezoceramics, the value of the maximum tangent of the inclination angle of the tangent line to the nonlinearity is about 1 nm/V for longitudinal piezoeffect and about 0.6 nm/V for transverse piezoeffect.

Let us consider the absolute stability criterion for the system in Figure 3 [4, 5] in the plane of the logarithmic amplitude frequency characteristic and the phase frequency characteristic.

In Figure 3 we have

$$L(\omega) = Q[\varphi(\omega)], L(\omega) = 20\lg\left| \nu_{ij}W_{ij}(j\omega)\right|. \tag{13}$$

The corrected logarithmic amplitude frequency characteristic is below the boundary curve in the form:

$$L(\omega) = 20\lg|\ 1/\cos\varphi|. \tag{14}$$

The characteristics of the correcting device for the control system are determined from the condition of the absolute stability of control system.

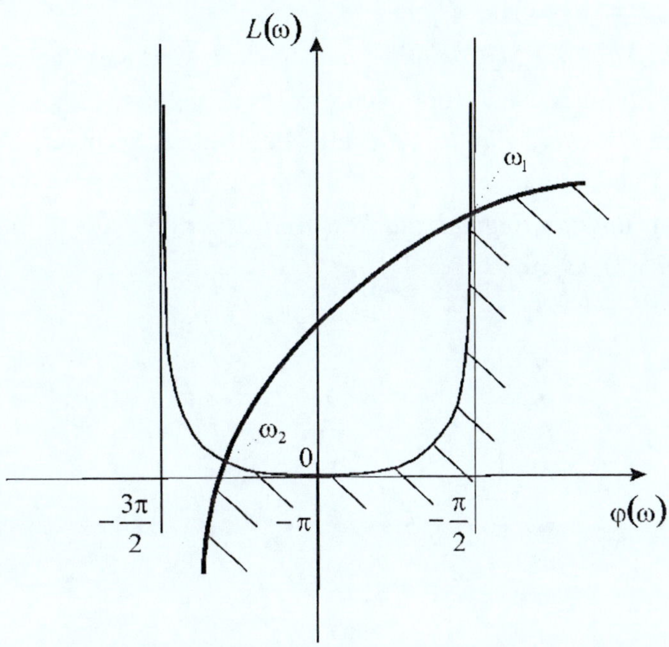

Figure 3. Choice of correcting device using the absolute stability criterion for control system of deformation of electromagnetoelastic actuator, shaded corrected logarithmic amplitude frequency characteristic from phase for which absolute stability condition is satisfied.

Conclusion

The stationary set of the control system of the electromagnetoelastic actuator deformation is the segment of the straight line. The obtained condition of the absolute stability on the derivative for the control system with the electromagnetoelastic actuator allows one to estimate and calculate the characteristics of the control system of the electromagnetoelastic actuator deformation.

References

[1] Schultz, J; Ueda, J; Asada, H. *Cellular Actuators*. Oxford: Butterworth-Heinemann Publisher, 2017, pp. 1–382.

[2] *Physical Acoustics: Principles and Methods*., Vol. 1. Part A. Methods and Devices; Mason, W.; Ed.; Academic Press, New York, 1964, pp. 1−515.

[3] Afonin, SM. In: *Piezoelectrics and Nanomaterials: Fundamentals, Developments and Applications*. Parinov, I. A.; Ed.; Nova Science Publisher, New York., 2015, 225-242.

[4] Afonin, SM. *Doklady Mathematics*. Pleiades Publishing, Springer, New York., 2006, vol. 74(3), 943–948.

[5] Afonin, SM. *Journal of Computer and Systems Sciences International*. Pleiades Publishing, Springer, New York., 2005, vol. 44(2), 266–272.

[6] Afonin, SM. *Doklady Physics*. Pleiades Publishing, Springer, New York., 2008, vol. 53(3), 137–143.

[7] Afonin, SM. *Russian Engineering Research*. Allerton Press, Springer, New York., 2015, vol. 35(2), 89-93.

In: Proceedings of the 2018 International Conference … ISBN: 978-1-53615-862-5
Editors: Ivan A. Parinov et al.

Chapter 38

Investigation of the Output Characteristics of Cantilever Type PEG with Proof Mass and Active Fixing under Low-Frequency Harmonic Loading

A. V. Cherpakov[1,2,*], I. A. Parinov[1], A. N. Soloviev[1,2] and E. V. Rozhkov[1]
[1]Southern Federal University, Rostov-on-Don, Russia
[2]Don State Technical University, Rostov-on-Don, Russia

Abstract

Experimental studies of the output parameters of a cantilever type piezoelectric generator (PEG) with active fixing are performed. The peculiarity of PEG is that the generator has two types of piezoelements: (i) elements disposed on the substrate forming bimorph and (ii) cylindrical piezoelements, which fix the cantilever plate. PEG can be used as a device for generating energy. The purpose of PEG is to transform mechanical oscillation energy from the environment into electrical energy. The results of the analysis of forced oscillations and output power, with the vibration excitation of the PEG substrate in the frequency range from 1 to 50 Hz, are present.

Keywords: energy harvesting, piezoelectric generator (PEG), cantilever type, bimorph piezoelectric elements, cylindrical piezoelectric elements, proof mass, active fixing, experimental modeling

---

* Correspdoning Author's E-mail: alex837@yandex.ru.

1. INTRODUCTION

Research and development of the systems for the generation of electric energy is quite relevant today. One such typical system includes elements in the form of piezoelectric generators (PEGs). For studies of such generators, there is a need to develop measurement systems for evaluating their real output properties. In the lot of publications, studies of PEG characteristics with the use of measuring installations and corresponding experimental techniques [1-4] are given. At the same time, in these and other researches, descriptions of experimental installations for measuring vibration parameters and recording PEG characteristics are mainly of cantilever type or, as in [2], for testing cantilever beams under three-point bending by using an Instron 5846 testing machine. To study the characteristics of cantilever type PEG, as a rule, serial vibrational setups are used. A specialized installation for testing piezoceramics elements and evaluating their properties is given in [4].

Among the various problems arising in the development of PEG, we can highlight the most significant, including the choice of energy-efficient piezoceramic materials (PCMs), the development of electrical circuits for the accumulation of electrical energy with minimal charge leakage, the search for geometric configurations and the assembly technology of the PEG sensitive element, which can provide maximum output power. Solutions of these problems use various approaches, which differ depending on the type and application of PEG. The reviews [5-12] contain research results on different PEG systems.

Experimental approaches to PEG studies are based on numerical-analytical modeling. In addition to analytical methods for describing PEG models, the main types of which are present in [13–15], the finite element method (FEM) is widely used. Its main advantage is the ability to perform numerical simulations of specific device models. This allows us to optimize the parameters of the model, to meet the specified performance, as well as to achieve maximum values of output power. This also allows one to solve the above-mentioned tasks. Examples of such studies are given below. The authors of [16] investigated the effect of various piezoceramics materials on energy efficiency and resonant characteristics of the cantilever bimorph sample. In paper [17], researchers investigated the possibility of creating a power source for non-destructive testing devices. In this treatment, the authors proposed a model of multimodal PEG operating due to vibrations, which are characteristic for the structures with sensors installed. In our previous studies [18, 19], based on the simulation of finite elements, we analyzed the influence of the proof mass and thickness of piezoelectric elements on the output characteristics of an stack-type PEG and studied harvesting the electric energy of the circuit.

2. PROBLEM FORMULATION AND DESCRIPTION OF MODEL PARAMETERS

The results of the investigations of an experimental approach were obtained for estimating the output parameters of a cantilever piezoelectric energy generator with active structures in the form of a bimorph. The piezoelectric transducer of the cantilever type for transformation of mechanical energy into an electrical one contains a cantilever beam, made of an elastic material, to which piezoelectric elements are glued on one side (unimorph) or on both sides (bimorph). One end of the cantilever beam is fixed at the base. On the free end of the cantilever, an additional proof mass is attached. Moreover, four piezoelectric elements, two upper and two lower in respect to the beam, having either opposite or identical directions of the polarization vector, are installed. The planes of their electrodes are pressed down with the help of the elements of mobile base to the conductive layers of thin elastic pads metallized on one side. The scheme of the PEG is shown in Figure 1. The construction materials of the base and Substrate are Duralumin, proof mass, is manufactured from steel.

Thin symmetrical piezoelectric elements (PEs) are polarized in thickness and glued to the console base. The geometric dimensions of the PEG are shown in Figure 1b. As materials of PEes, we used piezoceramic plates of a material PCR-7M, and piezocylinders were fabricated from a material PZT-19. Description of material parameters is given in [20]. The electrical scheme of junction of the PEG with active load is shown in Figure 1a. The value of the proof mass can vary from 3 to 25 grams (in the experiment, we used $M = 17.6$ g). The dimensions of the piezoplates were: $l_p \times b_p \times h_p = 50 \times 10 \times 0.45$ mm^3, piezo cylinder: $R \times H = 10 \times 10$ mm^2, substrate: $l \times b \times h = 135 \times 13.2 \times 1.5$ mm^3, $l_m = 127$ mm.

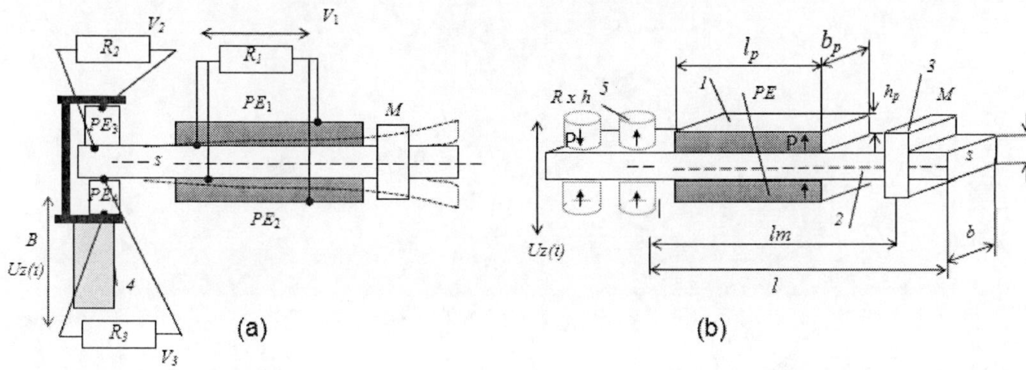

Figure 1. (a) Electric scheme of PEG at active load and (b) structural scheme of PEG with proof mass: 1 – PE; 2 – substrate; 3 – proof mass; 4 – place of PEG fixing (with movable base B); 5 – piezoelectric cylindrical element.

2.1. Description of Model

The studies were carried out on a vibration setup presented in Figure 2. Harmonic oscillations of PEG were investigated. The laser displacement sensors optoNSDT (3) and RF603 (6) measured displacements. The sensor (3) was used to record the movements of the PEG base. The sensor RF603 (6) was used to measure the movement of the proof mass attached to the free end of the PEG cantilever. The following parameters were recorded: the voltage applied to the plugs of the base of the vibrating Table 1, the output voltage from PEs (8) and (9) of PEG passing through the electrical path (5) to the ADC, and oscillation signals of the base from the sensors optoNSDT (3) and RF603 (6).

Figure 3 schematically shows a measuring system for the research of output parameters of PEG under vibration excitation. The setup works as follows. In the case of stationary vibration excitation of the vibration table, the oscillations and output characteristics of PEG were investigated and obtained in the frequency response of PEG. The first resonant frequency of the generator was ~ 22 Hz. Then, a signal in the form of a sinusoid with frequencies from 1 to 50 Hz was excited on the sound card of the computer using software [21, 22]. The vibrating table (2) with PEG fixed on it was driven by using the amplifier (9). In this case, all piezoelements, according to the scheme in Figure 1, were loaded with active resistance $R = 1$ MΩ. With the help of matching devices (4), (6), (8), the signals, generated by PEG and laser sensors of displacements, passed to the ADC (7) and were recorded by using the software of a computer (11).

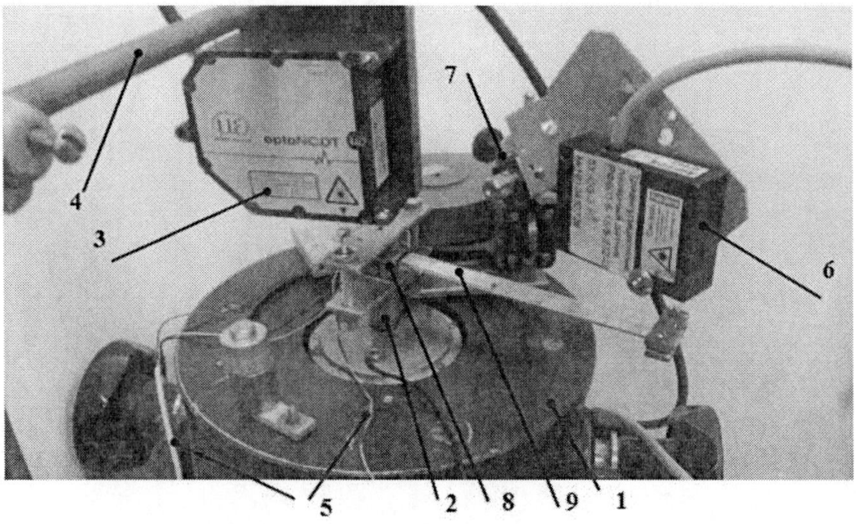

Figure 2. Vibration setup with PEG: 1 – vibrating table; 2 – PEG; 3 – laser sensor of displacements optoNSDT; 4 – mount of the laser sensor 3; 5 – electric tract PEG; 6 – RF603 triangulation laser sensor of displacements; 7 – support column of the sensor (6); 8 – PEs of cylindrical type, located at the PEG base; 9 – PEs in the form of plates.

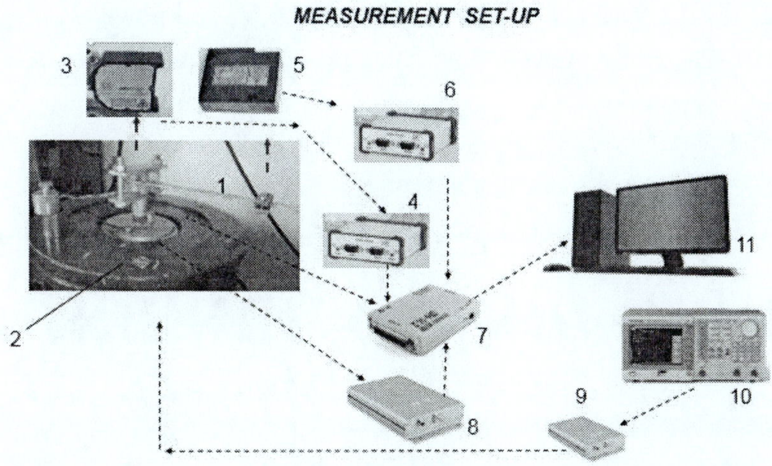

Figure 3. Measurement setup: 1 – PEG; 2 – vibration exciter; 3 – optical sensor of linear displacements optoNSDT; 4 – controller of the optical sensor 3; 5 – optical sensor of linear displacements RF603; 6 – controller of the optical sensor 5; 7 – external ADC/DAC module L-Card14-440; 8 – matching device of the acceleration sensor; 9 – power amplifier; 10 – set of the signal generator AFG3022 or sound card; 11 – computer.

3. RESULTS AND DISCUSSION

Studies were conducted using software [20, 21], which allows generating a signal, by sweeping and recording it on a computer. The examples of signals, obtained from displacement sensors (1, 2), terminals of connecting a vibrating table (3), and piezotransducers (4 – 6) are shown in Figure 4. Excitation was carried out at frequencies from 1 to 50 Hz. The dwell time of each frequency was equal to 3 s and hold time also was equal to 3 s.

Figure 5 shows an example of the signal obtained from 6 channels, respectively. The signal is characterized by two parameters: P_1 is the time of turning on the vibrator and *P*2 is the time of turning off the vibrator; they can be seen from the voltage at the terminals (see graph 3). The graph is divided into three sections. The figure shows that t_{excit} is the excitation time of oscillations, t_{garm} is the time of the established harmonic oscillations, and t_{damp} is the time of attenuation. For this case, $t_{excit} = 1.75$ s, $t_{garm} = 1.3$ s, $t_{damp} = 1.5$ s.

Figure 6a shows the voltage applied to the clamps of the vibration table and its amplitude of oscillations during harmonic excitation at frequencies from 1 to 50 Hz. The analysis shows that the amplitude of vibrations of the vibrating table changes, thereby changing the vibrations, excited at the base of the generator. When a resonance of about 22 Hz is reached, the oscillations of the end of the generator almost increase by a factor of 10 (Figure 6b). Figure 7 shows the output characteristics of the voltage *U* on the electrodes and of the power *P*, respectively, on piezoplates (*bim*), upper piezocylinders (*PCup*) and lower piezocylinders (*PCdn*). The analysis shows that at resonance, the voltage is U_{bim} = 8.12 V, U_{PCup} = 4.57 V, U_{PCdn} = 2.6 V, and power P_{bim} = 66×10^{-6} W, P_{PCup} = 20.9×10^{-6} W, P_{PCdn} = 6.8×10^{-6} W with a load of 1 MΩ.

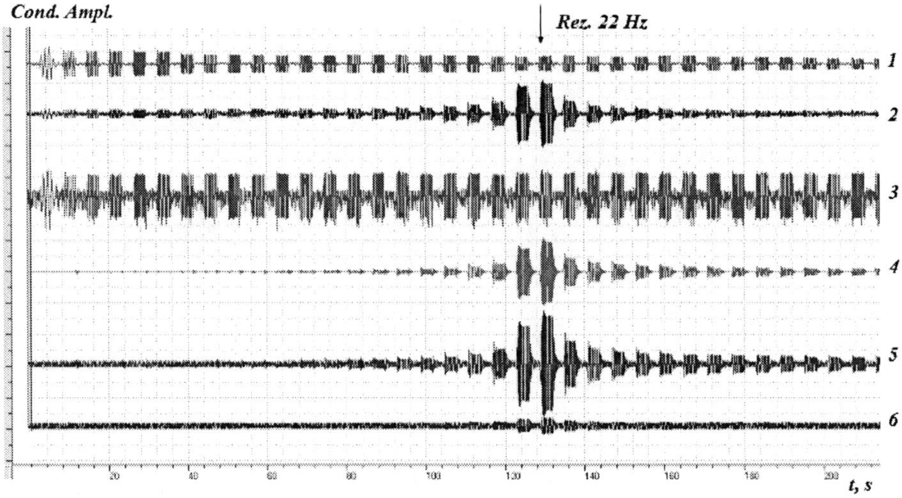

Figure 4. Testing the operation of the PEG construction under excitation by sweeping with frequency searching from 1 to 35 Hz at 1 Hz increments: 1 – conditional amplitudes of displacement of the generator base (vibration table); 2 – conditional amplitudes of movement of the load at the edge of cantilever; 3 – voltage on the electrodes of the vibration table; values of voltage, obtained on electrodes: (4) of piezoplates on the cantilever, (5) of upper piezo-cylinders and (6) of lower piezo-cylinders.

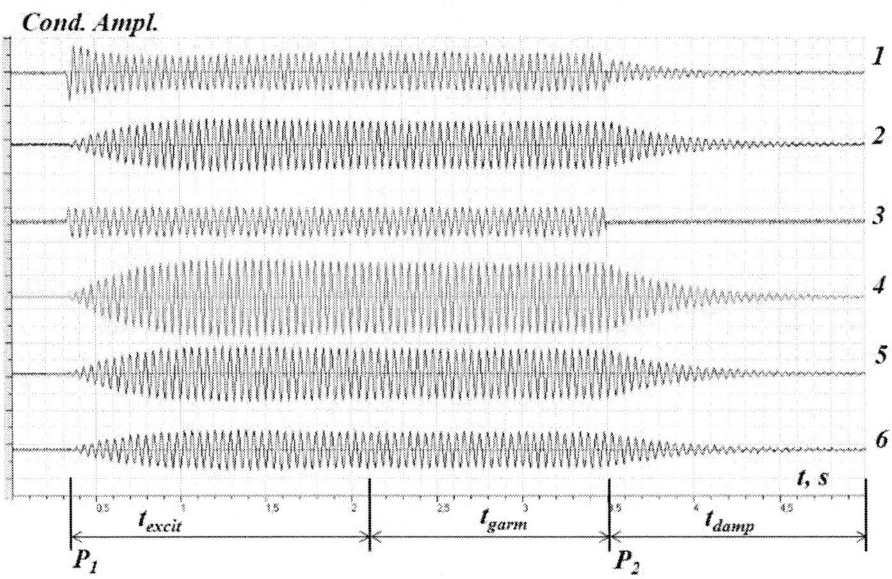

Figure 5. t_{excit} is the time of oscillation initiation, t_{garm} is the time of steady-state harmonic oscillations, t_{damp} is the decay time, P_1 is the switch on point of the vibrator of the vibrating table, P_2 is the switch off point of the vibrator of the vibrating table: 1 – conditional amplitudes of displacement of the generator base (vibration table); 2 – conditional amplitudes of movement of the load at the edge of cantilever; 3 – voltage on the electrodes of vibration table; values of voltage, obtained on electrodes: (4) of piezoplates on the cantilever, (5) of upper piezo-cylinders and (6) of lower piezo-cylinders.

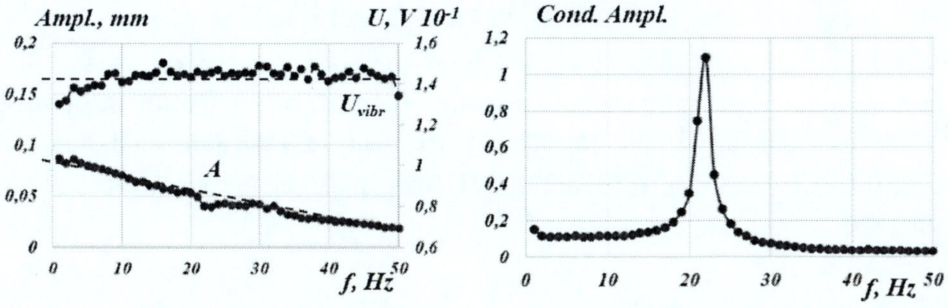

(a) (b)

Figure 6. (a) Amplitudes (A) of moving the vibrating table depending on the voltage U_{vibr} applied to its electrodes; (b) dependence of the movement of the load at the edge of the PEG cantilever on the excitation frequency of the harmonic oscillations of the generator base, mounted on the vibrating table.

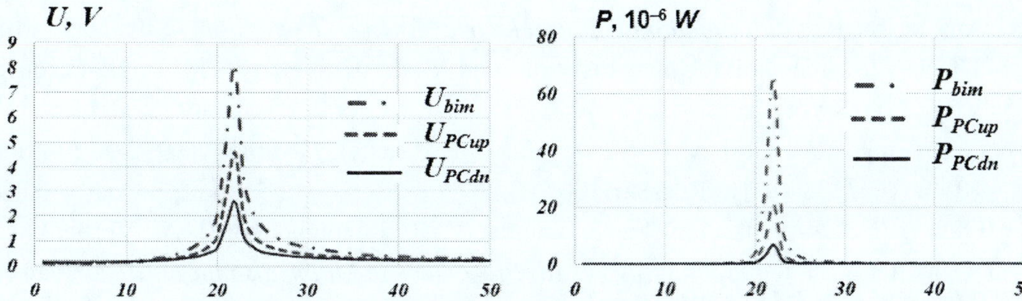

(a) (b)

Figure 7. Dependence of voltage (U_{bim}, U_{PCup}, U_{PCdn}) and power (P_{bim}, P_{PCup}, P_{PCdn}) with harmonic excitation of PEG at frequencies from 1 to 50 Hz; the indices point the following: *bim* – cantilever piezoplates, *PCup* – upper piezocylinders, *PCdn* – lower piezocylinders.

CONCLUSION

We considered full-scale experimental modeling of cantilever type PEG with a proof mass and active base of piezocylinders. The proof mass was disposed in the region of the free edge of the generator cantilever at a distance of l_m = 127 mm. The studies have shown that the first resonant frequency of PEG is equal to 22 Hz. We performed vibration tests of PEG by sweeping at frequencies from 1 to 50 Hz. The maximum power obtained on piezoplates with an active load of 1 kΩ at the resonance frequency with an output voltage $U_{bim} = 8.12$ V was $P_{bim} = 66 \times 10^{-6}$ W. For upper piezocylinders with a load of 1 MΩ, these characteristics were $U_{PCup} = 4.57$ V, $P_{PCup} = 20.9 \times 10^{-6}$ W. For lower piezocylinders, $U_{PCdn} = 2.6$ V, $P_{PCdn} = 6.8 \times 10^{-6}$ W.

A more detailed analysis of the output power of PEG requires calculation for different locations of proof mass and taking into account the other properties of the substrate material and geometric sizes of piezoelectric elements.

Acknowledgments

The research was partially supported by the Russian Foundation for Basic Research (grant No. 19-08-00365-a); the Russian Department of Education and Science No. 11.6767.2017/BCh, and No. 3.5378.2017/VU (organization of the implementation of researches).

References

[1] Alper Erturk, Daniel J. Inman. *Piezoelectric Energy Harvesting*. John Willey and Sons, USA. 2011.

[2] Anton, S. R. *Multifunctional Piezoelectric Energy Harvesting Concepts*. PhD Thesis Virginia Polytechnic Institute and State University. 2011. Blacksburg. Virginia. USA.

[3] Golovnin, V. A., Gornev, E. S., Dayneko, A. V., et al. *Vestnik TvGU. Ser. Physics*. 2010. *Is. 11*, 33–46 (In Russian).

[4] Gusev, A. A., Avvakumov, E. G., Isupov, V. P., Reznichenko, L. A., Verbenko, I. A., Miller, A. I., Cherpakov, A. V. In: *Piezoelectric Materials and Devices*. Ivan A. Parinov (Ed.), Nova Science Publishers, New Yourk. 2011, 189 – 234.

[5] Anton, S. R., Sodano, H. *Smart Mater. Struct.* 2007, *vol. 16*, K1 – K21.

[6] Priya, S. *J. Electroceramics*. 2007, *vol. 19*, 167 – 184.

[7] Cook-Cherault, K. A., Thamb, N., Sastry, A. M. *Smart Mater. Struct.* 2008, *vol. 17*, 043001.

[8] Kim, Heung Soo, Joo-Hyong, Kim, Jaehwan, Kim. *International Journal of Precision Engineering and Manufacturing.* 2011, *vol. 12(6)*, 1129 – 1141.

[9] Caliò, R., Rongala, U. B., Camboni, D., Milazzo, M., Stefanini, C., de Petris, G., Oddo, C. M. *Piezoelectric Energy Harvesting Solutions. Sensors,* 2014, *vol. 14(3)*, 4755 – 4790.

[10] Shevtsov, S. N., Soloviev, A. N., Parinov, I. A., Cherpakov, A. V., Chebanenko, V. A. *Piezoelectric Actuators and Generators for Energy Harvesting*. Innovation and Discovery in Russian Science and Engineering. Springer Cham, Switzerland, 2018.

[11] Soloviev, A. N., Parinov, I. A., Cherpakov, A. V., Chebanenko, V. A., Rozhkov, E. V. *Materials Physics and Mechanics.* 2018, *vol. 37(2)*, 168 – 175.

[12] Cherpakov, A. V., Chebanenko, V. A., Parinov, I. A., Chang, S. H., Jani, M. A. In: *Proc. Int. Symp. "Physics of Lead-free Piezoactive and Relative Materials" (Analysis of Current State and Prospects of Development), LFPM-2016,* 12-15 September, 2016, Tuapse, 2016, *vol. 2*, 265 – 269 (In Russian).

[13] Erturk, A., Tarazaga, P. A., Farmer, J. R., Inman, D. J. *Journal of Vibration and Acoustics,* 2009, *vol. 131(1)*, 011010.

[14] Elvin, N. G., Elvin, A. A. *Journal of Intelligent Material Systems and Structures*, 2009, *vol. 20(1)*, 3–9.

[15] Liao, Y., Sodano, H. A. *Smart Materials and Structures,* 2009, *vol. 18(4)*, 045011.

[16] Nechibvute, A., Chawanda, A., Luhanga, P. *ISRN Materials Science,* 2012.

[17] Zhang, Y., He, W. In: *The 15th International Symposium on: Smart Structures and Materials & Nondestructive Evaluation and Health Monitoring*, 2008, 69340Z – 69340Z.

[18] Soloviev, A. N., Parinov, I. A., Duong, L. V., Yang, C. C., Chang, S. H., Lee, J. C. Y. In: *Physics and Mechanics of New Materials and their Applications*. Ivan. A. Parinov, Shun-Hsyung Chang (Eds.). Nova Science Publishers, New York. 2013, 335 – 352.

[19] Shevtsov, S., Akopyan, V., Rozhkov, E., Chebanenko, V., Yang, C. C., Jenny Lee, C. Y., Kuo, C. X. In: *Physics, Mechanics and Applications,* Ivan A. Parinov, Shun-Hsyung Chang, Vitaly Yu. Topolov (Eds.). Springer Proceedings, 2016, *vol. 175*, 534 – 543.

[20] Panich, A. A., Marakhovskii, M. A., Motin, D. V. *Engineering Journal of Don,* 2011, No. 1 (In Russian).

[21] Parinov, I. A., Soloviev, A. N., Cherpakov, A. V. *Program "VIBROGRAF" for Registration, Visualization and Processing of Design Oscillations.* Certificate of Registration of the Computer Program, RUS 2016612309, 02.24.2016 (In Russian).

[22] Parinov, I. A., Cherpakov, A. V., Rozhkov, E. V., Soloviev, A. N., Chebanenko, V. A. *Program Signal Generator SGENERATOR.* Certificate of Registration of the Computer Program, RUS 2018610408, 11.13.2017 (In Russian).

In: Proceedings of the 2018 International Conference … ISBN: 978-1-53615-862-5
Editors: Ivan A. Parinov et al.

Chapter 39

CHARACTERISTICS OF UV SURFACE ACOUSTIC WAVE SENSORS BASED ON ZINC OXIDE

V. G. Dneprovski**, E. M. Kaidashev and G. Ya. Karapetyan
I. I. Vorovich Mathematics, Mechanics and Computer Science Institute,
Southern Federal University, Rostov-on-Don, Russia

ABSTRACT

The characteristics of UV surface acoustic wave (SAW) sensors based on zinc oxide obtained by different researchers are considered. In particular, SAW sensors are considered, both based on the SAW delay line or SAW filter, as well as based on the SAW generator. In our case, the elements sensitive to ultraviolet are either zinc oxide films, or ZnO nanorods and nanowires. A comparative analysis of such sensors has been carried out. Shift of all the main SAW device characteristics (resonant frequency, phase angle and insertion loss) of SAW sensors may be simultaneously used for the detection of UV, greatly enhancing the sensing accuracy. The largest photosensitivity (450 nW/cm^2) is obtained at using an oscillator circuit. The Rayleigh waves provide greater photosensitivity than the Lamb waves: 111.3 ppm/(mW/cm^2) and 55.8 ppm/(mW/cm^2), respectively.

Keywords: surface acoustic wave (SAW) devices, substrate, interdigital transducers (IDTs), piezoelectric, ZnO thin films, zinc oxide nanowires

---

* Corresponding Author's E-mail: valdnepr@ya.ru.

1. INTRODUCTION

In this paper, we consider the characteristics of ultraviolet (UV) surface acoustic wave (SAW) sensors based on the analysis of the work of various researchers. The influence on these characteristics of the substrate material of the device, the use of ZnO films, nanorods or nanowires, and annealing of ZnO films are considered. Methods of deposition of ZnO thin films, nanowires, and nanoparticles for SAW ultraviolet sensors are present in [1]. The principle of operation of these sensors is based on the acoustoelectronic effect that occurs when an ultraviolet radiation is applied to a zinc oxide film or to zinc oxide nanoelements. ZnO films have good photosensitivity due to change in the acoustoelectronic interaction, which arises from the variation of the sheet resistivity and carrier concentration in the ZnO under the influence of UV radiation. The semiconductor films of zinc oxide have band-gap energy of about 3.3 eV. In view of this, the photosensitivity of the zinc oxide lies in the UV range for radiation with a wavelength shorter 380 nm.

The changes in the velocity, phase, and amplitude of the SAW signal can be used to monitor the physical variations of the sensing layer, surface, and interfaces of the multi-layer film and substrate. For example, W. Water et al. detected the insertion loss shifts at the center frequency of the Love wave in ZnO/90°-rotated ST-cut (42°45`) quartz and ZnO nanorods/ZnO/90°-rotated ST-cut (42°45`) quartz devices under 365 nm light illumination [2].

2. UV SENSITIVITY OF THE SAW DEVICES

Designs of SAW devices can be divided into two main groups, which will be discussed below. In the various designs, the piezoelectric substrates, photosensitive ZnO films, interdigital transducers (IDTs), non-piezoelectric substrates, piezoelectric ZnO films, commercial SAW filters, flexible piezoelectric substrates, transparent non-piezoelectric substrates, and protective layers are used. To improve the photosensitivity of ZnO sensors nanorods, nanofibers, nanoparticles of zinc oxide and ultrathin gold layers are precipitated [3].

2.1. Sensitivity to Ultraviolet Radiation at Using ZnO Films

The UV sensing measurements of ZnO/$LiTaO_3$ structures were carried out in [4]. Particularly, they showed that at the power density of 350 μW/cm^2, the amplitude of the Love mode of SAW UV sensor, decreased up to −6.4 dB with the frequency shift of ~150 kHz under 254 nm illumination. S. Kumar et al. made SAW devices using ZnO films with a central frequency of 41.2 MHz [5]. The substrates were from glass or fused quartz. The downshift in frequency (~45 kHz) and a change in insertion losses (~1.1 dB) of the device show the promise of ZnO for fabrication of a low cost SAW UV sensor.

The paper [6] reports the fabrication of ZnO thin film based SAW UV sensors on glass substrates and the effect of post-deposition thermal annealing on the sensing performance. When the ZnO film annealed at 400 °C was used for sensors, the UV radiation induced resonant frequency shift increased more than 20 times with a response speed much better than in the case of ZnO films with a lower temperature of annealing.

The flexible ZnO/polyimide based UV light SAW sensors were fabricated and their performance was investigated in [7]. In [8], qualitative and quantitative analysis of SAW UV photodetectors with multiple optical sensing areas had been demonstrated.

In [9], as impedance changing under the action of UV radiation, it was used a set of parallel metal electrodes with IDT. The film of zinc oxide was deposited on it by using pulsed laser deposition. The developed device was a delay line with two unidirectional IDTs on $LiNbO_3$ of *YX*-128°-cut. Parameter S_{11} of transmitting-receiving IDT was measured. Its frequency dependence had an irregularity in the pass band of the IDT because of an incident on the reflected SAW. Since the electron concentration in the film increases under the action of UV radiation, it leads to a change in the impedance of the film as at a direct alternating current by changing its conductivity and permittivity. Measurements showed that under the action of UV radiation (λ = 325 nm, p = 280 mW/cm^2) and electric resistance of IDT with ZnO film, DC decreased in 6 times and the capacity increased in 3 times. Moreover, the SAW reflection coefficient from the reflective IDT, computed by the Fourier transformation of the frequency dependence of the parameter S_{11}, decreased in 1.6 times.

2.2. Sensitivity to Ultraviolet Radiation at Using ZnO Nanorods and Nanowires

The Love wave device was constructed using pure ZnO film (1.0 µm thick), and it did not respond to UV radiation [2]. The shifts of insertion loss reached 6.4 and 5.3 dB for the Love wave device using nanorod arrays synthesized on 1.0 µm ZnO film after exposure to UV for 30 s. The UV sensitivity of the device was markedly enhanced by the nanorod arrays that grew on the guiding layer. The obtained ZnO nanorods synthesized for 6 h were 0.54 µm in length, longer than those synthesized for 3 h (0.37 µm).

W. Peng et al. confirmed experimentally that the ZnO nanowire layer showed enhanced UV photoelectric response characteristics compared to the sputtered ZnO film layer. Figure 1 presents the frequency *versus* time curve for detector with ZnO nanowire layer showing a good repeatability of the characteristics of the sensor [10].

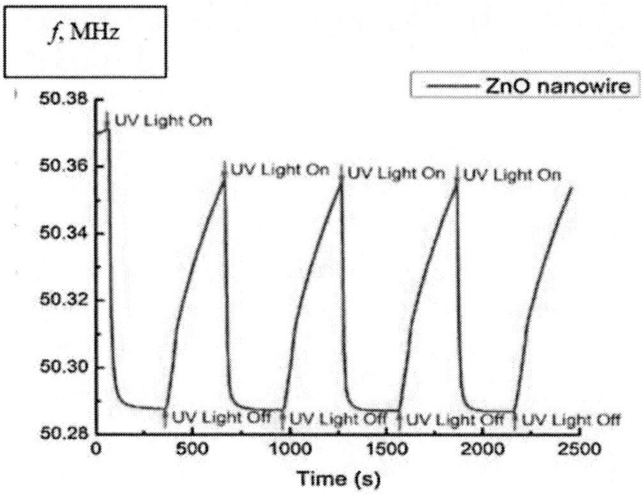

Figure 1. Frequency *versus* time curve of detector with ZnO nanowire layer shows repeatability [10].

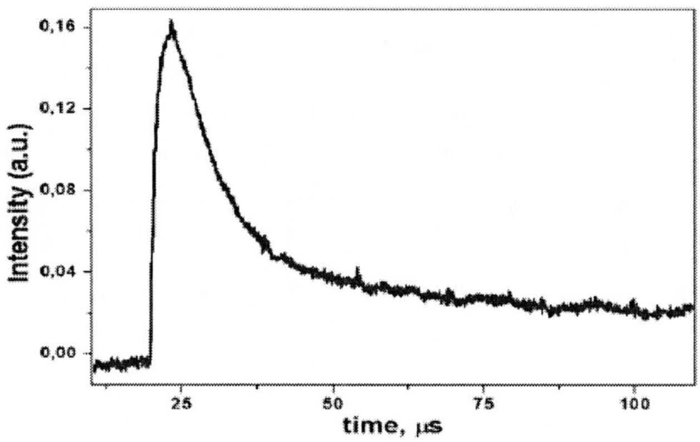

Figure 2. Temporal response of MSM detectors based ZnO nanorods excited by Nd-YAG laser (λ = 355 nm, τ_p=10 ns) [13].

It is easy to see that, in the case of a layer of nanowires, the detector had much higher reproducibility and a better response rate. W. Peng et al. had used a PEDOT - poly(3,4-ethylenedioxythiophene) surface coating for improvement in the efficiency of surface acoustic wave ultraviolet detector with ZnO nanowires [11].

A new approach to enhance the sensitivity of existing SAW UV-visible photo-detectors based on the monitoring of the multiple reflections of the acoustic waves and the Fourier transform of the frequency response (S_{21}) is demonstrated. By using this concept, it is possible to monitor the UV-visible radiation in a wide intensity range from very high to ultralow. In [12], a strategy is present to obtain an ultrafast SAW UV photo-detector with millisecond response by tuning the deposition conditions of the ZnO film and using the multiple-reflections concept.

Photoelectric characteristics of two types of metal-semiconductor-metal (MSM) structures based on ZnO nanorod arrays and two Schottky barriers were studied [13]. The nanorods with low and high amount of deep levels, related to oxygen defects, were designed by catalyst-free CVD from Zn vapor (T = 600°C) and by catalyst-free carbothermal synthesis (T = 950°C). The nanorods grown at lower temperature in oxygen rich gas flow show much less oxygen defect than those grown at an elevated temperature. Detectors with a high amount of deep defect levels show better performance under green illumination, while the defect-free structures show higher sensitivity to UV radiation. Defect free structures demonstrate fast photocurrent rise and recovery times of 1 μs and 10 μs, respectively.

The temporal response (Figure 2) of MSM structure for ZnO nanorods,, produced by the method of thermal evaporation, was received by laser pulses excitation with 10 ns pulse length (Nd-YAG laser, λ = 355 nm) at a bias 5V [13]. Rise time was ~ 1 μs. The fall time at full-width at half-maximum was ~ 10 μs. Due to good durability to radiation, high photosensitivity and performance, such photodetectors, could find wide application in many areas where the control of UV radiation needs.

3. UV Sensitivity of the SAW Oscillators

P. Sharma and K. Sreenivas presented the SAW UV sensor in which variable gain generators had a $ZnO/LiNbO_3$ SAW filter [14]. The amplitude of the oscillator output was found to decrease linearly with increasing UV radiation intensity. The sensitivity to UV radiation was dependent on the gain of the amplifier. UV radiation intensity of ~10 $\mu W/cm^2$ could be detected easily. Results [14] show its potential use in the measurement of very low (~10 $\mu W/cm^2$) and high intensity of UV radiation, with the possibility of wireless operation. S. Kumar et al. used the 71 nm thick ZnO overlayer in the $ZnO/LiNbO_3$ structure and tested the sensor for its photoconductivity in the presence of UV radiation [15]. The gain of the high frequency SAW oscillator had been controlled by the power supply.

In [16], the high-precision ultraviolet (UV) detector, combining ZnO nanostructures and SAW oscillator with dual delay lines, is present. For the delay lines, lithium niobate substrates of 128° YX-cut with zinc oxide nanostructures deposited on it were used. The center frequency of the SAW device was at 145 MHz. The maximum frequency shift over 40 kHz was observed under 365 nm illumination for the several on-off cycles, indicating that the ZnO nanorods based detector was sensitive to UV radiation with good repeatability. Moreover, the frequency shifts, reached to value of 19 kHz after 365 nm, were turned on for 10 s [16].

D. Dasgupta and K. Sreenivas used the 36 MHz surface acoustic wave delay line based oscillator to study the effect of acoustoelectronic interaction due to photogenerated charge carriers in ZnO film under UV illumination (λ = 365 nm, 20 – 100 $\mu W/cm^2$). The design aspects for developing a delay line based SAW oscillator are specified [17]. The observed linear downshift in frequency was 2.2 to 19.0 kHz with varying UV intensity (20 – 100 μW/cm^2).

Conclusion

For the SAW UV sensor, based on zinc oxide, there are used SAW filters, SAW delay lines, or SAW oscillators. The shift of all main characteristics (resonant frequency, phase angle and insertion loss) of the SAW sensor may be simultaneously used for the detection of UV, greatly enhancing the sensing accuracy.

One of the main characteristics of SAW UV sensors is photosensitivity. The largest photosensitivity (450 nW/cm^2) is obtained at using oscillator circuit [15, 18]. The Rayleigh waves provide greater photosensitivity than the Lamb waves: 111.3 ppm/(mW/cm^2) and 55.8 ppm/(mW/cm^2), respectively [19]. In turn, the Sezaw waves provide greater photosensitivity than the Rayleigh waves [20]. With that, it is important to note that the use of nanorods, nanowires, and nanoparticles significantly increases the photosensitivity, repeatability, and response speed of SAW UV sensors. Particularly, photosensitivity of ZnO nanowire is 0.4483 kHz/(μW/cm^2) and polycrystal ZnO film 0.05982 kHz/(μW/cm^2) [10]. In the SAW UV sensors, equidistant IDTs or slanted IDTs are used. The slanted IDTs permit to locate the area of UV radiation on the surface of the SAW device. Photosensitivity of the zinc oxide lies in the UV range for radiation with wavelengths shorter 380 nm.

Acknowledgments

This work was financially supported by the basic part of State task No.16.5405.2017/8.9 BCh of the Ministry Education and Science of Russian Federation and the grant No. 16-08-00740 of the Russian Foundation for Basic Research.

References

[1] Dneprovski, V. G., Kaidashev, E. M., Karapetyan, G. Ya. In: *Proc. of the 2016 Intern. Conf. on "Physics, Mechanics of New Materials and Their Applications,"* Ivan A. Parinov, Shun-Hsyung Chang, Muaffaq A. Jani (Eds.), Nova Science. Publishers, New York. 2017: 105 – 110.

[2] Water, W., Jhao, R. Y., Ji, L. W., Fang, T. H., Chen, S. E. *Sens. Actuators A Phys.,* 2010, *vol. 161:* 6–11.

[3] Dneprovski, V. G., Kaidashev, E. M., Karapetyan, G. Ya. *Abstracts & Schedule of the 2017 Intern. Conf. on "Physics and Mechanics of New Materials and Their Applications" (PHENMA 2017),* Jabalpur, India. 2017: 47 – 48.

[4] Pang, H. F. et al. *Sens. Actuators A. Phys.,* 2013, *vol. 193:* 87 – 94.

[5] Kumar, S., Kim, G.-H., Sreenivas, K., Tandon, R. P. *J. Electroceramics,* 2009, *vol. 22:* 198 – 202.

[6] Wang, W., et al. *J. Appl. Phys.,* 2014, *vol. 104:* 212107 (4 p.).

[7] He, X. L., et al. *J. Micromech. Microeng.*, 2014, *vol. 24*: 055014 (8 p.).

[8] Wang, W. S., Tsai, C. J., Ma, C. C. *J. Appl. Phys.,* 2013, *vol. 114:* 064911 (9 p.).

[9] Karapetyan, G. Ya., Kutepov, M. E., Kaydashev, V. E., Kaidashev, E. M. *Abstracts & Schedule of the 2017 Intern. Conf. on "Physics and Mechanics of New Materials and Their Applications" (PHENMA 2017),* Jabalpur, India. 2017: 98 – 99.

[10] Peng, W., He, Y., Wen, C., Ma, K. *Sens. Actuators A. Phys.*, 2012, *vol. 184:* 34 – 40.

[11] Peng, W., et al. *Sens. Actuators A. Phys.*, 2013, *vol. 199:* 149 – 155.

[12] Karapetyan, G. Ya., Kaidashev, V. E., Zhilin, D. A., Minasyan, T., Abdulvakhidov, K. G., Kaidashev, E. M. *Smart Materials and Structures,* 2017, *vol. 26(3):* 035029.

[13] Zhilin, D. A., Lyanguzov, N. V., Pushkarev, V. I., Nikolaev, A. L., Kaidashev, E. M. *Journal of Physics Conference Series,* 2014, *vol. 541:* 012038.

[14] Sharma, P., Sreenivas, K. *Appl. Phys. Lett.,* 2003, *vol. 83:* 3617 – 3619.

[15] Kumar, S., Sharma, P., Gulia, V., Sreenivas, K. *Low Intensity Ultraviolet Light Detection Using a ZnO/LiNbO$_3$ SAW Oscillator.* http://een.iust.ac.ir/profs/Sadr/Papers/int3.6.pdf.

[16] Wang, W. S., Wu, T. T., Chou, T. H., Chen, Y. Y. *Nanotechnology*, 2009, *vol. 20:* 135503 (5 p.).

[17] Dasgupta, D., Sreenivas, K. *J. Appl. Phys.,* 2011, *vol. 110:* 044502 (8 p.).

[18] Kumar, S., Sharma, P., Sreenivas, K. *Semicond. Sci. Technol.,* 2005, *vol. 20:* L27 – L30.

[19] Zhou, J., et al. *J. Appl. Phys.*, 2013, *vol. 114:* 044502 (8 p.).

[20] Nuri, W., Emanetoglu, J., Zhu, Y., Chen, J., Zhong, Y., Chen, Y., Lu, Y. *Appl. Phys. Lett.,* 2004, *vol. 85:* 3702 – 3704.

In: Proceedings of the 2018 International Conference … ISBN: 978-1-53615-862-5
Editors: Ivan A. Parinov et al.

Chapter 40

UV PHOTODETECTOR BASED ON ZNO FILM AS EXTERNAL IMPEDANCE FOR SAW SENSOR

G. Ya. Karapetyan**[*]**, ***M. E. Kutepov,***
T. A. Minasyan and E. M. Kaidashev
Laboratory of Nanomaterials,
Southern Federal University, Rostov-on-Don, Russia

ABSTRACT

The UV sensor, based on surface acoustic waves (SAWs) with a sensitive element taken out from the acoustic channel, is investigated. We proposed a new design of UV sensor, based on the change of the SAW transmission through the interdigital directional coupler, which is loaded on the impedance of the photosensitive zinc oxide film, taken out from the acoustic channel. The Fourier transform of the sensor frequency responses makes it possible to obtain the impulse response of the sensor, where the temporal position of the SAW pulses, reflected from the IDT, is seen. This makes it possible to increase the sensitivity of UV sensors. The design of the sensor allows a measurement of the UV intensity not only by the change in the amplitude of the wave, reflected from the IDT, but also by the change of its phase. The obtained results can be used to create wireless passive UV sensors.

Keywords: SAW device, UV light sensor, ZnO films, interdigital transducer (IDT), interdigital system, external impedance

[*] Corresponding Author's E-mail: gkarapertyan@sfedu.ru.

1. INTRODUCTION

Zinc oxide films have recently been increasingly used for SAW sensors of pressure, displacement, magnitudes of magnetic induction, and the intensity of ultraviolet radiation (UV), as well as for the detection of various chemical substances [1–6]. The structures with zinc oxide films, located between the interdigital transducers (IDTs), were investigated in [3, 4] to determine the most suitable film parameters for the sensor. Only frequency responses were considered in upper listed reports, which do not allow one to visually obtain information about the SAW reflected from the IDT during UV illumination of the film. In our paper [5], we propose the use of the Fourier transform of frequency responses, which makes it possible to obtain the impulse response of the sensor, where the temporal position of the SAW pulses reflected from the ITD is seen. This made it possible to increase the sensitivity of UV sensors, since reflected SAW can pass under the film more than once. However, in references [1–5], the zinc oxide film is located directly in the acoustic channel in the pathway of SAW propagation. This leads to an increase in the SAW damping, especially at high frequencies. Moreover, propagation of the SAW through the nanorods array leads to even greater acoustic wave damping [6]. Thus, with the increase in the SAW damping, the sensitivity of the UV sensors is drastically decreased. Therefore, the creation of a sensor with a UV sensitive film outside the acoustic channel is an actual task.

2. EXPERIMENTAL DETAILS

It is shown that the external load, connected to transmitting-receiving IDT influence S_{11}, is much more than in the case of a reflective one. By considering it, we place coupler, which consists of two unidirectional IDTs, exciting SAW in opposite directions (Figure 1) between reflective IDTs located in different acoustic channels. Then, for the IDT located in the lower acoustic channel, the IDT in the upper channel of the coupler can be considered transmitting-receiving or *vice versa*. Sensor design, shown in Figure 1, allows one to change not only the SAW reflection coefficient, but also the coefficient of SAW transmission from one acoustic channel to another, which leads to the increase of the external impedance influence on the S-parameters of the sensor. To measure the parameter S_{21}, it is necessary to connect both transmitting-receiving and reflective IDTs to a network analyzer. The coupler transfer coefficient can be significantly changed by tuning capacitance C, and inductance L. Figure 2(a) shows that the coefficient of the coupler transfer during SAW transition from one acoustic channel to another change by 23 dB. The influence of SAW reflected from IDT on S_{11} is clearly seen in Figure 2(b), where the frequency of pulsations doubles (dashed curve) in comparison with the pulsations is caused by SAW reflections from the coupler (solid curve). On the Fourier transformation of the S_{11} frequency dependence shown in Figure 2(c), it is clearly seen that SAW reflected from the coupler is significantly suppressed when the coupler is open, wherein coupler poorly reflects SAW and transfers it to another acoustic channel. Since the delay lines are identical in different acoustic channels, the secondary reflection of SAW from the coupler coincides with the primary reflection of the SAW from the reflective IDT, which is evident in Figure 2(c). Next, IDS with zinc oxide film was connected to the coupler, see Figure 3(a). Using the tuning inductor and capacitor, the coupler transfer coefficient was made maximum in the absence of UV. When UV was on, the impedance of IDT with the film changed, which led to a change of the transfer coefficient, see Figure 3(a).

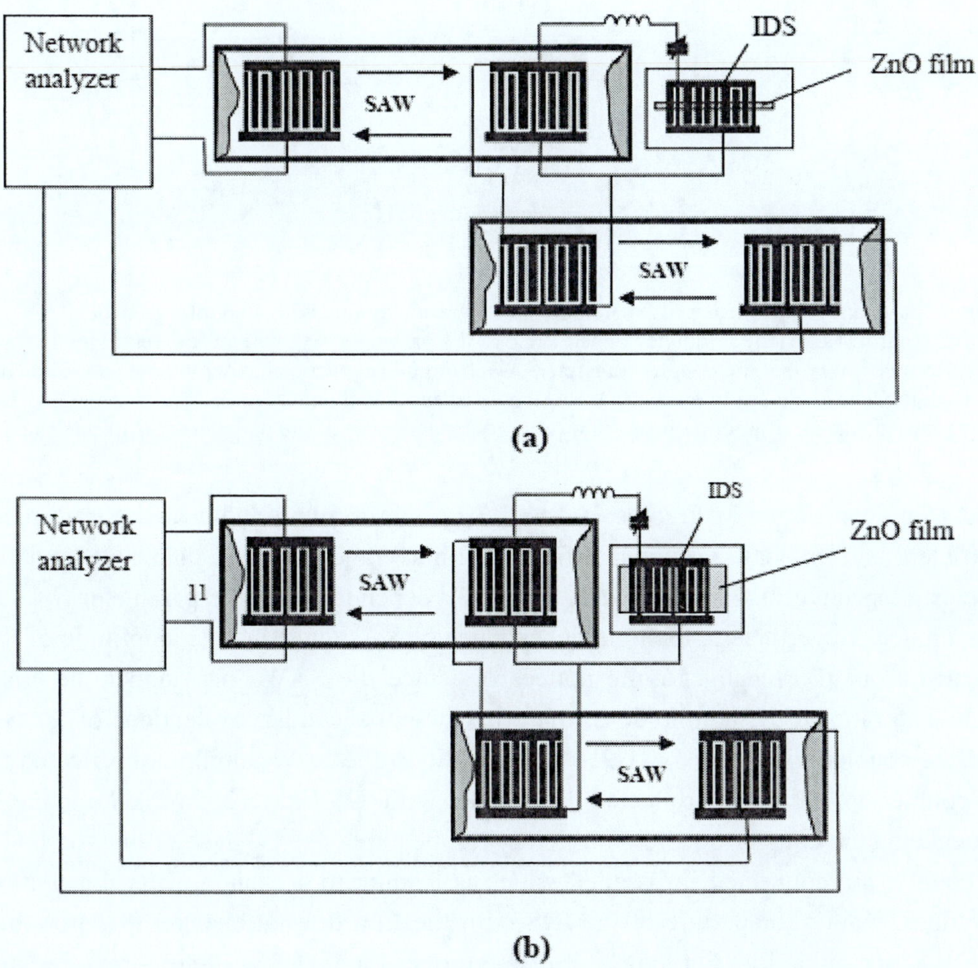

Figure 1. UV sensor with a coupler in the form of two unidirectional IDTs: (a) UV sensor with two acoustic channels, where the IDS with the film is connected to the IDT, which transfers SAW from one acoustic channel to another; (b) UV sensor with two acoustic channels, where the IDS with a film in the form of a narrow band is connected to the width of the IDT that converts the surfactant from one acoustic channel to another.

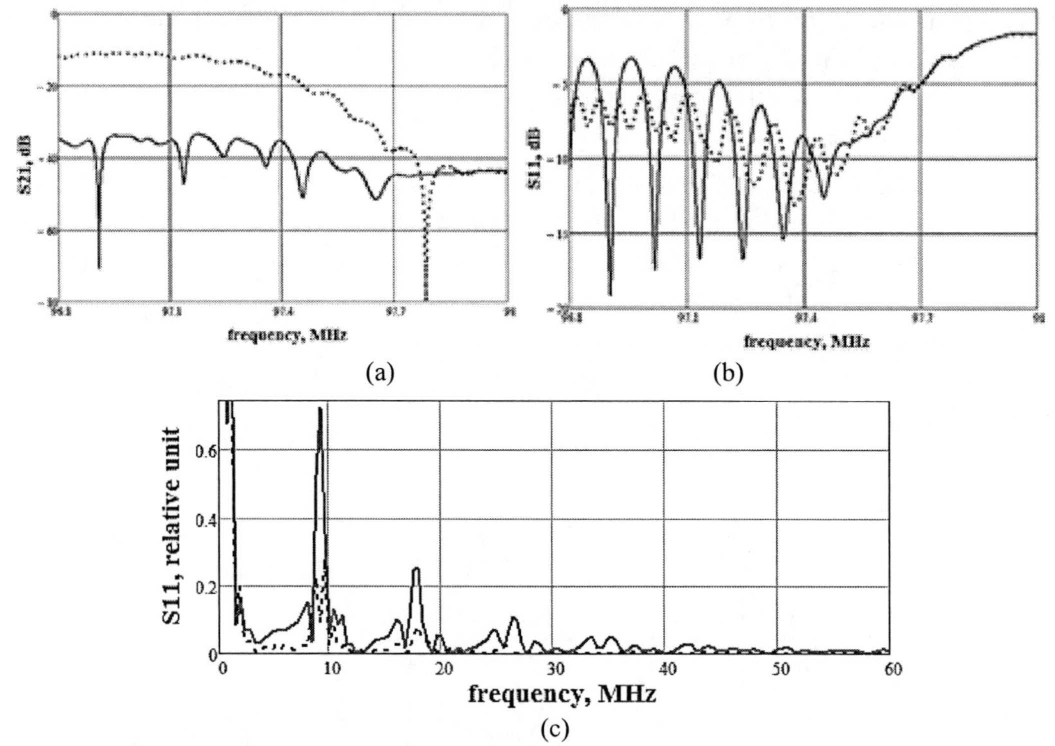

Figure 2. Frequency dependence of S-parameters of SAW delay line with a coupler as two unidirectional IDTs: (a) the frequency dependence of the parameter S_{21}, when the external load is chosen in such a way that the coupler transfers SAW from one acoustic channel to another (solid curve) and when the load is chosen in a way, that SAW pass to another channel with attenuation of more than 20 dB (dashed curve); (b) the frequency and (c) time dependences of the parameter S_{11} for the same cases.

As one can notice in Figures 3(a) and 3(c), the absolute value of the transmission coefficient and its phase vary insignificantly (on average not more than 1 dB and 1 – 2 degrees, respectively). Moreover, the frequency dependence of the parameter S_{11} varies not so much. Nevertheless, the time dependence of S_{11} shows that the amplitude of SAW reflected from IDT changes quite noticeably, since the SAWs pass under the coupler twice. The ratio of the amplitude of the primary and secondary reflections of the SAW from the coupler without UV is 1.91, and the ratio of the SAW amplitudes reflected from the coupler to the SAW reflected from IDT with UV is 1.12. Weakening the UV influence on the coupler transfer coefficient can be explained by the fact that the IDTs of the coupler are connected in parallel, which is leading to a change of its impedance in two times, while the aperture of the IDS with the film did not change. It is possible to deposit a zinc oxide film not to the entire aperture of the IDT, but in the form of a narrow strip with a width of 0.5 mm perpendicular to the electrodes of the IDT, as shown in Figure 2. In this case, impedance of IDS with the film will change significantly, and its influence on SAW transition from one acoustic channel to another will increase. As can be seen from Figures 3(b) and 3(d), the use of zinc oxide film in the form of a narrow strip increases the influence of UV on the transition of SAW from one acoustic channel to another. Attenuation decreases by approximately 2.5 dB, and the phase changes by approximately 28° as a 3 mm diameter UV beam with 3.2 mW/cm^2 intensity and 325 nm wavelength is on.

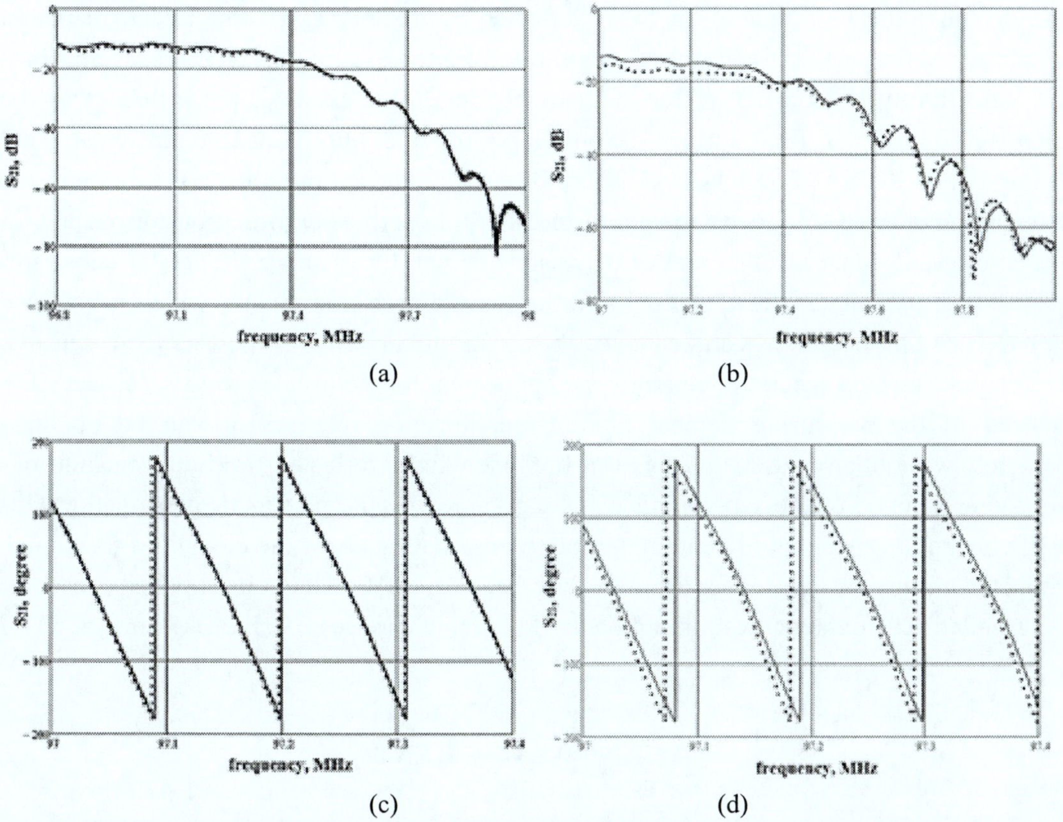

Figure 3. Frequency dependence of the absolute value and phase of the S_{21} parameter of SAW delay line with a coupler in the form of two unidirectional IDTs: IDS with a wide zinc oxide film (a, c) and IDS with the narrow strip film (b, d); solid curve corresponds to the case with UV, dashed curve corresponds to the case without UV.

CONCLUSION

We have carried out measurements of the amplitude and phase of the SAW, transferred through a coupler, consisting of unidirectional IDTs and exciting SAW in opposite directions. When the additional impedance that contains the zinc oxide film sensitive to UV is connected to the coupler, located between the reflective and transmitting-receiving IDT, the transfer coefficient of the coupler changes. This leads to 28°-change in the phase and 2.5 dB-change of the amplitude of SAW passing under the coupler.

The use of the Fourier transform of the frequency responses makes it possible to obtain the impulse response of the sensor, where the temporal position of the SAW pulses reflected from the IDT is seen. It allows one to increase the sensitivity of the UV sensors. The SAW based sensor with a sensitive element removed from the acoustic channel, proposed in this paper, can be used to measure the intensity of UV radiation not only by the change in the amplitude of the SAW reflected from the IDT, but also by the change in its phase. At the same time, it differs favorably from the existing SAW UV sensors, since UV-sensitive film is outside of the acoustic channel. This excludes the influence of the photosensitive ZnO film surface morphology (or the photosensitive array of ZnO nanorods) on the SAW propagation. Therefore, such a design provides no increase in the attenuation of the SAW due to the photosensitive film surface deformations, especially in the microwave range. It also makes it possible to use films on single-crystal substrates with lattice parameters close to ZnO and the type of crystal structure, which may allow one to vary film parameters and enhance the sensitivity and time response of UV-sensors. If the transmitting-receiving antenna is connected to the transmitting-receiving IDT instead of the measuring device, the information from the sensor can be obtained remotely without any wires. Interrogator impulse will be reflected from the transmitting-receiving antenna of the sensor and will be obtained by the reader of interrogator. This will change S_{11},, which will contain complete information about the change of the signal, reflected from the sensor under the influence of UV. Thus, its intensity can be determined. The obtained results can be used to create wireless passive UV-sensors.

Acknowledgment

This research work is supported by Russian Education and Science Ministry, the project No. 16.5405.2017/8.9.

References

[1] Aurel Marcu, Cristian Viespe. (2015) *Sensors and Actuators B: Chem., vol. 208*, 1–6.

[2] Lu Yicheng, Chen Ying, Zhang Zheng, *Multifunctional Biosensor Based on ZnO Nanostructures*, US patent 7989851, 02.08.2011.

[3] Wen-Che Tsai, Hui-ling Kao, Kun-Hsu Liao, Yu-Hao Liu, Tzu-Ping Lin, Erik S. Jeng. (2015) *Optics Express, vol. 23(3)*, 2187.

[4] Kumar, Sanjeev, Kim, Gil-Ho, Sreenivas, K., Tandon, R. P. (2009) *J. Electroceram., vol. 22*, 198–202.

[5] Ballantine, D. S. Jr., White, R. M., Martin, S. J., Ricco, A. J., Frye, G. C., Zellers, E. T., Wohltjen, H. (1997) *Acoustic Wave Sensors: Theory, Design, and Physico-Chemical Applications*. Academic Press.

[6] Karapetyan, G. Ya., Kaydashev, V. E., Zhilin, D. A., Minasyan, T. A., Abdulvakhidov, K. G., Kaidashev, E. M. (2017) *Smart Materials and Structures, vol. 26(3)*, 1 – 6.

In: Proceedings of the 2018 International Conference … ISBN: 978-1-53615-862-5
Editors: Ivan A. Parinov et al.

Chapter 41

SIMULATION OF THE WORK OF FLAT JACK IN LARGE DISPLACEMENTS

A. S. Yudin*

I. I. Vorovich Institute of Mathematics, Mechanics and Computer Science, Southern Federal University, Rostov-on-Don, Russia

ABSTRACT

The work model of the shell of the flat jack type at the deformations close to limit is considered. A semi-inverse method is applied, in which the functional of pressure is stated to a stationary value.

Keywords: shell, large deformations, mathematical modeling, semi-inverse method

1. INTRODUCTION

To solve nonlinear problems of large shell deformations, the semi-inverse method is effectively used when the shape of the meridian of the obtained shell and the bond reaction are *a priori* given. This approach was used in the works of the author in relation to the problems of dome-shaped shells extraction [1] and cylindrical shell blowing [2].

In the present work, we continue the investigation of the possibilities to apply this method to the shells of other geometries. The design scheme of the simulation of flat

* Corresponding Author's E-mail: yudin@math.sfedu.ru.

jacks (FJs) used for lifting and leveling of heavy structures is considered. FJ can be considered a kind of mechanism for generating large forces and displacements.

The device is formed by two coaxial parallel round plates, which are conjugated on the contours with a torus cut along the inner diameter. The design is also symmetrical with respect to the horizontal plane, equidistant with respect to the plates. The torus meridian has a composite geometry and is formed by arcs of smooth curves, in particular, circles. At the points of conjugation of elements of the torus, there are discontinuities of curvature with change of sign.

Here, the problem of mathematical modeling of the flat jack in the range of large movements is posed up to the maximum possible. This requires the application of geometrically and physically nonlinear equations. It is necessary to take into account the changes in the metrics of the shell due to the appearance of large deformations and inelastic behavior of the material.

The applied algorithm reduces the problem to the derivation of some functional by a constant, depending on the vector of variable parameters. The constant is equal to the value of uniform pressure, under the action of which the shell is deformed to a given level of characteristic displacement. With a small number of parameters, they can be selected interactively.

2. Equations and Method

The paper uses a geometrically and physically nonlinear mathematical model based on the equations of shell theory for large displacements, angles of rotation and deformations and the relations determining the properties of the material, in which the plasticity diagram for logarithmic deformations is used.

Let us present a summary of the constitutive equations of the model.

Kinematic relations:

$$\varepsilon_1^\zeta = (\varepsilon_1 + \zeta\kappa_1),\ \kappa_1 = (\Phi_o' - \delta_3\Phi')/\alpha_o,$$
$$\varepsilon_1 = (w'\sin\Phi + u'\cos\Phi)/\alpha_o + \cos(\Phi - \Phi_o) - 1, \tag{1}$$

$$\varepsilon_2^\zeta = (\varepsilon_2 + \zeta\kappa_2),\ \varepsilon_2 = u/r_o,\ \kappa_2 = (\sin\Phi_o - \delta_3\sin\Phi)/r_o.$$
$$\gamma = \delta_1^{-1}(\gamma_o + \zeta\gamma_1),\ \gamma_1 = \varepsilon_3'/\alpha_o,$$
$$\gamma_o = (w'\cos\Phi - u'\sin\Phi)/\alpha_o - \sin(\Phi - \Phi_o). \tag{2}$$

$$K_1 = \Phi'/\alpha_o,\ K_2 = (\sin\Phi)/r_o. \tag{3}$$

All functions depend on the curvilinear coordinate $\alpha_1 = \xi$. Here the prime denotes the derivative with respect α_1; α_o is the Lamé coefficient of the original shell; Φ_o is the angle between the normal to the reference shell, and the z-axis; Φ is the angle between the normal to the deformed membrane and the z-axis; $\delta_j = 1 + \varepsilon_j$, $j = 1,2,3$. We assume that the incompressibility condition is fulfilled: $\delta_1\delta_2\delta_3 = 1$, $\delta_3 = \delta_1^{-1}\delta_2^{-1}$.

Equilibrium equation:

$$(r_o\bar{V}^o)' + \alpha_o r_o p_w^o = 0\,, \tag{4}$$

$$(r_o\bar{H}^o)' - \alpha_o\bar{N}_2^o + \alpha_o r_o p_u^o = 0\,, \tag{5}$$

$$(r_o M_1^o)' - \alpha_o M_2^o\cos\Phi - \alpha_o r_o\delta_1 Q^o = 0\,. \tag{6}$$

$$p_w^o = \delta_1\delta_2 p_w\,,\ \ p_u^o = \delta_1\delta_2 p_u\,, \tag{7}$$

Internal forces $\bar{V}^o$ and $\bar{H}^o$ are oriented, respectively, vertically (along the axis of symmetry) and horizontally (on the radius of the cylindrical coordinate system), where Q^o is the shear force:

$$\bar{V}^o = \bar{N}_1^o\sin\Phi + Q^o\cos\Phi\,,\ \ \bar{H}^o = \bar{N}_1^o\cos\Phi - Q^o\sin\Phi\,; \tag{8}$$

Quantities $\bar{N}_1^o$, $\bar{N}_2^o$ are the generalized forces, associated with tangential internal forces and moments M_1^o, M_2^o:

$$\bar{N}_1^o = N_1^o + \delta_1^{-1}(K_1M_1^o + K_2M_2^o)\,,\ \ \bar{N}_2^o = N_2^o + \delta_2^{-1}(K_1M_1^o + K_2M_2^o)\,,$$
$$N_1^o = \delta_2 N_1,\ \ N_2^o = \delta_1 N_2,\ \ Q^o = \delta_2 Q\,,\ \ M_1^o = \delta_2 M_1,\ \ M_2^o = \delta_1 M_2\,. \tag{9}$$

Internal pressure is considered as a load. Taking into account that the positive direction of the normal is directed inside the shell, we have:

$$p_w = -p\cos\Phi,\ \ p_u = p\sin\Phi\,. \tag{10}$$

The constitutive equations of the Davis-Nadai type are accepted in the form:

$$N_1^o = \bar{B}_1(\bar{\varepsilon}_1 + 0.5\bar{\varepsilon}_2)\,,\ \ N_2^o = \bar{B}_2(\bar{\varepsilon}_2 + 0.5\bar{\varepsilon}_1)\,,$$
$$M_1^o = \bar{D}_1\ \ \bar{\kappa}_1 + 0.5\bar{\kappa}_2\ \ ,\ \ M_2^o = \bar{D}_2\ \ \bar{\kappa}_2 + 0.5\bar{\kappa}_1\ \ ; \tag{11}$$

$$\bar{\varepsilon}_1 = \ln(\delta_1),\ \bar{\varepsilon}_2 = \ln(\delta_2),\ \bar{\kappa}_1 = \kappa_1/\delta_1,\ \bar{\kappa}_2 = \kappa_2/\delta_2; \tag{12}$$

$$\bar{B}_1 = \bar{B}/\delta_1,\ \bar{B}_2 = \bar{B}/\delta_2,\ \bar{D}_1 = \delta_3^2\delta_2\bar{D},\ \bar{D}_2 = \delta_3^2\delta_1\bar{D},$$
$$\bar{B} = (4/3)\Lambda(\bar{\varepsilon})h_o,\ \bar{D} = (1/9)\Lambda(\bar{\varepsilon})h_o^3. \tag{13}$$

Here $\Lambda(\bar{\varepsilon}) = C\bar{\varepsilon}^{\eta-1}$ is the secant modulus of material plasticity diagram, where $\bar{\varepsilon} = (2/\sqrt{3})(\bar{\varepsilon}_1^2 + \bar{\varepsilon}_1\bar{\varepsilon}_2 + \bar{\varepsilon}_2^2)^{0.5}$ is the intensity of the logarithmic strain.

The material diagram can be identified by the characteristic values, which are usually given in reference books. This is the modulus of elasticity E, the elastic limit σ_{02}, the limit of elastic deformation $\varepsilon_{02} = \sigma_{02}/E$, temporary stress intensity σ_B, and temporal intensity of logarithmic deformations $\bar{\varepsilon}_B$. Based on these data, the parameters of the exponential approximation of the nonlinear material properties diagram are found as $\sigma(\bar{\varepsilon}) = C\bar{\varepsilon}^{\eta}$, where

$$\eta = [\ln(\sigma_B) - \ln(\sigma_{02})]/[\ln(\bar{\varepsilon}_B) - \ln(\varepsilon_{02})],\ C = \sigma_B/\bar{\varepsilon}_B^{\eta}.$$

We integrate equations (4) and (5). Taking into account (8), we obtain:

$$r_o(\bar{N}_1^o\sin\Phi + Q^o\cos\Phi) = p \cdot I_{11} + C_V, \tag{14}$$

$$r_o(\bar{N}_1^o\cos\Phi - Q^o\sin\Phi) = I_{21} - p \cdot I_{22} + C_H, \tag{15}$$

where C_V and C_H are integration constants, associated with vertical and horizontal forces on the contour $\xi = \xi_0$; I_{11}, I_{21}, I_{22} are integrals with variable upper limits:

$$I_{11} = \int_{\xi_0}^{\xi} \alpha_o r_o \delta_1\delta_2\cos\Phi d\xi,\ I_{21} = \int_{\xi_0}^{\xi} \alpha_o \bar{N}_2^o r_o d\xi,\ I_{22} = \int_{\xi_0}^{\xi} \alpha_o r_o \delta_1\delta_2\sin\Phi d\xi. \tag{16}$$

From Equations (14) and (15), we express Q^o. To do this, we perform the operation:

$$r_o^{-1}[(14) \cdot \cos\Phi - (15) \cdot \sin\Phi].$$

We obtain:

$$Q^o = r_o^{-1}[(p \cdot I_{11} + C_V)\cos\Phi - (I_{21} - p \cdot I_{22} + C_H)\sin\Phi]. \tag{17}$$

By substituting expression (17) for Q^o in Equation (6) and integrating it, we get:

$$r_o M_1^o = I_{31} + p \cdot (J_{11} + J_{22}) + C_V J_{31} - J_{32} - C_H J_{33} + C_M, \tag{18}$$

where

$$J_{11} = \int_{\xi_0}^{\xi} \alpha_o \delta_1 \cos\Phi \cdot I_{11} d\xi,\ J_{22} = \int_{\xi_0}^{\xi} \alpha_o \delta_1 \sin\Phi \cdot I_{22} d\xi,\ J_{31} = \int_{\xi_0}^{\xi} \alpha_o \delta_1 \cos\Phi d\xi.,$$

$$J_{32} = \int_{\xi_0}^{\xi} \alpha_o \delta_1 \sin\Phi \cdot I_{21} d\xi,\ J_{33} = \int_{\xi_0}^{\xi} \alpha_o \delta_1 \sin\Phi d\xi, \tag{19}$$

C_M is a integration constant, associated with the moment reaction on the contour.

The pressure p does not depend on ξ and is taken out beyond the integral sign. From (18) it follows:

$$p = \frac{r_o M_1^o + J_{32} - J_{31} - C_M - C_V J_{31} + C_H J_{33}}{J_{11} + J_{22}} \tag{20}$$

3. Design Scheme

The numerators and denominators of the functional (20) depend on ξ. The further problem is to derive the pressure p through a value close to a constant value with acceptable accuracy. The control levers here are the parameters of the shell shape approximation and the integration constant. This can be performed in an interactive numerical experiment. At the same time, of course, physical relations (11)–(13), as well as the condition of incompressibility of the material, are involved.

We introduce the notation of the main parameters of the shell in the initial state, Figure 1: $h_p = h_1$ is the thickness of the plate; r_3 is the outer radius of the middle surface of the torus; 2δ and $2\delta_1 = 2\delta + h_1$ are the distances between the plates and their median surfaces; $r_p = r_1$ is the radius; R_2 is the radius of the arc (2), the generatrix of the convex part of the torus (section 2); $r_{O1} = r_1$, and $r_{O2} = r_3 - R_2$ are the coordinates of the centers of circles, arcs of which form the toroidal part; ϕ_1 and ϕ_2 are the coverage angles of arcs of circles; $r_2 = r_{O2} - R_2 \cdot \sin(\phi_2 - \pi/2)$ and $z_2 = R_2 \cdot \cos(\phi_2 - \pi/2)$ are the coordinates of the point of junction between the arcs; the arc radius is 1, forming the concave part of the torus (plot 1) and the interfacing plate with part 2 of the torus is calculated as $R_1 = 0.5[(z_2 - \delta_1)^2 + (r_2 - r_1)^2]/(z_2 - \delta_1)$. The thickness of the shell of the torus changes from h_1, at the point $r = r_p$ to $k_h h_1$, $k_h < 1$, at the point $r = r_3$.

The equations of the sections of the middle surface of the composite shell are as follows: plate, $Z_p(r) = \delta_1$; the torus in sections 1 and 2, $Z_1(r) = R_1 + \delta_1 - [R_1^2 - (r - r_1)^2]^{1/2}$, $Z_2(r) = [R_2^2 - (r - r_{O2})^2]^{1/2}$. A single surface $Z_o(r)$ is formed by a superposition of piecewise defined functions, given by the corresponding equations on its interval, and equal to zero outside of this interval.

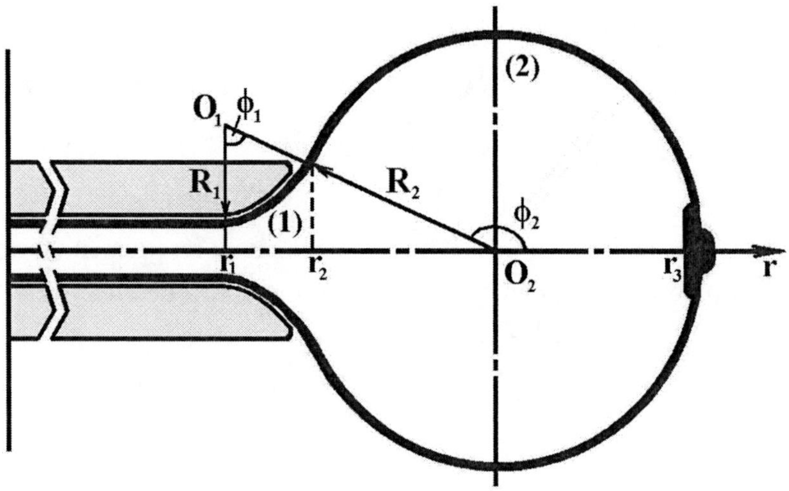

Figure 1. Construction scheme.

When using the shell for its intended purpose, it is placed in a package between the inserts, which are round plates of thick plywood. With a significant load resistance of the structure being lifted, the liners prevent the metal plate from bulging, leaving it flat in the normal lifting mode.

From the structural features of the shell, we note the presence of a sufficiently rigid ring on the outer contour, on which the upper and lower parts are connected by welding. Therefore, when considering one (upper) half, we can assume that the fixing of the shell is stiff.

Let us consider the plate separately under the action of the tensile load on its contour $r = r_1$. We use the homogeneity of the stress-strain state and the condition of incompressibility of the material. During the operation of the system, plastic deformation appears, but not so large that it was necessary to use the logarithmic relative elongations. Taking these factors into account, it is relatively easy to obtain a connection between the boundary load N_1 and the displacement u_1 arising on the plate contour.

The formula looks like:

$$N_{1p} = C \cdot \left(\frac{2u_1}{r_1} \right)^{\eta - 1} \cdot \frac{2u_1 h_1}{r_1 + u_1} \tag{21}$$

where C and η are material constants. For the material of steel 3 type: C = 441.01286, η = 0.10741. The plate works on stretching in its plane and effectively resists deformation.

In the calculation scheme, we take into account the displacement of both the contour of the plate (u_1) and the external diameter of the structure (u_3).

We are interested in the height and weight of the load close to the maximum possible jacks. The maximum height is determined by the movement of the inner contour of the toroidal part of the shell. Consider it separately, setting the power characteristics on the inner circuit. These values refer to the functional control parameters (20).

We assume that the point of joining the torus to the plate obtains the displacement w_p at the transition of the shell from the initial state (0) to the deformed one (d), Figure 2. To approximate the deformed surface, it is preferable to use a single equation that provides continuity of the meridian curvature change. It is acceptable to use an elliptical shape:

$$Z_d(r_d) = (b_{d2}/a_{d2})\sqrt{a_{d2}^2 - (r_d - r_1)^2}\ . \tag{22}$$

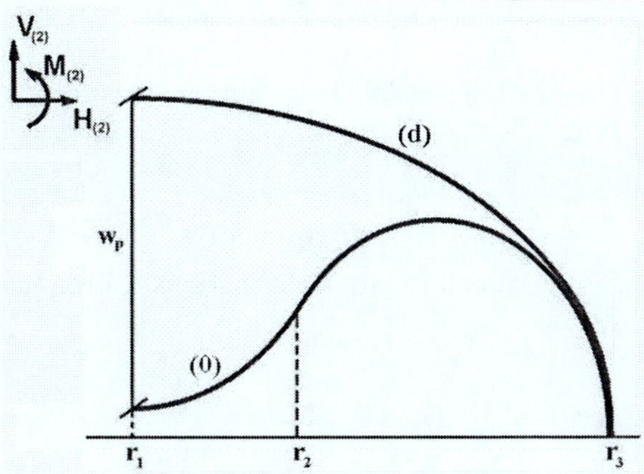

Figure 2. Shell meridian before (0) and after (d) deformation.

Since the change in the shape of the torus zone occurs by bending, it can be assumed that the elongation along the meridian is small: $\varepsilon_1 \approx 0$. Then, the lengths of the meridian arcs of the shell are preserved. The arc length before deformation is $S_0 = S_1 + S_2$, where $S_1 = R_1\phi_1$, $S_2 = R_1\phi_2$.

The length of the arc of the meridian of the deformed shell is:

$$S_d = \int_{r_1}^{r_3} \sqrt{1 + (d(Z_d(r_d))/dr_d)^2}\, dr_d \tag{23}$$

The parameters of the curve (22) are chosen by the condition $S_d \approx S_0$. Also, the coordinates of s and s_d points coincide when moving in an arc, which means that s_d is Lagrangian. Therefore, it is further advisable to move to the independent coordinate s.

The polar radius of the initial median surface of the convex torus r_o as a function of the current point of the meridian arc is given as

$$r_{01}(s) = r_1 + R_1\sin[(s - s_1)/R_1],\ s\in[s_1, s_2], \tag{24}$$

$$r_{02}(s) = r_2 + R_2(k_r - \cos[\pi/2 - \xi_{12} + (s - s_2)/R_2],\ s\in[s_2, s_3], \tag{25}$$

where $k_r = (r_3 - r_2)/R_2 - 1$, $s_1 = r_1$, $s_2 = s_1 + S_1$, $s_3 = s_2 + S_2$. Functions (24) and (25), given piecewise, can be combined into $r_0(s)$ using logical operators. When counting the arc, the Lamé coefficient is $\alpha_0 = 1$. The equation of the meridians, depending on the s will be $Z_{0s}(s) = Z_0(r_0(s))$, $Z_{ds}(s) = Z_d(r_0(s))$.

$$\Phi_0(s) = -\operatorname{arctg}(\frac{d}{ds}Z_{ds}(s)),\ \Phi(s) = -\operatorname{arctg}(\frac{d}{ds}Z_{ds}(s)) \tag{26}$$

From the relations (1), (2) it is possible to obtain the equation:

$$u' = \alpha_o[(1+\varepsilon_1)\cos\Phi - \gamma_o\sin\Phi - \cos\Phi_o)], \tag{27}$$

which is a condition of for compatibility of deformations. Considering that $\varepsilon_1 \approx 0$, $\gamma_0 \approx 0$, we have a simplified version:

$$u' = \alpha_0(\cos\Phi - \cos\Phi_o). \tag{28}$$

Since the initial and deformed forms of the shell are given, this ratio is integrated:

$$u(s) = \int_{r_1}^{s} \alpha_0(s)[\cos(\Phi(s)) - \cos(\Phi_0(s))]ds + u_1 \tag{29}$$

Then the relative elongation $\varepsilon_{22}(s) = u(s)/r_0(s)$ is determined. The incompressibility of the material $(1 + \varepsilon_{22}(s))(1 + \varepsilon_{33}(s)) = 1$ implies $\varepsilon_{33}(s) = (1 + \varepsilon_{22}(s))^{-1} - 1$. In this case, the intensity of deformation is $\varepsilon = (2/\sqrt{3})\varepsilon_{22}$.

Numerators and denominators in the formula (20) for p are given by functions of s. Controlling the parameters of the approximation of the shell shape and the integration constants, it is necessary to deduce the pressure p by a value close to the constant value with acceptable accuracy. This can be done interactively in a numerical experiment.

4. Numerical Experiment

Let us consider a structure having dimensions (here and further in mm): $h_1 = 3$, outer diameter $D = 520$, $r_3 = (D - h_1)/2 = 258.5$; distance between plates $2\delta = 2$, $\delta = 1$; $r_1 = 408.65/2 = 204.325$, $r_2 = 223.63$, $z_2 = 12.005$. The angles of coverage of arcs of circles: $\phi_1 = 52°$, $\phi_2 = 142°$.

Consider the lifting height of the plate $w_p = 28.55$, which makes it possible to fulfill the condition of preserving the lengths of the arcs. Then, the vertical half-axis of the ellipse $b_d = w_p + \delta + h_1/2 = 31.05$, and horizontal one $a_d = (r_3 - r_1) + (u_3 - u_1) = 59.175$.

When the constant value is reached, p is the pressure that corresponds to the specified blow-up level. In this case, the pressure was about 10 MPa.

The load rating is not carried out by simple calculation. To do this, it is necessary to involve techniques based on experimental data. This is due to the specific operation of the flat jack with insert. With a change in the value of the jack's stroke, the radius of the torus and the area of contact of the jack with the support insert change. Therefore, the load capacity of the jack decreases with increasing stroke. According to the technical method, by taking into account these factors, the load capacity for this internal pressure is estimated at 1422 kN.

Acknowledgments

The work was carried out within the project No 11.6767.2017/BCh of the basic part of state task of Russian Ministry of Science and High Education.

References

[1] Yudin, AS. *Stability and Oscillations of Structurally Anisotropic and Artificial Shells of Revolution*. Southern Federal University Press, Rostov-on-Don., 2011, pp. 1 – 362 (In Russian).

[2] Yudin, AS; Yudin, SA; Sigaeva, TV. In: *Advanced Materials - Studies and Applications*, I. A. Parinov, S. H. Chang, S. Theerakulpisut (Eds.). Nova Science Publishers, New York., 2015, 341 – 351.

In: Proceedings of the 2018 International Conference … ISBN: 978-1-53615-862-5
Editors: Ivan A. Parinov et al.

Chapter 42

UTILIZATION OF GLASS WASTE AS A RAW MATERIAL MAKING OF ELECTRICAL COMPONENTS AND ENGINEERING CONSTRUCTION

Ahmad Habib[1,*], Bambang Dwi Sulo and Agus Darwanto

[1]Department of Informatics Engineering, 17 Agustus 1945 University, Surabaya, Indonesia

[2]Department of Electronic Engineering, Islam Malang University, Malang, Indonesia

ABSTRACT

Glass is an inorganic material that cannot decompose, so in sufficient quantities, its existence as a waste will affect the environmental conditions. To reduce this impact, glass waste can be transformed to electrical components, such as fuse housing, bushing, insulator, and bus bar support. In this study, glass waste is made into a glass powder and then mixed with epoxy resin with a ratio of 3:1, then printed according to the desired shape with the cold printing method. Once tested electrically and mechanically, the molding characteristics have a penetrating voltage of 17kV/mm and mechanical strength of $5kgf/mm^2$.

With reference to these characteristics, the objects or components can be made, related to electrical and construction, engineering components and used in mechanical and civil constructions.

Keywords: glass waste, epoxy resin, electrical and mechanical characteristics

---

* Corresponding Author's E-mail: habib@untag-sby.ac.id.

1. Introduction

Glass is a material naturally destroyed similar to an organic material, unless there is an obstacle in its processing. This processing is made only by recycling the glass into original materials or goods, namely glass, glass bottle, or glass sheet. In this processing, the necessary facilities and infrastructures are complex and costly because they must provide process equipment equivalent to the development of the glass industry [1].

Starting from the idea that there are some products, which from the viewpoint of raw materials can be replaced by glass, the use of glass waste as raw materials of new products can be realized. Some hypothetical analysis of some components of engineering construction can be made for glass (waste) material.

By processing of the systems, based on glass waste, raw materials can be used for the manufacture of the corresponding components of electrical constructions such as insulators, bushing, bus bar support, and fuse housing, as well as some other construction components as mechanical as building.

Due to the goods that will be obtained, that have high economic value and environmental cleanliness there is a decrease in the negative impact of glass waste. On the significant utilization of glass wastes, it has not been announced yet, so that they generally have a negative impact on the environment. The places of glass waste producers include hospitals, hotels, households, industrial and tourist areas [4].

2. Problem Formulation

2.1. Glass

Glass is a construction material manufactured from inorganic materials such as quartz sand. The nature of the glass is hygroscopic, transparent and has a melting point of 1200°C. Glass has a penetrating voltage of 100kV/mm rendering the same strong effect as a mechanical stress of 5kgf/mm^2. Due to this nature, glass wastes can be utilized for various construction purposes [3].

2.2. Processing

Glass waste processing can be performed by two methods of hot and cold influences. In this research, the processing is performed by using the cold method [2, 5].

Diagrammatically, the process of making electrical components and construction from glass-based materials is shown in Figure 1.

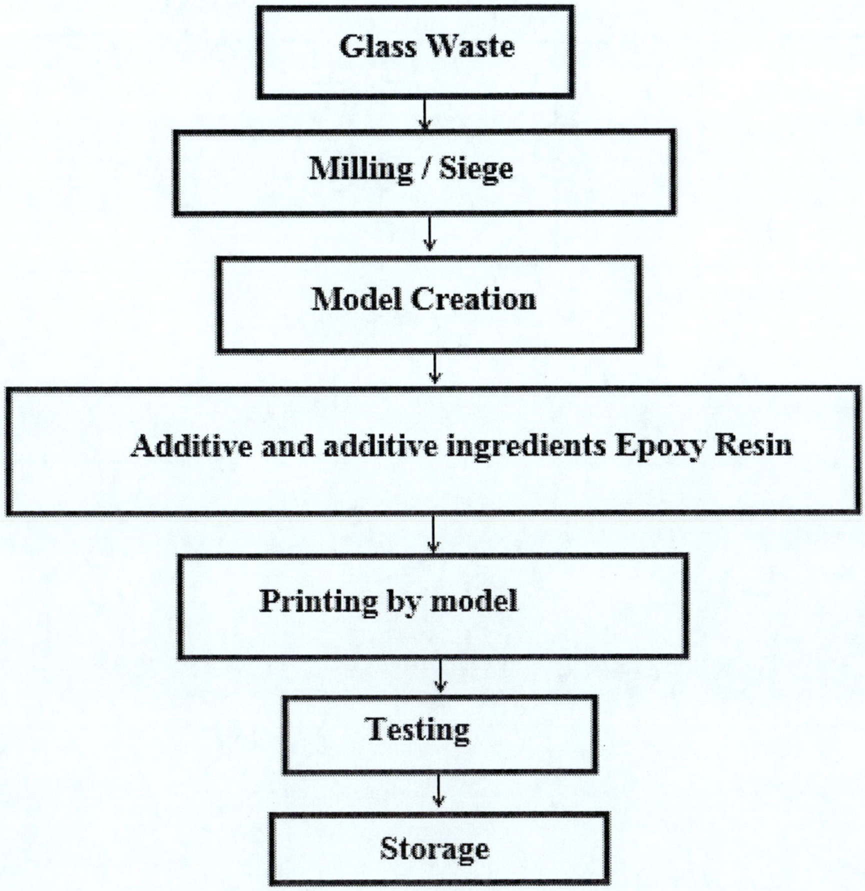

Figure 1. Flowchart of research method.

2.3. Equipment

Equipment used in this research include: (i) disc mill; (ii) cutoon screen; (iii) silicon rubber; (iv) Teflon; (v) wax.

3. Volume Fraction Optimization

3.1. Strong Insulation Testing

Testing the insulation value is defined by the circuit, presented in Figure 2.

The breakdown test results are present in Table 1. The test results, by using the circuit for measurement of insulation value (see, Figure 2), show the equivalent value of insulation resistance equal to 17kV/mm.

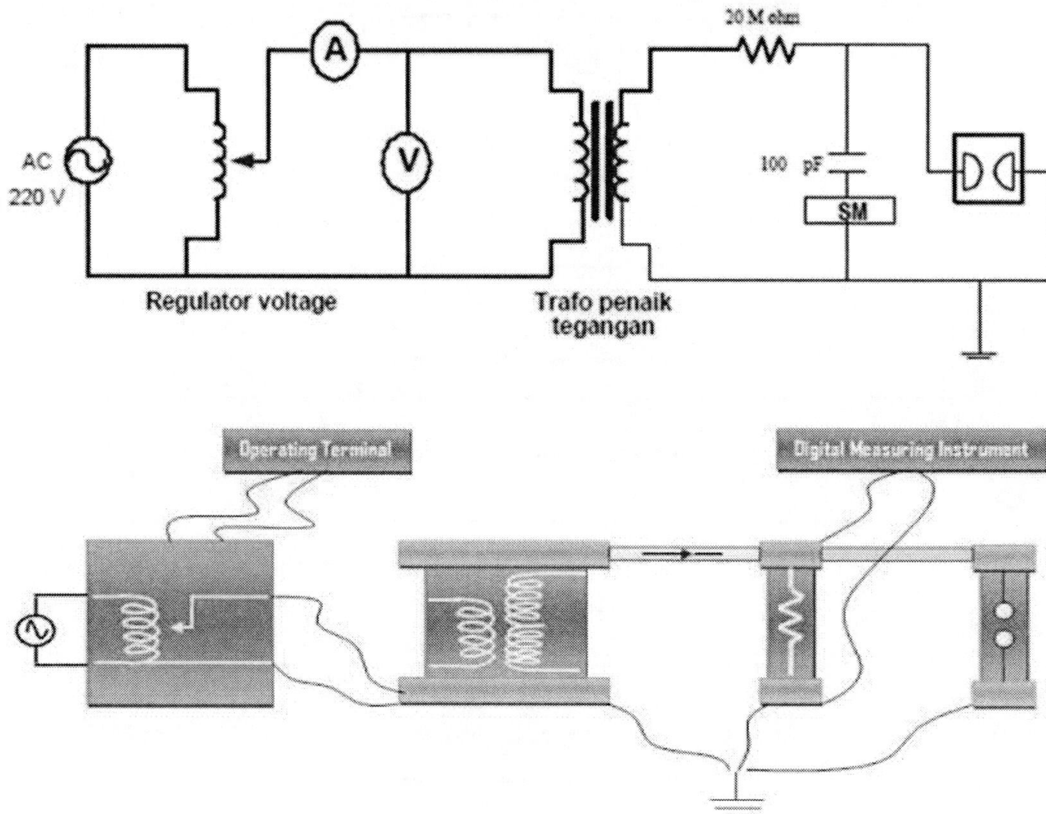

Figure 2. Circuit for measurement of insulation value.

Table 1. Breakdown test results

Test Number	Translucent Voltage (kV)	Average Translucent Voltage (kV)	Leakage Current (μA)	Dielectric Strength (kV/mm)
1	28	35.8	2	17.9
2	49		2	
3	34		2	

3.2. Mechanical Strong Testing

The mechanical test is performed by using a specimen for tensile test (see Figure 32 with a thickness of 1/8 "material).

The results of strain testing are present in Table 2 and allow us to obtain the equivalent tensile stress of 5kgf/mm^2.

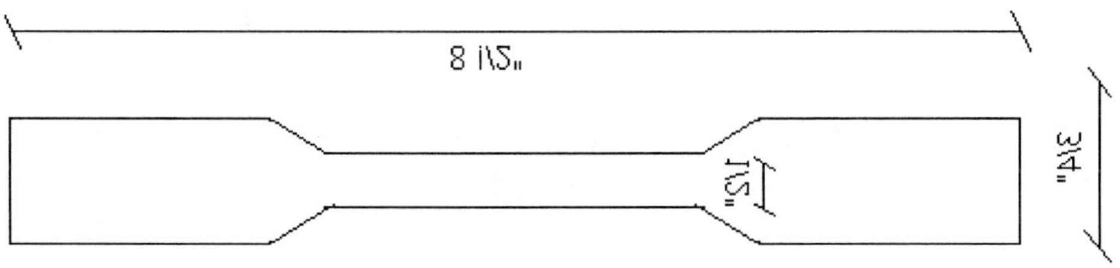

Figure 3. Sizes of sample for tensile test.

Table 2. Results of strain testing

Test Number	Maximum Tensile Stress (MPa)	Strain (%)
1	16.79	6.35
2	9.86	5.13
3	8.80	4.14

The above strain tests allow us to obtain equivalent tensile stress, therefore showing that the composition of the glass waste powder mixed with the resin provides sufficient insulation strength because it has a leakage current value of 2μA at a voltage of 48kV. So, the usefulness of this material as a medium voltage isolator is technically expedient. Moreover, in the terms of mechanical strength, the use of this material for bushing, isolator and bus bar support is also adequate.

One of the drawbacks of this material is that the influence of a direct fire during a few minutes can damage the shape of the construction. Being of use in the open air, this material is hygroscopic and does not lead to the growth of moss that is conductive.

Conclusion

From this research, it can be concluded that the glass waste transformed to powder and then mixed with epoxy resin can be used as an electrical insulator with a dielectric value of 17kV/mm and has a tensile strength of 5kgf/mm^2.

By using the mixtures of both materials by substituting wood, their application for manufacturing roofs or similar construction is appropriate because, due to these materials, we obtain termite-free construction, which is also escaped of other environmental and weather influences.

REFERENCES

[1] Fitgerald, A. E.; Djoko Achyanto, *Electric Machines*, 4th ed., 1997 (In Indonesian).
[2] Heimsat, C. *Toward an Accountable Design Process*, New York Pantheon Books, New York, 1971.
[3] EL-Wakil, M. M. *Power-Plant Technology*, McGraw-Hill, New York, 1984.
[4] Thomas F. Froumann, *Design Source and Resource*, UNR, New York. 1990.
[5] Kenneth F. Bates, *Basic Design*, The New Word Pub. New York, 1979.

In: Proceedings of the 2018 International Conference … ISBN: 978-1-53615-862-5
Editors: Ivan A. Parinov et al.

Chapter 43

THEORETICAL STUDY OF THE EXTRUSION PROCESS OF BIOMASS

D. V. Rudoy, I. A. Khozyaev, A. N. Soloviev, V. I. Pakhomov, Yu. A. Ivanov and A. S. Babadzhanyan
Don State Technical University, Rostov-on-Don, Russia

ABSTRACT

An improvement of theoretical models of screw extruders with adjustable vegetable oil processing of multicomponent mixtures with respect to temperature denaturation of proteins is present, developed on the base of thixotropic media and using the law of Oswald de Ville. The model of vegetable oil processing the protein-rich multi-component mixtures takes into account regulation parameters and limitations of denaturation processes in the conversion of protein thermal exposure.

Keywords: extrusion, biomass, thixotropic medium, Oswald de Ville power law

1. DETERMINATION OF THE PRODUCTIVITY OF EXTRUDER

The main characteristics of the screw extruder is performance and the pressure developed in the chamber within of which the mixing and forming granules in the holes of the matrix occur [1, 2].

In the extruder, it is necessary to highlight the main areas of material processing (Figure 1):

1) the loading area (flow area); the role of the loading zone consists in passing raw materials to the press, grab it, displace into the case and tighten.
2) the zone of compression (plasticizing), the densification of the product.
3) the output area (dosing), where the material becomes viscous-fluid state, and comes out of the extruder through the formed holes of the matrix.

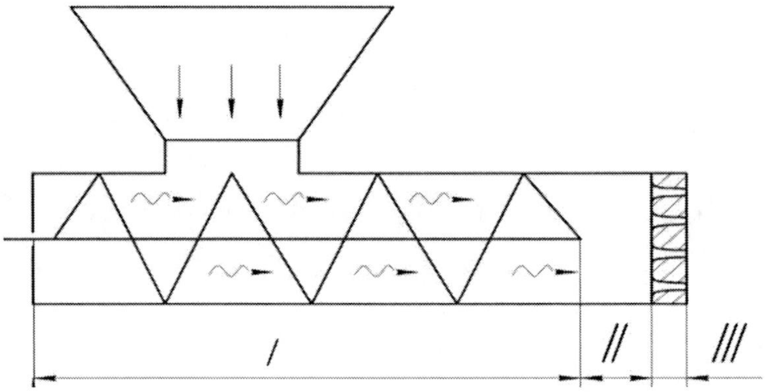

Figure 1. Stages of pressing the mixture in a screw extruder.

The geometric sizes of augers can greatly affect the output performance of the zone and depend on them to define its performance. Screw channel during rotation changes its position in respect to the cylinder, and the material moves along the screw, which can be seen as the mirror image of the screw [12]. The overall performance of the extruder is determined as (see Figure 2 [3, 4, 13])

$$Q = Q_1 - Q_2 - Q_3, \tag{1}$$

where Q_1 is the volumetric capacity of the longitudinal flow, kg/h; Q_2 is the volumetric capacity of reversed flotation, kg/h; Q_3 is the volumetric output of the flux leakage in the gap between the screw and the cylinder, kg/h.

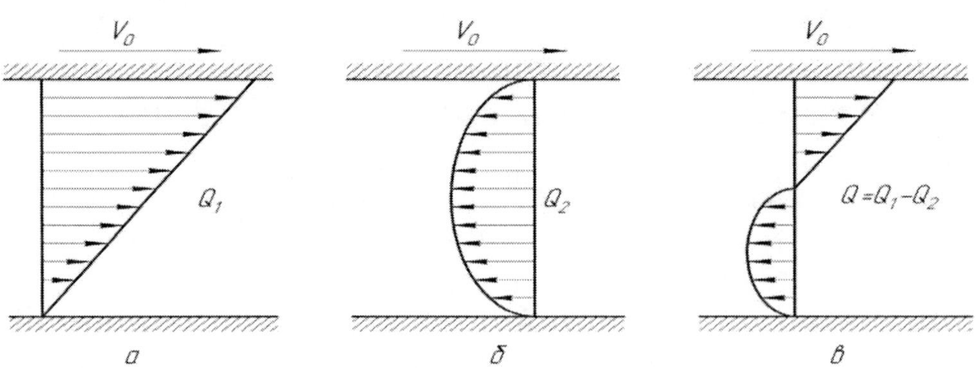

Figure 2. Formation of the resulting flow performance.

The volumetric capacity Q_1 is the result of a distribution of forces between the surfaces of the rotating screw and the stationary cylinder. We suppose that these surfaces are "wet" biomass. The volumetric c capacity Q_2, appearing along the inter-turn channel screw, is the result of the pressure drop of the mixture and such reverse flow. The volumetric capacity Q_3 is defined between the auger and cylinder. Due to the gap between the screw and the cylinder not having significance, the value of Q_3 is neglected. That is:

$$Q = Q_1 - Q_2, \tag{2}$$

The pressure dropping Δp is determined by the total pressure drop Δp_0 [5] as

$$\Delta p_0 = p_1 - p_2, \tag{3}$$

where Δp_0 is the total pressure drop, PA; p_1 is the flow pressure before entry into the holes of the matrix, PA; p_2 is the pressure at the outlet of the matrix PA.

Determine the pressure drop along the channel length corresponding to one step of the helix of the screw [5] as

$$\Delta p_1 = \frac{t \cdot (p_1 - p_2)}{L} = \frac{\pi \cdot D \cdot tg\varphi}{L} \cdot (p_1 - p_2), \tag{4}$$

where Δp_1 is the pressure drop along the channel, PA; t is the pitch helix of the screw, m; D is the diameter of the screw, m; L is the length of channel, m; φ is the length of channel, m.

Figure 3 outlines the key dimensions of the screw elements and scan coils.

In Equation (4), it is necessary to substitute the value of the differential pressure between points A and B.

It is assumed that if the pressure dropping in the channel of the screw changes in linear law, then the pressure increment Δp between points A and b can be calculated by multiplying the ratio of Δp_1 to the distances along the spiral channel.

From Figure 3 we obtain:

$$\frac{\Delta p_1}{\Delta p} = \frac{\pi \cdot D}{K}, \tag{5}$$

where $K = \pi \cdot D - A \cdot F$.

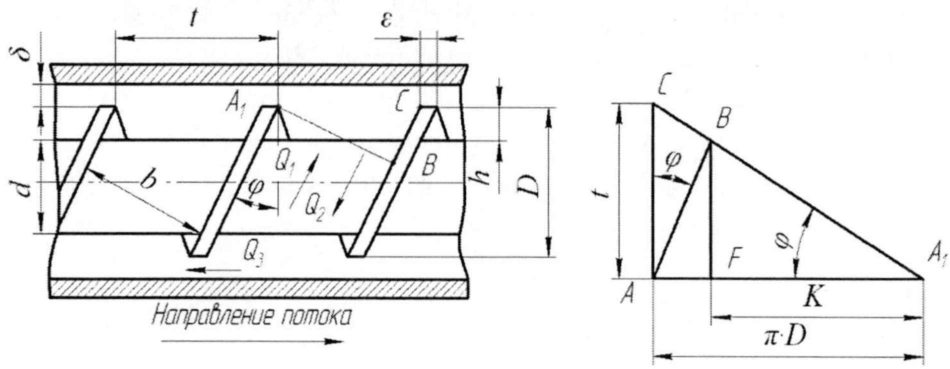

Figure 3. Designation of the dimensions of the screw elements.

Then we obtain:

$$A \cdot F = A \cdot B \cdot \sin\varphi, \tag{6}$$

where $A \cdot B = t \cdot \cos\varphi$.

Since $t = \pi \cdot D \cdot tg\varphi$, then $K = \pi \cdot D \cdot \cos^2\varphi$ and we obtain:

$$\Delta p_0 = \frac{\Delta p_1 \cdot K}{\pi \cdot D} = \Delta p_1 \cdot \cos^2\varphi, \tag{7}$$

Substituting the value of Δp_1 from equation (4) and the value of Δp_0 from equation (3) into this equation, we find the pressure difference (8) as

$$\Delta p = \frac{\pi \cdot D \cdot tg\varphi}{L} = \Delta p_0 \cdot \cos^2\varphi, \tag{8}$$

The performance of the two-screw extruder is regarded as forced flow along the channel axis of a helix of the screw; the flow is inextricably linked with the conditional reversed flotation.

Both streams are defined by the following differential equation [5]:

$$\frac{\partial^2 v}{\partial y^2} + \frac{\partial^2 v_z}{\partial x^2} = \frac{1}{\eta} \cdot \frac{\partial p}{\partial z}, \tag{9}$$

where η is the viscosity, Pa·s; x, y, z are coordinates; v is the particle velocity of material, m/s.

This is a uniform differential equation, describing the flow of viscous material in the presence of friction forces.

The flow velocity of the fluid at any point varies with the depth and position of the point along the length of the channel. Such a two-dimensional velocity distribution leads to the use of Equation (9), which is difficult to integrate, even for a particular case, because of the complex shape of the flow.

Equation (9) can be simplified, if we accept that the velocity distribution does not depend on the channel walls, i.e., for the calculation of the screw extruder, it can apply a special case as the movement of mass between infinite parallel plates. Therefore, it is possible to use a one-dimensional law of distribution of velocities [3–5].

When the ratio of channel width to depth is equal to 10 or more, and the error is not more than 10%, this error is negligible; hence, the equation of one-dimensional velocity distribution can be used for practical calculations of a screw extruder [4, 5].

The flow equation has the following differential form [5]:

$$\frac{\partial^2 v}{\partial y^2} = \frac{1}{\eta} \cdot \frac{\partial p}{\partial z}, \tag{10}$$

By integrating twice Equation (10), we obtain the velocity equation:

$$v = \frac{1}{\eta} \cdot \frac{\partial p}{\partial z} \cdot \frac{y^2}{2} + K \cdot y + K_1, \tag{11}$$

Boundary conditions to determine the constants of integration K and K_1, have the form: $v = 0$ at $y = 0$ and $y = hv = v_0$ (12). Then

$$K = \frac{v_0}{h} - \frac{1}{\eta} \cdot \frac{\partial p}{\partial z} \cdot \frac{h}{2}, \tag{12}$$

By substituting the constant of integration the speed equation reduces to:

$$v = \frac{v_0 \cdot y}{h} + \frac{(y^2 - hy)}{2 \cdot \eta} \cdot \frac{dp}{dz}, \tag{13}$$

The first term in this equation is the speed of longitudinal flow or forced flow, the second term corresponds to speed of reversed flotation.

The equation for the volume velocity obtained by integration the follows equation:

$$Q = \int_0^h b \cdot v \cdot dv, \tag{14}$$

where b is the channel width, m.

The channel width is determined as

$$b = (t - \varepsilon) \cdot \cos\varphi, \tag{15}$$

where t is the step helix, m; φ is the angle of the helix, grad; ε is the thickness of the spiral auger, m.

Performing the integration and making the substitution, we obtain:

$$Q = \frac{b \cdot h}{2} \cdot v_0 - \frac{b \cdot h^3}{12 \cdot \eta} \cdot \frac{dp}{dz}, \tag{16}$$

The first term at the right hand of the equation represents the volume velocity of the longitudinal forced flow, the second term defines the flow rate of reversed flotation.

The speed of rotation is defined as [13, 16, 17, 21, 22]

$$v = \frac{\pi \cdot D \cdot n}{60}, \tag{17}$$

and the component of the peripheral velocity is:

$$v_0 = \frac{\pi \cdot D \cdot n}{60} \cdot \cos\varphi, \tag{18}$$

By substituting the values of v_0 and b into Equation (16), we obtain [13]:

$$Q = \frac{\pi \cdot D \cdot n \cdot h \cdot (t - \varepsilon) \cdot \cos^2\varphi}{120} - \frac{h^3 \cdot (t - \varepsilon) \cdot \cos\varphi}{12 \cdot \eta} \cdot \frac{dp}{dz}, \tag{19}$$

where z is the axis along the screw channel of the worm, it replaces the longitudinal axis of the auger and mark as x;

$$dz = \frac{dx}{\sin\varphi}, \tag{20}$$

The volumetric capacity of flow in this case will be expressed as

$$Q = \frac{\pi \cdot D \cdot n \cdot h \cdot (t-\varepsilon) \cdot \cos^2\varphi}{120} - \frac{h^3 \cdot (t-\varepsilon) \cdot \sin 2\varphi}{24 \cdot \eta} \cdot \frac{dp}{dx}, \tag{21}$$

We introduce the following designations:

$$Q = \frac{\pi \cdot D \cdot n \cdot h \cdot (t-\varepsilon) \cdot \cos^2\varphi}{120} = \alpha, \tag{22}$$

$$\frac{h^3 \cdot (t-\varepsilon) \cdot \sin 2\varphi}{24 \cdot \eta} = \beta, \tag{23}$$

and define

$$Q = \alpha \cdot n - \beta \cdot \frac{dp}{dx}, \tag{24}$$

The resulting equation is solved relatively of pressure gradient and producing an integrating in both parts, we obtain:

$$(\alpha \cdot n - Q \cdot \int_0^{t_0} dx = \beta \cdot \int_{p_2}^{p_1} dp, \tag{25}$$

After integrating, we have:

$$(\alpha \cdot n - Q) \cdot L = \beta \cdot (p_1 - p_2), \tag{26}$$

Let us express α via β and by using Equations (22) and (23), we obtain the equation of the volumetric capacity of the flow [13, 16, 17, 21, 22] as

$$Q = \frac{\pi \cdot D \cdot n \cdot h \cdot (t-\varepsilon) \cdot \cos^2\varphi}{120} - \frac{h^3 \cdot (t-\varepsilon) \cdot \sin 2\varphi}{24 \cdot \eta \cdot L} \cdot \Delta p, \tag{27}$$

Formula (27) has the following dimension:

$$[\mathrm{kg}/\mathrm{h}]=[\mathrm{m}]\cdot[\min]^{-1}\cdot[\mathrm{m}]^{2}-[\mathrm{m}]^{2}\cdot\left([\mathrm{kg}]\cdot[\mathrm{m}]^{-1}\cdot[\mathrm{s}]^{-2}\cdot[\mathrm{s}]\cdot[\mathrm{m}]\right)^{-1}\cdot[\mathrm{kg}]\cdot[\mathrm{m}]^{-1}\cdot[\mathrm{s}]^{-2}, \qquad (28)$$

To check the theoretical model, we find the pressure necessary for extruding the material from the formula (27) as

$$\Delta p=\frac{\frac{\pi\cdot D\cdot n\cdot h\cdot(t-\varepsilon)\cdot\cos^{2}\varphi}{120}-Q}{\frac{h^{3}\cdot(t-\varepsilon)\cdot\sin 2\varphi}{24\cdot\eta\cdot L}}. \qquad (29)$$

2. Determination of the Productivity of Twin-Screw Extruder

We have shown earlier the advantages of the twin-screw extruder. The creation of such structures has led to a gradual convergence of the screws, and then the overlap of the screw flights. Separate design multistep extruders showed that the most effective is the design of the twin-screw extruder with screws [4, 12, 14, 15, 20].

The advantages of these extruders compared to a single screw are:

1) final products are obtained from raw materials in one pass than through the extruder;
2) high pressures of extrusion is obtained in the processing of biomass;
3) uniform output of material is extruded without pulsation, which is beneficial to the quality of the product;
4) there is additional mixing of all components of biomass.

The flow process of biomass in this type of extruders is more complex than in single-screw extruders. It has not been studied and calculations of these extruders were mainly based on empirical data. However, this problem can be solved [1, 7–11].

Figure 4 shows the layout of overlapping the augers of the twin-screw extruder. So, the extruded volume is taken by a circular area filled with ground.

The performance of the two-screw extruder is defined as [13, 16, 17, 21, 22]:

$$Q=(V_1+V_2-V_{nep})\cdot m\cdot n, \qquad (30)$$

where Q is the performance of twin-screw extruder, kg/h; V_1 is the volume, formed by the first auger, bounded on one side by the turns of the screw, and on the other hand, – by the cylindrical surfaces of the housing of the extruder and the shaft of the screw, m^3; V_2 is the volume, formed by the second auger, bounded on one side by the turns of the screw, and on the other hand, – by the cylindrical surfaces of the extruder housing and the screw shaft, m^3; V_{nep} is the amount of overlapping of the two screws, m^3; m is the number of threads of the screw; n is the frequency of rotation of the auger, s^{-1}.

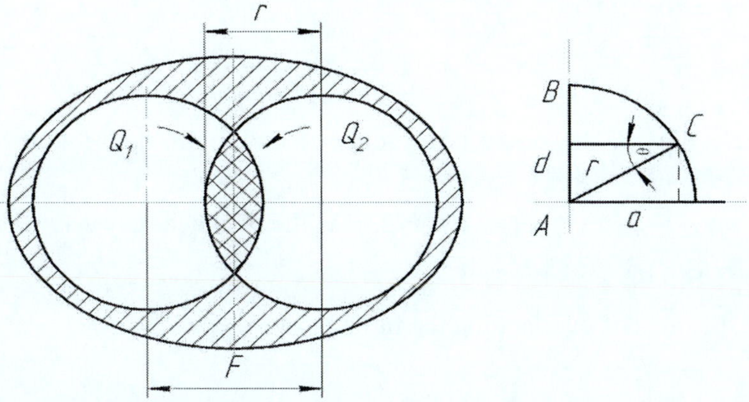

Figure 4. Design diagram of two-screw extruder.

The volume, formed by the first auger, is determined as

$$V_1 = S_{ш1} \cdot t, \tag{31}$$

where V_1 is the volume, formed by the first screw, m^3; $S_{ш1}$ is the area of the revolution of first screw, m^2; t is the distance between coils, m.

The volume, formed by the second auger, is determined as

$$V_2 = S_{ш2} \cdot t, \tag{32}$$

where V_2 is the volume, formed by the second endless screw, m^3; $S_{ш2}$ is the area of revolution of the second screw, m^2; t is the distance between coils, m.

The amount of overlapping of the two augers is determined as

$$V_{nep} = 2 \cdot S_{nep} \cdot \varepsilon, \tag{33}$$

where V_{nep} is the amount of overlapping the two screws, m^3; S_{nep} is the area of overlapping the turns of the two screws, m^2; ε is the thickness of the spiral auger, m.

Let us determine the area of overlapping the screw flights as

$$S_{nep} = 4 \cdot S_{1nep}, \tag{34}$$

where S_{1nep} is the area of overlapping one section of ceiling, m^2.

From Figure 4, we find the overlapping area of one section as

$$S_{1nep} = \frac{r^2 \cdot (\frac{\pi}{2} - \theta) - f \cdot a}{2}, \tag{35}$$

where r is the radius of the screw (the radius of the circle, describing the part of the overlapping), m; $\theta = arc\cos\frac{a}{r}$, grad; $f = F/2$, F is the distance between the centers of the circle, m; $a = \sqrt{r^2 - d^2}$.

Therefore, the area of overlapping the turns of two screws is:

$$S_{nep} = r^2 \cdot (\frac{\pi}{2} - 2\theta) - 2f \cdot a, \tag{36}$$

By substituting (31)–(33) into (30), we obtain the performance of the two-screw extruder, taking into account the overlapping area of the screws as

$$Q = (2\pi \cdot r^2 \cdot t - 2 \cdot \varepsilon \cdot (r^2 \cdot (\frac{\pi}{2} - 2\theta) - 2f \cdot a)) \cdot m \cdot n, \tag{37}$$

Then we define the shape of the holes in the die along its length to obtain a uniform pressure and maximum throughput.

3. Determining the Optimal Input Radius of One Hole

Taking into account the rheological parameters of the considered biomass and geometrical parameters of the forming matrix, we can obtain the following expression to determine the performance [11–13, 16, 17, 21, 22]:

$$Q \approx Q_M,$$

where Q is the output of the extruder, kg/h; Q_M is the performance matrix, kg/h.

$$Q_M = Q_{отв} \cdot n_{отв}, \tag{38}$$

where $Q_{отв}$ is the performance of one of the holes, kg/h; $n_{отв}$ is the number of holes in the matrix.

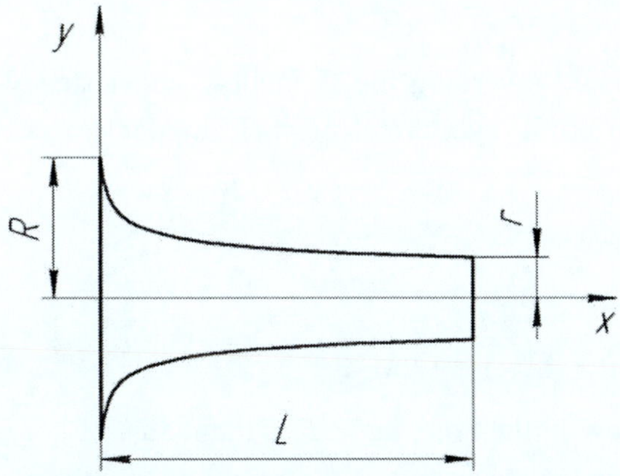

Figure 5. Cross-section of hole with maximum throughput.

It has been stated that in the process of pushing the product, as well as increasing the quality of the pellets, it is necessary to have a constant pressure in the polymer along the entire length of the hole [12, 13, 16, 17, 21, 22]. As shown by the analysis of the work of granulation, when using cylindrical, conical, etc. shapes of holes, it is not possible. However, similar problems were solved in the outflow of bulk solids from bins [23] and it was determined that the shape of the hole with the greatest bandwidth (see Figure 5) without loss of pressure corresponds to the curve equation [12]:

$$y = \frac{R}{\sqrt[4]{1 + 2 \cdot \eta^2 \cdot g \cdot \dfrac{R^4}{Q_{np}^2} \cdot x}}, \tag{39}$$

where R is the radius of the input section of the conical hole, m; g is the acceleration of gravity, m/s^2; Q_{np} is the limit consumption, kg/h; η is the viscosity, Pa·s; x is the coordinate along x-axis.

For the considered length, $x = L$, Equation (39) takes the corresponding form:

$$y = \frac{R}{\sqrt[4]{1 + 2 \cdot \eta^2 \cdot g \cdot \dfrac{R^4}{Q_{np}^2} \cdot L}}, \tag{40}$$

Based on the known small radius at the point O, at known length $y = r$, we obtain:

$$r = \frac{R}{\sqrt[4]{1 + 2 \cdot \eta^2 \cdot g \cdot \frac{R^4}{Q_{np}^2} \cdot L}}, \tag{41}$$

According to the law, governing the fluid flow under steady-state flow of viscous incompressible fluid in a thin cylindrical pipe of circular cross-section (Poiseuille law) [6], we obtain:

$$Q = \frac{\pi \cdot R^4 \cdot \Delta p}{8 \cdot \eta \cdot L}, \tag{42}$$

where Q is the fluid flow in the hole, kg/h; R is the radius of hole, m; Δp is the pressure difference at the input and at the output of the hole, PA; η is the viscosity of fluid, Pa·s; L is the hole length, m.

Since the considered mixture is the pseudoplastic non-Newtonian fluids, in order to describe the flow of biomass in the extrusion process, we used the Oswald de Ville equation of power law.

Accordingly, the viscosity, taking into account the rheological properties of biomass, takes form [3, 8, 18, 19]:

$$\eta = \mu \gamma^{n-1}, \tag{43}$$

where η is the viscosity of biomass, Pa·s; μ is the rheological parameter that characterizes the consistency of the material, kg/(m·s); γ is the strain rate, s^{-1}; n is the index of the flux.

We believe that the performance in Equation (41) is equal to the performance in (42) at the initial time t_0.

From the formula (41) for the output data and knowing the radius, we express the input radius as

$$r^4 = \frac{R^4}{1 + 2 \cdot \eta^2 \cdot g \cdot \frac{R^4}{Q_{np}^2} \cdot L}, \tag{44}$$

or we have

$$R^4 = \frac{1}{\frac{1}{r^4} - 2 \cdot \eta^2 \cdot g \cdot \frac{L}{Q_{np}^2}}, \tag{45}$$

By substituting (42) into (45), we have:

$$R^4 = \frac{1}{\frac{1}{r^4} - 2 \cdot \eta^2 \cdot g \cdot \frac{L}{\left(\frac{\pi^2 \cdot R^8 \cdot \Delta p}{8^2 \cdot \eta^2 \cdot L^2}\right)^2}}, \tag{46}$$

or we have

$$\frac{R^8}{r^4} - R^4 = \frac{128 \cdot \eta^4 \cdot L^3 \cdot g}{\Delta p^2 \cdot \pi^2}, \tag{47}$$

Taking into account that $R^4 = R_1$, we obtain from (47) the following quadratic equation:

$$R_1^2 - r^4 \cdot R_1 - \frac{128 \cdot \eta^4 \cdot L^3 \cdot r^4 \cdot g}{\Delta p^2 \cdot \pi^2} = 0, \tag{48}$$

with the solutions:

$$R_{1\,1,2} = \frac{r^4 \pm \sqrt{\frac{r^8 + 496 \cdot \eta^4 \cdot L^3 \cdot r^4 \cdot g}{\Delta p^2 \cdot \pi^2}}}{2} = 0, \tag{49}$$

Due to $R_1 > 0$, and $R = \sqrt[4]{R_1}$, we obtain the radius equation for the hole of maximum throughput as

$$R = \sqrt[4]{\frac{r^4 + \sqrt{\frac{r^8 + 496 \cdot \eta^4 \cdot L^3 \cdot r^4 \cdot g}{\Delta p^2 \cdot \pi^2}}}{2}}. \tag{50}$$

4. Determination of the Velocity of Material Particles, Forced through Hole

The compaction process of the mixture before extrusion, was investigated by several authors [3, 14, 15].

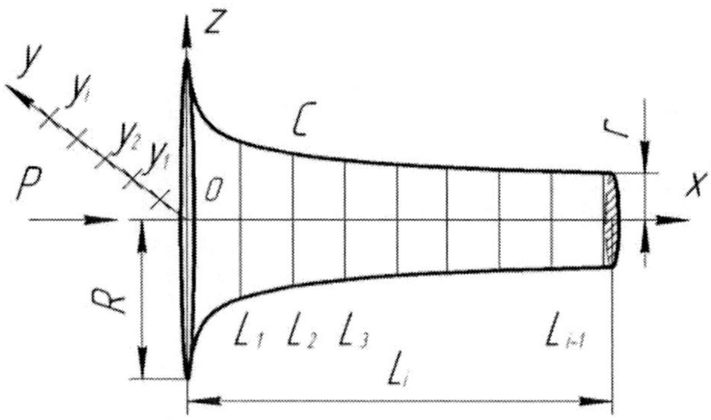

Figure 6. Diagram of holes with maximum throughput.

Let us assume that the pressure p_1, corresponding to the start of extrusion, is known, as well as density of the material ρ and the law of change of viscosity. Taking into account that the extrusion process is isometric and steady state, we use the Navie–Stokes equations [4, 5, 9]:

$$\frac{\partial p}{\partial x} = 0, \frac{\partial p}{\partial y} = 0, \tag{51}$$

$$v \cdot \frac{\partial v}{\partial z} = -\frac{1}{\rho} \cdot \frac{\partial p}{\partial z} + \frac{\eta}{\rho} \cdot (\frac{\partial^2 v}{\partial x^2} + \frac{\partial^2 v}{\partial y^2} + \frac{\partial^2 v}{\partial z^2}), \tag{52}$$

$$\frac{\partial v}{\partial z} = 0, \tag{53}$$

where v is the speed of material particles along the z-axis, m/s; p is the pressure along the axis of the material, PA.

Figure 6 shows a diagram of holes with maximum throughput. From Equation (52), it follows that $v = v(x, y)$, and from (51), we have $p = p(x)$.

The system of Equations (51) –(53) is reduced to the following equation:

$$\eta \cdot (\frac{\partial^2 v}{\partial x^2} + \frac{\partial^2 v}{\partial y^2}) = \frac{dp}{dz}, \tag{54}$$

The left part of Equation (54) is a function of x and y, and the right part is a function of z. Therefore, the left and right side of Equation (61) are separate constants. Hence:

$$\frac{dp}{dz} = const = -\frac{p_1 - p_2}{L} = -\frac{\Delta p}{L}, \tag{55}$$

where $\Delta p = p_1 - p_2$ is the pressure in the system, that is extrusion, PA.

We reduce Equation (54) to a linear equation in partial derivatives of second order in the plane Oxy:

$$\frac{\partial^2 v}{\partial x^2} + \frac{\partial^2 v}{\partial y^2} = -\frac{\Delta p}{L}, \tag{56}$$

Boundary condition for Equation (56) is the equality to zero of the speed of the material particles on the walls of the holes:

$$v|_C = 0, \quad 0 \le z \le L, \tag{57}$$

where C is the contour equation of the normal to the z-axis of the cross-section, which defines the cylindrical surface bounding the hole.

Consider a special case of the contour C. Let the contour C is a circumference in the plane Oxy:

$$x^2 + y^2 = R^2, \tag{58}$$

The solution of Equation (56), satisfying the boundary condition (64) has the form:

$$v = A \cdot (R^2 - L^2 - r^2), \tag{59}$$

Substituting Equation (59) into (56) and fine constant A:

$$v''_{xx} + v''_{yy} = \frac{\Delta p}{\eta \cdot L}, \tag{60}$$

$$-2A - 2A = \frac{\Delta p}{\eta \cdot L}, \tag{61}$$

Then we obtain:

$$A = \frac{\Delta p}{\eta \cdot L}, \tag{62}$$

So finally, the speed of the material particles is forced through hole is calculated as

$$\nu = \frac{\Delta p}{\eta \cdot L} \cdot (R^2 - L^2 - r^2), \tag{63}$$

Conclusion

The developed theory of operation of the screw extruder is based on the consideration of the extruded biomass as a thixotropic medium. The relationships between stress, strain and viscosity is described by the Oswald de Ville power law.

The developed model extruding multicomponent mixtures for a rational choice of constructive-technological parameters provide a high intensity of heat at a uniform pressure, plague bloom mode, and are limited to protein denaturation.

References

[1] Guskov, K. P., Machikhin, Yu. A. *Rheology of Food Nasses*. Food Industry, Moscow 1970, pp. 1–208 (In Russian).

[2] Kosoy, V. D., Vinogradov, Ya. I., Malyshev. A. D. *Engineering Rheology of Biotechnological Media*. GIORD, Saint-Petersburg, 2005, pp. 1–648 (In Russian).

[3] Tadmor, 3, Gogos. K. *Theoretical Foundations of Polymer Processing*. Chemistry, Moscow. 1984, pp. 1–632 (In Russian).

[4] Shenkel, G., *Screw Presses for Plastics.* A. Ya. Chapito (Ed.). Goskhimizdat, Leningrad. 1962, pp. 1–467 (In Russian).

[5] Gruzdev, I. E., Mirzoev, R. G., Yankov, V. I. *Theory of Screw Devices.* Leningrad State University Press, Leningrad. 1978, pp. 1–144 (In Russian).

[6] Fisher, E. *Plastic Extrusion.* S. I. Gdalina (Ed.). Chemistry, Moscow. 1970, pp. 1–288 (In Russian).

[7] Klemens Kohlgruber, James Lindsay White, Helmut Potente. *Co-Rotating Twin Screw Extruders*. Berghaus - Hanser Gardner, 2007, pp. 1–362.

[8] Rauwendaal, C. *Polymer Extrusion 3rd ed* / C. C. Rauwendaal, / - *Publisher. - New York*, 1994, pp. 1–568.

[9] Rauwendaal, C., Gardner, H. *Polymer Extrusion.* Hanser Publisher, Munich, 2006, 580.

[10] Sowitt, Ed. R. *Extrusion Cooking Technology.* New York, 1984, 212.

[11] Steffe, J. F. *Rheological Methods in Food Process Engineering*. 2nd ed. Freeman Press, East Lansing. 1996, 417.

[12] Rudoy, D. V., Owners, I. A. In: *Development of Innovative Technologies and Technical Means for the Agro-industrial Complex. Proc. 8th Intern. Sci.-Pract. Conf.* 2013, 145–151.

[13] Rudoy, D. V., Ore, D. V. In: *Proc. 7th Int. Sci.-Pract. Conf., 17th Int. Agricultural Exhibition "Interagromash-2014.*" 2014, 139 - 142.

[14] Rudoy, D. V. *Bulletin of Kazan State Agrarian University.* 2014, *No. 3(33)*, 95–97 (In Russian).

[15] Pakhomov, V. I., Rudoy, D. V., Tupolskikh, T. I., Soloviev, A. N., Braginets, S. V., Bakhchevnikov, O. N. *Technologies and Equipment for the Extrusion of Plant Materials.* Ministry of Education and Science of the Russian Federation, Don State Technical University. Rostov-on-Don, 2018 (In Russian).

[16] Pakhomov, V. I., Rudoy, D. V., Ivanov, Yu. A. In: *Mathematical Modeling and Biomechanics in a Modern University.* Abstracts of the XIII All-Russian School-Seminar. Southern Federal University, Don State Technical University. 2018, 66 (In Russian).

[17] Pakhomov, V. I., Braginez, S. V., Bahchevnikov, O. N., Rudoy, D. V. In: *Abstracts and Schedule of the 2017 International Conference.* 2017, 171-172.

[18] Khozyaev, I. A., Rudoy, D. V. In: *Proc. 9th Int. Sci.-Pract. Conf., 19th Int. Agricultural Exhibition "Interagromash-2016."* 2016, 107–109.

[19] Khozyaev, I. A., Rudoy, D. V. In: *Proc. 9th Int. Sci.-Pract. Conf., 19th Int. Agricultural Exhibition "Interagromash-2016."* 2016, 109–112.

[20] Rudoy, D. V. *Research of Technological Process and Determination of Rational Parameters of a Screw Extruder for the Production of Mixed Feeds.* PhD Thesis (Technical sciences). Don State Technical University. Rostov-on-Don, 2015 (In Russian).

[21] Khozyaev, I. A., Rudoy, D. V. In: *Proc. 8th Int. Sci.-Pract. Conf. "Innovative Developments for the AIC."* V. F. Khlystunov, V. B. Rykov, A. I. Pakhomov, V. S. Gazalov, S. V. Brahineu, A. P. Kushnarev (Eds.). 2013, 145–151 (In Russian).

[22] Braginets, S. V., Bakhchevnikov, O. N., Alferov, A. S. *Rural Mechanicizer*. 2018, *No. 2*, 28–29 (In Russian).

[23] Gachev, L. V. *Motion of Bulk Materials in Pipes and Bunkers.* Mashinostroenie, Moscow. 1968, pp. 1–184 (In Russian).

PART 5. INDUSTRY AND MANAGEMENT

In: Proceedings of the 2018 International Conference … ISBN: 978-1-53615-862-5
Editors: Ivan A. Parinov et al.

Chapter 44

VARIABLE SELECTION IN DATA ENVELOPMENT ANALYSIS USING STEPWISE MODELING APPROACH: A CASE STUDY OF TOURISM SECTOR IN INDONESIA

Erni Puspanantasari Putri and Danaipong Chetchotsak***
Department of Industrial Engineering, Faculty of Engineering,
Khon Kaen University, Khon Kaen, Thailand

ABSTRACT

Data Envelopment Analysis (DEA) presents itself as a non-parametric method to measure the relative effectiveness of some similar objects, as a rule called Decision Making Units (DMUs), which transforms the number of input variables to the number of output variables. One of the important research goals in DEA is the selection of available input and output factors. Besides the selection of the DEA numerical procedure, choosing input and output variables is another decisive treatment, which should be taken into account. To regret, this issue has been treated insignificantly, but the most consideration was directed on the development of models. Often in the studies, there was subjective selection of parameters, or effectiveness of calculation was based on other data, results, and criteria. In this paper, we shall use the stepwise modeling method to perform the choosing variables in DEA. This approach is directed to develop DEA models, which insert only variables with the greatest influence on the DEA results. The backward approach of the stepwise numerical procedure begins with treatment of all-possible input and output variables in the DEA model. At each step, one variable is deleted from the model due to analyses of the effectiveness factors of the DMUs. Theoretically, the numerical procedure can be continued until only one input and one output variable are present in the treatment. From a practice, ending of calculations can be considered on the base of the decision criterion to develop an optimal DEA model. In this paper, we shall illustrate this approach based on data of a case study of the touristic sector of Indonesia.

* Corresponding Author's E-mail: erniputri@untag-sby.ac.id.

Keywords: variable selection, DEA, stepwise modeling approach

1. INTRODUCTION

DEA is a technique of linear programming, related to multiple efficiency factors in an integrated model. The measurement of multiple efficiency is related to input and output variables. Input variables are the factors usually subjected to minimization, for example, costs, man-power, materials used, etc. Output variables are the factors usually subjected to maximization, for example, profit, revenue, products, etc. Input and output parameters are classified and chosen before using the DEA method [1]. The DEA method is particularly available for the researcher studying the effectiveness of transforming multiple input variables into multiple output variables [2]. In certain cases, the research could use many variables violating the rules. It will lead to a great difference between DMUs. It is necessary to delete some factors due to the existence of a correlation relationship between them [3]. By applying the DEA method, we can determine a number of the used input and output variables, which is greater than the number of DMUs. It will cause a value of DMU effectively at a great number of quantities. This result is not available [4]. One of the ways to overcome this problem consists in increasing the number of DMUs. This method is not practical due to the restricted number of DMUs. Another possible way to resolve this problem consists in reducing the number of input and output variables [5]. This paper will illustrate a case study of a classification of hotels in Indonesia in the Indonesian tourist sector in order to choose the variables in the DEA method by using the Stepwise Modeling Approach.

2. METHODS

2.1. Data Envelopment Analysis (DEA) Method

The Data envelopment analysis (DEA) method was developed by Charnes, Cooper and Rhodes (CCR) (1978) on the base of Farrell ideas, consisting in the evaluation of technical effectiveness and effective frontiers [6]. DEA is a "data-oriented" method to estimate the performance of a set of peer objects called Decision Making Units (DMUs), which transform multiple input variables into multiple output variables. The determination of a DMU is universal and versatile. In the last year, there were a great number of DEA applications used for the estimation of various types of objects, relating to different activities in various contexts, and touching many different countries. The DEA applications have used different DMUs to estimate the performance of various objects, namely, hospitals, universities, towns, regions, industrial enterprises, and so on, including countries, states, etc. [7].

In particular, the DEA is applied when the researcher studies the effectiveness of transforming multiple input variables into multiple output variables. Charnes et al. presented DEA as a model of mathematical programming based and used with data of observations. This demonstrates a new method to obtain empirical evaluations of extremal relationships, namely the production functions and/or effective surfaces, and describe production possibilities, which are the original points of modern economics [2]. As an effective frontier technique, the DEA method defines the ineffectiveness of a specific DMU by comparing it with certain DMUs treated as effective [8]. By using mathematical programming, the DEA method states a weighting system, allowing input and output variables, each of which should be cumulated and have the effectiveness factors computed. It is required as more than one set of weights. Rather, the DEA method, at repeating solutions, determines a set of weights for every DMU. The weights are those that are most favorable for the considered unit. They state for it the highest effectiveness estimation, if weights are not negative and when these weights, applied to any unit, do not lead to an effectiveness factors exceeding 1.0 (in a range from zero to one with 1.0 indicating an effective DMU) [9, 10].

2.2. Variable Selection in DEA Method

The DEA method uses multiple input and output variables for analysing the effectiveness, but it does not provide control in the choosing of these variables. The researchers tend to restrict themselves by applying some methods. If the number of variables used is very great, it will directly influence on the values of effectiveness. It will reduce the values of the power effectiveness, which will cause all effective values of DMU. Restricted data of DMU will cause an unreasonable, very great number of variables. It will directly influence their application in computational procedure. In favorable conditions, the number of DMUs should be three times greater compared to the number of input and output parameters [11, 12].

In the choosing variables, DEA method does not state rules. It depends on the estimation, policy, and user experience and qualification. Therefore, all researchers will have various opinions. In the choosing variables, some of the problems arisen are: (i) data are not appropriate; (ii) high dimensions of sampling in the production process; and (iii) input and output data are improper [13].

Choosing the appropriate input and output variables is an interesting research task of the DEA method. In the processing, several arising problems are the following: (i) a procedure used for the choosing; (ii) the correlation analysis; and (iii) the classification input and output variables. Charnes et al. in their seminal work showed that effectiveness is the ratio of output to input weighted variables. Other researchers developed this approach further. Choosing the variables is a central point for any formulation [14]. Choosing the input and output variables is another important treatment, which should be taken into account by the researchers together with the technique of choosing the DEA model [15].

2.3. Stepwise Modeling Approach

The fundamental stepwise procedure in the Stepwise Modeling Approach will use the backward approach in modelling DEA parameters. The approach begins with treatment of existing input and output parameters of the DEA method. At every stage, one of the variables is deleted from the numerical procedure by analysing the effectiveness factors of DMUs. Theoretically, the numerical procedure can be continued until only one input and one output parameters are present in the treatment. Due to practice requirements, the ending of calculations can be considered on the base of the decision criterion to develop an optimal DEA model. The stages of the backwards approach are as follows:

(i) *Start Stage.*

Perform a DEA analysis including the total set of j input parameters and k output parameters. Fix the effectiveness factors for every of DMUs in this treatment (set E^*).

(ii) *Stage 1.*

Perform a series of DEA procedures $i = 1,. . ., j + k$, deleting one of input variables and one of output variables at a moment of every treatment. For every analysis:

a. Fix the effectiveness factors for every of DMUs (set $E_{1,i}$) for each of i-treatments.
b. Compute, for every of DMUs, the differences in the corresponding DMU effectiveness factors ($E^* - E_{1,i}$).
c. Compute the mean difference in values of effectiveness (based on series of i differences).

Select one input or output parameter, which is deleted by choosing the parameter with the minimal mean difference among the above-mentioned effectiveness factors. At least one input and one output parameters should remain in the analysis. If only one input or output parameter is remained at this stage, then this parameter cannot be deleted and another parameter must be considered based on the numerical procedure of choosing. For the parameter deleted, mark the DEA results as E_1^*; E_1^* is based on the effectiveness factors of the DMUs for the keeping of input and output parameters.

(iii) *Stage n + 1.*

Repeat every stage by treating a series of DEA procedures $i = 1, \ldots, j + k - n$. With the kept $j + k - n$ input and output parameters, we compare results $E_{n+1,i}$ and E_n* (the effectiveness factors at the previous stage) and select the parameter, which is deleted on the base of minimal mean difference in effectiveness factors.

(iv) *End Stage.*

The computational procedure is finished, when the model has only one input and one output parameters [16].

3. Results and Discussion

Table 1. Data of number of accommodation, average worker, visitor per day, average duration of stay of guests and room occupancy rate for classified hotels by province in 2016

No.	Province	X1	X2	X3	X4	X5	Y1	Y2	Y3	Y4	Y5
1	Aceh	20	1,503	2,579	47.6	0.6	704	45	1.99	1.67	46.03
2	Sumatera Utara	116	10 732	16 418	79.5	0.9	4,921	683	2.03	1.52	54.19
3	Sumatera Barat	61	3,653	6,015	44.6	0.7	1,985	397	1.73	1.45	54.32
4	R i a u	58	5,206	7,832	63.9	0.7	2,503	88	3.29	1.66	48.51
5	J a m b i	31	2,187	2,885	51.5	0.7	777	6	3.22	1.84	48.20
6	Sumatera Selatan	64	4,781	6,800	57.3	0.8	3,520	53	3.34	1.70	52.81
7	Bengkulu	9	523	786	47.2	0.8	307	3	2.36	1.49	60.45
8	Lampung	18	1,954	2,847	88.3	0.8	2,705	58	3.48	1.67	50.43
9	Kep Bangka Belitung	42	2,528	3,653	41.5	0.7	1,034	13	2.15	1.79	39.97
10	Kepulauan Riau	88	11 393	15,836	107	0.8	3,795	3 458	1.96	1.69	49.41
11	DKI Jakarta	232	39 806	55,293	161.4	0.9	29,116	7 755	2.26	1.75	57.89
12	Jawa Barat	314	29 053	46,013	73.1	0.8	16,144	1 951	3.09	1.71	55.14
13	Jawa Tengah	217	17 246	26,878	55.9	0.7	14,322	440	2.56	1.41	46.76
14	DI Yogyakarta	94	9,256	14,432	77.8	0.8	8,419	649	2.06	1.56	58.55
15	Jawa Timur	158	17,014	26,393	90.3	0.8	10,609	1 536	2.97	1.63	59.57
16	Banten	58	5,939	8,844	87.6	0.9	3,352	628	1.87	1.56	53.88
17	B a l i	317	33,599	49,549	140.9	1.3	7,872	18,760	3.17	2.34	61.74
18	Nusa Tenggara Barat	75	4,525	6,721	69.3	1.1	1,782	1 162	2.02	1.67	42.75
19	Nusa Tenggara Timur	23	1,646	2,652	59.9	0.8	693	102	2.92	1.97	53.65
20	Kalimantan Barat	36	3,821	5,264	69.5	0.7	2,249	101	2.79	1.64	51.11
21	Kalimantan Tengah	12	1,051	1,548	74.3	0.8	543	14	2.31	1.49	55.76
22	Kalimantan Selatan	49	3,584	5,254	61.1	0.8	2 516	81	2.51	1.49	43.10
23	Kalimantan Timur	51	5,468	8,048	88.1	0.8	4 755	147	2.80	1.57	50.06
24	Kalimantan Utara	10	461	710	41.9	0.9	236	6	1.49	1.49	41.31
25	Sulawesi Utara	35	2,576	3 752	61.9	0.8	1 006	245	3.77	2.03	71.12
26	Sulawesi Tengah	8	766	1 182	87.3	0.9	396	3	2.28	1.93	59.07
27	Sulawesi Selatan	93	6,475	9 420	51.4	0.7	5 744	200	2.85	1.57	44.68
28	Sulawesi Tenggara	18	1,295	1 959	56.1	0.8	796	19	2.48	1.65	40.43
29	Gorontalo	8	471	750	39.5	0.7	166	6	2.14	1.31	47.64
30	Sulawesi Barat	7	374	587	37.7	0.7	157	-	4.41	1.54	45.71
31	M a l u k u	22	1,110	1,566	36.8	0.7	371	18	3.24	1.97	37.50
32	Maluku Utara	3	313	492	88.7	0.8	86	1	2.53	2.39	45.86
33	Papua Barat	15	1,014	1,477	51.2	0.8	435	12	2.23	2.18	49.95
34	Papua	25	1,684	2,524	59.6	0.9	1,464	48	3.49	1.91	60.80

3.1. Input and Output Data

The data, used in this study, are the data of classified hotels in Indonesia by a province in 2016 as shown in Table 1 [17]. The data consist of five types as follows: (i) number of accommodation; (ii) average worker; (iii) visitor per day; (iv) average duration of stay of guests; and (v) room occupancy rate. Based on these types, it can then be determined the input and output variables as seen in Table 2.

Table 2. Input and output variables

Input Variable (Xi)		Output Variable (Yi)	
X1	Number of Accommodations for Establishment	Y1	Number of Domestic Guests per Day
X2	Number of Accommodations for Room	Y2	Number of Foreign Guests per Day
X3	Number of Accommodations for Bed	Y3	Average Duration of Stay of Indonesia Guests
X4	Average Worker per Establishment	Y4	Average Duration of Stay of Foreign
X5	Average Worker per Room	Y5	Room Occupancy Rate

3.2. Processing of Input and Output Data

The Input Oriented DEA Envelopment Method is used in this research to calculate the efficiency score for each DMU. The standardization of data is shown in Table 3. There are two input and one output data standardized by percentage, namely: (i) number of accommodations for room (X2); (ii) number of accommodations for bed (X3); and (iii) number of domestic guests per day (Y1).

Table 3. Standardization of data

DMU	Province	X2	X3	Y1	DMU	Province	X2	X3	Y1
DMU1	Aceh	0.65	0.74	0.52	DMU18	Nusa Tenggara Barat	1.94	1.94	1.32
DMU2	Sumatera Utara	4.61	4.73	3.63	DMU19	Nusa Tenggara Timur	0.71	0.76	0.51
DMU3	Sumatera Barat	1.57	1.73	1.47	DMU20	Kalimantan Barat	1.64	1.52	1.66
DMU4	R i a u	2.23	2.26	1.85	DMU21	Kalimantan Tengah	0.45	0.45	0.40
DMU5	J a m b i	0.94	0.83	0.57	DMU22	Kalimantan Selatan	1.54	1.51	1.86
DMU6	Sumatera Selatan	2.05	1.96	2.60	DMU23	Kalimantan Timur	2.35	2.32	3.51
DMU7	Bengkulu	0.22	0.23	0.23	DMU24	Kalimantan Utara	0.20	0.20	0.17
DMU8	Lampung	0.84	0.82	2.00	DMU25	Sulawesi Utara	1.11	1.08	0.74
DMU9	Kep Bangka Belitung	1.08	1.05	0.76	DMU26	Sulawesi Tengah	0.33	0.34	0.29
DMU10	Kepulauan Riau	4.89	4.56	2.80	DMU27	Sulawesi Selatan	2.78	2.72	4.24
DMU11	DKI Jakarta	17.08	15.94	21.49	DMU28	Sulawesi Tenggara	0.56	0.56	0.59
DMU12	Jawa Barat	12.47	13.26	11.92	DMU29	Gorontalo	0.20	0.22	0.12
DMU13	Jawa Tengah	7.40	7.75	10.57	DMU30	Sulawesi Barat	0.16	0.17	0.12
DMU14	DI Yogyakarta	3.97	4.16	6.21	DMU31	M a l u k u	0.48	0.45	0.27
DMU15	Jawa Timur	7.30	7.61	7.83	DMU32	Maluku Utara	0.13	0.14	0.06
DMU16	Banten	2.55	2.55	2.47	DMU33	Papua Barat	0.44	0.43	0.32
DMU17	B a l i	14.42	14.28	5.81	DMU34	Papua	0.72	0.73	1.08

Table 4. Stepwise process and efficiency score for classified hotels by province in 2016

	Start	Step1	Step2	Step3	Step4	Step5	Step6	Step7	END
Remaining inputs	X1 X2 X3 X4 X5	X2 X3 X4 X5	X2 X3 X4 X5	X3 X4 X5	X3 X4 X5	X4 X5	X4 X5	X5	X5
Remaining outputs	Y1 Y2 Y3 Y4 Y5	Y1 Y2 Y3 Y4 Y5	Y2 Y3 Y4 Y5	Y2 Y3 Y4 Y5	Y3 Y4 Y5	Y3 Y4 Y5	Y4 Y5	Y4 Y5	Y5
Variable Dropped	X1	Y1	X2	Y2	X3	Y3	X4	Y4	
DMUs	Efficiency Scores								
DMU1	1	1	1	1	1	1	1	1	1
DMU2	0.7713	0.7713	0.7562	0.7562	0.7389	0.7389	0.7389	0.7389	0.7389
DMU3	1	1	1	1	1	1	1	0.9515	0.9515
DMU4	0.9782	0.9782	0.9574	0.9574	0.9558	0.9558	0.8854	0.8854	0.8854
DMU5	0.9971	0.9971	0.9947	0.9947	0.9947	0.9947	0.9552	0.9371	0.8819
DMU6	0.9232	0.9326	0.8632	0.8632	0.8632	0.8632	0.8480	0.8176	0.8176
DMU7	1	1	1	1	1	1	1	0.8937	0.8937
DMU8	1	1	0.8557	0.8557	0.8557	0.8557	0.7938	0.7938	0.7938
DMU9	0.9668	0.9668	0.9599	0.9599	0.9599	0.9599	0.9599	0.9048	0.8571
DMU10	0.9131	0.9110	0.9110	0.9110	0.7837	0.7837	0.7837	0.7837	0.7837
DMU11	1	1	1	1	0.7717	0.7717	0.7717	0.7717	0.7717
DMU12	1	1	0.9347	0.9347	0.8646	0.8646	0.8408	0.8408	0.8408
DMU13	1	1	0.9137	0.9137	0.8965	0.8965	0.8655	0.8655	0.8655
DMU14	1	1	0.8918	0.8918	0.8748	0.8748	0.8748	0.8748	0.8748
DMU15	0.9712	0.9712	0.9381	0.9381	0.8859	0.8859	0.8849	0.8849	0.8849
DMU16	0.7902	0.7902	0.7518	0.7518	0.7362	0.7362	0.7362	0.7362	0.7362
DMU17	1	1	1	1	1	1	1	1	0.5579
DMU18	0.7736	0.7736	0.6537	0.6537	0.5985	0.5985	0.5985	0.5455	0.5455
DMU19	0.9310	0.9310	0.9214	0.9143	0.9137	0.9098	0.9098	0.8924	0.8259
DMU20	0.9584	0.9616	0.9424	0.9424	0.9414	0.9414	0.9150	0.9150	0.9150
DMU21	0.9447	0.9259	0.9173	0.9173	0.9173	0.8470	0.8470	0.8470	0.8470
DMU22	0.8437	0.8496	0.7788	0.7788	0.7769	0.7769	0.7620	0.7500	0.7500
DMU23	0.9467	0.9352	0.8206	0.8206	0.8177	0.8177	0.7902	0.7902	0.7902
DMU24	0.9353	0.9388	0.8996	0.8996	0.8986	0.8883	0.8883	0.6667	0.6667
DMU25	1	1	1	1	1	1	1	1	1
DMU26	1	1	1	1	1	0.8047	0.8047	0.8047	0.7822
DMU27	1	1	0.9164	0.9164	0.9079	0.9079	0.8853	0.8571	0.8571
DMU28	0.8573	0.8625	0.8174	0.8174	0.8174	0.7893	0.7888	0.7500	0.7500
DMU29	1	1	1	1	1	1	1	0.8755	0.8755
DMU30	1	1	1	1	1	1	1	0.8571	0.8571
DMU31	1	1	1	1	1	1	1	0.9762	0.8571
DMU32	1	1	1	1	1	1	1	1	0.7500
DMU33	1	1	1	1	1	1	1	0.9471	0.7891
DMU34	1	1	0.9381	0.9265	0.9265	0.8878	0.8873	0.8062	0.7975

Table 5. Results of stepwise process

	Start	Step1	Step2	Step3	Step4	Step5	Step6	Step7	END
1. Number of DMU Efficient	18	18	12	12	11	9	9	4	2
2. Total of Efficiency Score	32.5016	32.4964	31.3338	31.3152	30.6975	30.3483	30.0132	28.9609	27.7911
3. Average Efficiency Score	0.9559	0.9558	0.9216	0.9210	0.9029	0.8926	0.8827	0.8518	0.8174
4. Average Change in Efficiency Score (ACES)		0.0002	0.0342	0.0005	0.0182	0.0103	0.0099	0.0310	0.0344

The process of Stepwise Modeling Approach for input and output variables are shown in Table 4. In the beginning Step (Start) there are original variables. There are five input variables (X1, X2, X3, X4, X5) and five output variables (Y1, Y2, Y3, Y4, Y5). In Step 1, X1 is a variable dropped; therefore, the remaining variables are four input variables (X2, X3, X4, X5) and five output variables (Y1, Y2, Y3, Y4, Y5). In the same way, Step 2 until Step 7 will also drop one input variable or one output variable. The END of this step had the remaining variables of input (X5) and output (Y5). Therefore, X5 is the selected input variable and Y5 is the selected output variable. The efficiency scores for every step were variants. It had an impact on the number of DMU efficient and inefficient.

The results of the stepwise process are shown in Table 5. Based on the number of DMUs efficient for each step are as follows: Step Start (18 DMUs), Step1 (18 DMUs), Step2 (12 DMUs), Step3 (12 DMUs), Step4 (11 DMUs), Step5 (9 DMUs), Step6 (9 DMUs), Step7 (4 DMUs) and Step END (2 DMUs). The number of DMUs inefficient for each step are as follows: Step Start (16 DMUs), Step1 (16 DMUs), Step2 (22 DMUs), Step3 (22 DMUs), Step4 (23 DMUs), Step5 (25 DMUs), Step6 (25 DMUs), Step7 (30 DMUs) and Step END (32 DMUs). Based on Total of Efficiency Score and Average Efficiency Score, Step Start has the biggest value, 32.5016 and 0.9559, respectively. The smallest number is at Step END, 27.7911 and 0.8174, respectively. Based on the Average Change in Efficiency Score (ACES), for Step 1, eliminating the input variable X1 provides the smallest ACE (0.0002) and is eliminated from the model.

18 DMUs are effective. In Stage 2, where there are 4 input and 5 output parameters, the output variable Y1 is eliminated with ACES (0.0342) producing in 12 efficient DMUs. In Stage 3, where there are 4 input and 4 output parameters, the input variable X2 is eliminated with ACES (0.0005) producing in 12 efficient DMUs. In Stage 4, where there are 3 input and 4 output parameters, the output variable Y2 is eliminated with ACES (0.0182) producing in 11 efficient DMUs. In Stage 5, where there are 3 input and 3 output parameters, the input variable X3 is eliminated with ACES (0.0103) producing in 9 efficient DMUs. In Step 6, where there are 2 input and 3 output parameters, the input

variable Y3 is eliminated with ACES (0.0099) producing in 9 efficient DMUs. In Step 7, where there are 2 input and 2 output parameters, the input variable X4 is eliminated with ACES (0.0310) producing in 4 efficient DMUs. In Step END, where there are 1 input and 2 output parameters, the input variable Y4 is eliminated with ACES (0.0344) producing in 2 efficient DMUs.

CONCLUSION

The results of variable selection in Data Envelopment Analysis using the Stepwise Modeling Approach are as follows: (i) selected variables are X5 (selected input variable) and Y5 (selected output variable); (ii) number of DMUs efficient is 2 DMUs and inefficient is 32 DMUs; (iii) total of efficiency score is 27.7911; (iv) average efficiency score is 0.8174; and (v) average change in efficiency score (ACES) is 0.0344.

ACKNOWLEDGMENTS

This work was supported by Research Unit on System Modeling for Industry (Grant No. SMI.KKU 572001/2558) Khon Kaen University, Thailand and International Relation Division of Khon Kaen University, Thailand that had provided first author the University Scholarship for ASEAN and GMS Countries' Personnel of Academic Year 2015.

REFERENCES

[1] Cook, W. D.; Zhu, J., *Data Envelopment Analysis: Modeling Operational Processes and Measuring Productivity*, 2008, 19-86.

[2] Adler, N.; Friedman, L.; Sinuany-Stern, Z., *European Journal of Operational Research*, 2002, *vol.140*, 249-265.

[3] Cooper, W.; Seiford, L.M.; Zhu, J., *Handbook on Data Envelopment Analysis*. London, Kluwer Academic Publishers, 2004.

[4] Cooper, W.W.; Seiford, L.M.; Tone, K., *Introduction to Data Envelopment Analysis and its Uses with DEA-Solver Software and References*. 2nd ed. USA, Springer, 2007.

[5] Tolooa, M.; Babaeeb, S., *Applied Mathematics and Computation*. 2015, *vol 270*, 527–533.

[6] Yun, Y. B.; Nakayama, H.; Tanino, T., *European Journal of Operational Research*, 2004, *vol.157*, 87-105.

[7] Cooper, W. W.; Seiford, L. M.; Zhu, J., *Handbook on Data Envelopment Analysis*, Springer, 2011.

[8] Avkiran, N. K., *Socio-Economic Planning Sciences*, 2001, vol. 35, 57-80.

[9] Memillan, M. L.; Datta, D., *Analyse de Politiques*, 1998, vol. 24(4), 487-511.

[10] Goksen, Y.; Dogan, O.; Ozkarabacak, B., *Procedia Economics and Finance*, 2015, vol. 19, 226-237.

[11] Madhanagopal, R.; Chandrasekaran, R., *International Journal of Data Envelopment Analysis and Operations Research*, 2014, vol. 1(2), 28-33.

[12] Dyson, R. G.; Allen, R.; Camanho, A. S.; Podinovski, V. V.; Sarrico, C. S.; Shale, E.A., *European Journal of Operational Research*, 2001, vol. 132(2), 245-259.

[13] Nataraja, N. R.; Johnson, A.L., *European Journal of Operational Research*. 2011, vol. 215, 662–669.

[14] Luo, Y.; Bi, G.; Liang, L., *Expert Systems with Applications*. 2012, vol. 39, 1118–1123.

[15] Paradi, J. C.; Vela, S.; Yang, Z. J., *Assessing Bank and Bank Branch Performance – Modeling Considerations and Approaches*. 2004.

[16] Wagner, J. M.; Shimshak, D. G., *European Journal of Operational Research,* 2007, vol. 180, 57–67.

[17] *Statistics Indonesia, 2018, Tourism,* https://www.bps.go.id/subject/16/pariwisata.html#subjekViewTab3.

In: Proceedings of the 2018 International Conference … ISBN: 978-1-53615-862-5
Editors: Ivan A. Parinov et al.

Chapter 45

DEVELOPMENT OF CREATIVE INDUSTRY IN INDONESIA BASED ON LOCAL CULTURE

Erni Puspanantasari Putri[1,2,*] and Muslimin Abdulrahim[2]
[1]Department of Industrial Engineering, Khon Kaen University (KKU), Khon Kaen, Thailand
[2]Department of Industrial Engineering, University of 17 Agustus 1945 (UNTAG), Surabaya, Indonesia

ABSTRACT

The purpose of this research is to study the development of the creative industry in Indonesia based on local culture. Indonesia has a diverse cultural heritage. It includes aspects of art, beauty, social, empathy, ceremony and others. It signifies the high creativity of Indonesian people in their specific skills and talents. Cultural diversity is also supported by ethnic diversity in Indonesian society. It shows that Indonesia has a very strong supporting factor in developing a creative economy [2]. Culture should be the basis for the development of creative industries. The uniqueness of each region must be maintained. Therefore, local culture-based potentials need to manifest themselves into the creative economy program. This is where the creative economy can begin. The creative industry developed is in line with the potential resources available in the local region. The creative economy is seen not only in the economic context, but also in the cultural dimension. The results of this research conclude that Indonesia has great potential to develop creative industries. This can be seen from two things, namely: the economic potential of tourism and a culture with high artistic taste. This shows that Indonesia has wide creativity. It is a major capital in the development of creative industries. Indonesia's creative economy can be more developed if combined with tourism. The combination can bring the centers of creative economy as a tourist attraction. The role of the creative industry will have a major impact on economic growth. Increased growth of micro, small

* Corresponding author: E-mail: erniputri@untag-sby.ac.id.

and medium enterprises (MSMEs) affect the establishment of several cities that are central to the creative industry. The city will be a potential area for the growth of industrial and tourism centers. The city will also contribute to the advancement of economic development.

Keywords: development, creative industry, local culture

1. Introduction

The fourth industrial revolution has begun. World economists predict that the global economy will increasingly depend on the creative industry sector. The global community's dependence on information technology led to an exponential growth of the creative industries, which include computer software, film, music, publications, entertainment, and fashion [1].

The development of the creative economy in each country built competence in accordance with existing capabilities in the country. There are three directions of creative industry development that focus more on the industry, based on: (i) creative and cultural business field as creative cultural industry; (ii) creative industry and (iii) intellectual property rights as copyright industry [2].

The creative economy industry in Indonesia has experienced exponential growth in the last three years. Based on national data, the creative industry sector has absorbed 15.9 million workers with a 7.3% contribution to gross domestic product (GDP) equivalent to IDR 852 trillion. On a global scale, the economic value of the creative industry goes even beyond the petroleum industry. Based on data released by the United Nations Trade and Development Conference (UNCTAD) in 2012, the creative industry contributes USD 2.2 trillion or 230% more than the value of OPEC oil exports. The creative economy has also become one of the important pillars of the Indonesian economy. According to BPS data on the creative economy in 2015, the creative economy has absorbed 15.9 million workers (13.90%) and has an export value of US $19.4 billion (12.88%) [1]. The creative industry must be improved optimally and continuously. Therefore, the creative industry can develop well and compete with other economic players.

2. Methods

2.1. Definition of Creative Economy and Creative Industry

The creative economy is a concept that puts creativity and knowledge as the main asset in driving the economy. This concept has sparked the interest of various countries to study the creative economy and make the creative economy the main model of economic development. The term creative economy has begun to be known globally since the publication of the book "The Creative Economy: How People Make Money from Ideas" in 2001 by John Howkins. The definitions of economy and creative industries in Indonesia, based on the Blueprint of National Creative Economy Development 2009 – 2015 (2008), are as follows:

(i) creative economy is defined as a new era of economics after the agricultural economy, industrial economy, and information economy, which intensify information and creativity by relying on the idea and knowledge of human resources as the main factor of production in its economic activity;
(ii) creative industry is defined as an industry, derived from the utilization of creativity, skills and talents of individuals to create welfare and employment through the creation and utilization of creative power and creativity of the individual [3].

2.2. Classification of Creative Industries

The creative economy has begun to develop since the launch of the Indonesia Design Power program in 2006. The program is increasingly intensified by Presidential Instruction No. 6 of 2009 on Creative Economy Development Year 2009 – 2015. This determination is taken to create jobs and reduce poverty. Thus, the creative industry in Indonesia is an industry derived from the utilization of creativity, skills and individual talents to create welfare and employment through the creation and utilization of creative power and creativity of the individual. The profile of the creative industry is grouped based on four measurement indicators, namely: (i) the value of gross domestic product, (ii) employment, (iii) corporate activity and (iv) impacts on other sectors. Based on this, the creative industries are classified into 14 creative industry sub-sectors, including: advertising, architecture, market and art, craft, design, fashion, film, video, photography, interactive games, music, performing arts, publishing and printing, computers and software, television and radio, as well as research and development [4].

2.3. Creative Industry Based on Local Culture

Indonesia has cultural diversity and uniqueness in each region. Culture should be the base for the development of creative industries. The uniqueness of each region must be maintained. Therefore, local culture-based potentials need to manifest themselves into the creative economy program. This is where the creative economy can begin. The creative industry developed is in line with the potential resources available in the local region. The creative economy is seen not only in the economic context, but also in the cultural dimension. Creative ideas that arise are cultural products. Cultural strategy is crucial to the direction of creative economic development. Creative industries that develop well can become the identity and pride of the nation. A well-developed creative industry will produce two positive effects as follows: (i) it can increase gross domestic product (GDP) and (ii) create employment so that unemployment and poverty can be gradually overcome [5].

2.4. The Key to Success in the Development of the Creative Industry

The creative industry sector can produce a leap of progress for the regional economy. This is because the industrial sector relies more on technological innovation. Special attention is required for the development of creative industries. The key successes in the creative industry development are as follows:

(i) Regions with rich natural resources must begin to move from raw material commodities to the manufacturing sector.

(ii) Regions that have not abundant natural resources should focus on developing an industrial sector, which is based on non-natural resources.

(iii) Investing in the preparation of human resources.

(iv) Improve the standards and competencies of workers engaged in the creative industry sector.

If this is achieved, indirectly the manufacturing industry in the country has a competitive edge. This is not only in terms of price but also value added. With the increasing quality of human resources is expected Indonesia can be a production base and not just a market base.

(v) Give special attention to investment in the cultural sector and strengthen its accessibility;

(vi) Improve the performance of the tourism sector.

This is caused by two things, namely: (1) The contribution of the tourism sector to the national economy is greater. As export revenues fell sharply due to uncertain global conditions, the tourism sector showed positive progress. (2) The tourism sector is the largest labor creation sector in Indonesia.

(vii) Improve the development of infrastructure based on visual and soft [6 – 8].

3. Result and Discussion

3.1. Potential of Creative Industry Development in Indonesia

Indonesia has a diverse cultural heritage. It includes aspects of art, beauty, social, empathy, ceremony and others. It signifies the high creativity of Indonesian people in specific skills and talents. Cultural diversity is also supported by ethnic diversity in Indonesian society. It shows that Indonesia has a very strong supporting factor in developing a creative economy [2]. Indonesia also has great potential in developing creative industries. This can be seen from two things, namely: the economic potential of tourism and culture with high artistic taste. This shows that Indonesia has a wide range of creativity. This is in accordance with the opinion of John Howkins, who states that it is a major capital in the development of creative industries. Indonesia's creative economy can be more developed if combined with tourism. The combination can bring the centers of the creative economy as a tourist attraction. The process of combining and integrating these two activities must be through development and preparation. In addition, it is necessary to have a master plan for the creative economy. This master plan includes the development of human resources, the development of promotion centers, packaging and sales [4].

3.2. Best City for Indonesia Creative Industry

Indonesia has been listed as the third most developed country in the world after China and India in 2016. Therefore, the role of the creative industry will have a major impact on economic growth. The increased growth of micro, small and medium enterprises (MSMEs) affects the establishment of several cities that are central to the creative industry. The city will be a potential area for the growth of industrial and tourism centers. The city will also contribute to the advancement of economic development. The nine best cities for creative industries in Indonesia consist of Pekalongan, Bandung, Special Region Yogyakarta, Denpasar, Makasar, Malang, Solo, Palembang and Pare. Each city develops its creative industry based on local culture [9].

Creative Industry of Pekalongan City

Pekalongan, is a city on the North Coast of Java Island (Pantura). The city occupies the first position for the development of creative industries in Indonesia. Pekalongan City is the first city in Indonesia to enter into the UNESCO Creative Cities Network in 2014 in the category of crafts and folk arts. There has been a built batik museum, as well as an international batik center. The city that was crowned as a batik city of this world turned out to beat Yogyakarta, and Solo as a producer of industrial batik. Pekalongan is not only creative in handicrafts, but also proves its creativity in the field of technology application for its government system (Smart City).

Creative Industry of Bandung City

In addition to the city of Pekalongan, the city of Bandung also entered into the network of UNESCO Creative Cities Network in 2015 in the design category. The city is intensively promoting development in the field of youth creativity. Bandung is also famous for various fashion industries. This is evident with the presence of the jeans industrial center in Cihampelas, shoes in Cibaduyut, the Factory Outlet around Dago, and some of the Best Malls in Indonesia. This is what helped made Bandung an industrial city for fashion in Indonesia.

Creative Industry of Special Region Yogyakarta

Yogyakarta is known as the city of tourism, cultural, history and the best first education in Indonesia. In addition, Yogyakarta is also known as a community city. This is because Yogyakarta has produced a diverse community of young people in the social, hobby and art fields. The existence of this community is increasing in number and growing rapidly. Yogyakarta is also rich with typical handicrafts, such as pottery in Kasongan and silver in Kota Gede. Another advantage of this city is to become one of the cities with the best technological developments in Indonesia. One of the contributing factors is the strong role of the youth community to make the gaming industry and startup world grow rapidly.

Creative Industry of Denpasar City

Denpasar, a city in the province of Bali, is of a friendly nature for the development of world tourism climate. Bali also occupies the position of the best tourist spot for some land marks, which became the icon for the Island of the Gods. This factor causes the creative industry to grow rapidly in Denpasar as the capital of the Bali province. The hotel and culinary industries that support the tourism potential in Bali are experiencing rapid growth. The industry has spread in Denpasar and areas outside Denpasar city, such as Sanur, Kuta and Ubud. Fashion and boutiques for surfing activities are scattered in almost every location of Bali Beach. Moreover, creative industries from handicrafts are also scattered in several corners of Trade Center, such as Art Market Sukowati.

Creative Industry of Makassar City

The city of Makassar and some areas in South Sulawesi have great potential for the growth of the creative industry. This potential has been managed hereditarily. Several areas outside Makasar city are also places for creative industries, such as: (i) the ship industry of Pinishi as a vessel of South Sulawesi pride located in Tana Beru; (ii) sculpture industry "Ke'Te Kesu" in Toraja Land; and (iii) cocoa and tea plantations with the cocoa industry in some areas of South Sulawesi. The advantage of Makassar as the best city for creative industry growth in Eastern Indonesia is further strengthened by the existence of international and national events, such as: Toraja Festival and Eight Festival (F8). The city is also considered successful in the use of applications on a good governance system.

Creative Industry of Malang City

Malang city has a predicate as a city of education. This is because the city is one of the largest educational cities in Indonesia. Malang also earned the nickname 'Paris Van East Java' because of the beautiful nature of the city of Paris in East Java. Malang and its surroundings are also famous for apple fruit plantations. Agro Industrial Area is one of the plantation areas of apple fruits. This area has become a potential for tourism. Not far from the city of Malang, precisely the Batu area is also famous with many tourist attractions, such as: East Java Park 1 and 2, Batu Night Spectacular (BNS), Eco Green, Secret Zoo Stone, Umbrella and others. Malang is also the host of various local television companies that became the development for the Creative Industry in Indonesia.

Creative Industry of Solo City

Solo city is known for the establishment of the batik industry, culinary industry and organizing the best event in Indonesia. Village of Laweyan and Kauman are a home industry for the Batik Solo industry. This village has existed from generation to generation from their ancestors. This is one of the reasons why Solo city is known as the city of batik. The existence of Danar Hadi Batik Museum, as the famous Solo Batik producer, is further strengthening the identity of Solo as the city of batik. In this place, hundreds of batiks are on display at the museum. Culinary industry is also scattered in Solo city. Some of the best culinary trademarks are tengkleng, liwet rice, solo sausage, surabi and spikoe. Organizing an international event is also further strengthening the existence of the city and became one of the cities with rapid growth of the creative industries. Some international events are Solo Batik International Carnival and Umbrella Festival in Taman Balekambang.

Creative Industry of Palembang City

Palembang city is known as Venice from Indonesia. The underlying factor is the potential of Musi River, and Ampera Bridge has contributed greatly to the tourism of Indonesia. This is where the international Musi Festival is held. Through the Songket weaving, wicker, miniature and culinary handicraft industry, Palembang has become a considerable city for the growth of the creative industry. Palembang city is also a Smart city because it successfully implemented a good working government system.

Creative Industry of Pare District

Pare city is part of the Kediri Regency. Pare city is known as the English Village and the Course Village. In the villages, there have been established hundreds of institutions of English courses and other courses. The courses are scattered in Tulungrejo and Pelem Villages. The institute combines the courses system with the dormitory system. Some homes in Pare are functional as home industries, such as course services, dormitory, bookstores and bicycle rentals. This condition makes Pare city one of the cities for creative industries in Indonesia.

3.3. The Role of Government in Developing Creative Industries

Most of the small and medium enterprises (SMEs) in Indonesia are engaged in the creative industry sector based on ideas and creative innovation. The industry uses resources to produce value for added products in the market. The creative industry is the future of hope for Indonesian economic development. Therefore, the government of Indonesia made various efforts to develop the creative industry optimally [1, 5]. The role of the Indonesia Government in developing creative industry is as follows:

(i) Government established "*Badan Ekonomi Kreatif*" (BEKRAF) in 2015. It is the Creative Economy Agency. This step is an effort to explore the potential that has not been explored in the industry. Furthermore, the agency is tasked with a strategic mission to promote innovation-based employment by creating an ecosystem conducive to its development.

(ii) Central and local governments should continue to assist and encourage creative industries as well as improving the welfare of the creative industry players in the region.

CONCLUSION

The Indonesia government seeks to develop the creative industry optimally based on local culture. Indonesia has cultural diversity and uniqueness in each region. Culture should be the basis for the development of creative industries. The uniqueness of each region must be maintained. Therefore, local culture-based potentials need to manifest themselves into the creative economy program. This is where the creative economy can begin. The creative industry developed is in line with the potential resources available in the local region. The creative economy is seen not only in the economic context, but also in the cultural dimension. Regions with rich natural resources must begin to move from raw material commodities to the manufacturing sector. Regions, which do not have abundant natural resources, should focus on developing an industrial sector, which is based on non-natural resources.

Indonesia also has great potential to develop creative industries. It can be seen from two things, namely: the economic potential of tourism and culture with high artistic taste. This shows that Indonesia has a wide range of creativity. Indonesia's creative economy can be more developed if combined with tourism. The combination can bring the centers of creative economy to be a tourist attraction. It will affect the establishment of cities as a central of creative industry. Therefore, the city will contribute to the advancement of economic development.

References

[3] Kompas, *Menyadari Peran Negara dalam Industri Kreatif* [*Realizing the Role of the State in the Creative Industry*], 2017, http://nasional.kompas.com/read/2017/12/05/06080321/menyadari-peran-negara-dalam-industri-kreatif.

[4] Indonesiakreatif, *Ekonomi Kreatif Berbasis Kebudayaan* [*Creative Economy based on Cultural*], 2018, http://indonesiakreatif.info/iknews/ekonomi-kreatif-berbasis-kebudayaan/.

[5] Indonesiakreatif, *Definisi Ekonomi Kreatif dan Industri Kreatif* [*Definition of Creative Economy and Creative Industry*], 2018, http://indonesiakreatif.bekraf.go.id/ikpro/programs/apa-itu-ekonomi-kreatif/.

[6] Hyastina, *Konsep dan Definisi Ekonomi Kreatif dalam Perspektif Indonesia* [*Concept and Definition of Creative Economy in Indonesian Perspective*], 2014, https:// hotgantina11sinaga.wordpress.com/ 2014/ 02/ 26/ konsep-dan-definisi-ekonomi-kreatif-dalam-perspektif-indonesia/.

[7] Siagian, *Industri Kreatif Berbasis Budaya Lokal* [*Creative Industry Based on Local Culture*], 2014, http:// medanbisnisdaily.com/ news/ read/ 2014/ 08/ 07/ 110038/ industri-kreatif-berbasis-budaya-lokal/.

[8] Lampungpro, *Jokowi Beberkan Kunci Keberhasilan Industri Kreatif dan Pariwisata* [*Jokowi Reveals Key to the Success of Creative and Tourism Industry*], 2017, https:// lampungpro.com/ post/ 3214/ jokowi-beberkan-kunci-keberhasilan-industri-kreatif-dan-pariwisata.

[9] Jppn, *Jokowi Beberkan Kunci Keberhasilan Industri Kreatif* [*Jokowi Reveals Key to the Success of Creative Industry*], 2017, https://www.jpnn.com/news/jokowi-beberkan-kunci-keberhasilan-industri-kreatif.

[10] Syukro, *Industri Kreatif: Kunci Sukses Indonesia Hadapi MEA 2015* [*Creative Industry: Indonesia's Key Success Face MEA 2015*], 2014, http://www.beritasatu.com/ nasional/ 172266-industri-kreatif-kunci-sukses-indonesia-hadapi-mea-2015.html.

[11] Pangluar, *9 Kota Terbaik Bagi Industri Kreatif Indonesia* [*The 9 Best Cities for Indonesia's Creative Industry*], 2016, http://www.thecolourofindonesia.com/2016/11/9-kota-terbaik-bagi-industri-kreatif.html.

In: Proceedings of the 2018 International Conference … ISBN: 978-1-53615-862-5
Editors: Ivan A. Parinov et al.

Chapter 46

REDESIGN WORK ENVIRONMENT WITH ERGONOMICS INTERVENTIONS TO DECREASE FATIGUE AND INCREASE PRODUCTION OUTPUT AT BATIK HOME INDUSTRY "CAHAYA SARI" IN SRAGEN, CENTRAL JAVA

Sajiyo*

Department of Industrial Engineering,
University of 17 Agustus 1945 (UNTAG), Surabaya, Indonesia

ABSTRACT

This research is a type of experimental research with the same subject to resolve problems faced by the craftsmen of batik on Batik home industry "CAHAYA SARI" in Sragen, Central Java. Problem solving is done by means of the redesigned lingkugan in the workplace ergonomics intervention. The population in this research is the whole batik home industry of craftsmen, namely 12 women, who were samples for this study and created total sampling. In this research that was conducted in relation to the work environment, the following ergonomics interventions were: (i) installation of exhoust to regulate circulation, humidity, and temperature. in indoor workplaces; (ii) engineering taking into account lighting needs for indoor workplaces. This experiment was done as follows: the original batik craftsmen worked in the workplace conditions (prior to intervention ergonomics) for 12 days, then we performed a washing out for 1 day every 6 days of the experiment. Next, the new conditions of the work place (after the intervention of ergonomics) were presented for adaptation for 12 days; after adapting to the new conditions of the workplace, batik craftsmen were told to work on the venue for the

* Corresponding Author's E-mail: sajiyo@untag-sby.ac.id.

following 12 days. To get the research data, we carried out research of the following variables measured: (i) air temperature, wind speed, humidity air, and light intencity of (ii) pulse rate work and rest as an indicator of the level of fatigue, and objective data on retrieval fatigue subjectively by using questionnaire fatigue. The results of measurements of the environment after the intervention of ergonomics is as follows: (i) increasing the itensity of light from 160.45 Lux to 360.40 Lux, (2) lower of the air humidity from 72.15% to 57.40%, (3) lower of the temperature from 30.01°C to 26.43°C, (4) increasing wind speed from 0.12 m/s to 0.18 m/s. The results of the measurement of fatigue after performed intervention ergonomics are as follows: lowering work fatigue subjectively from extremely tired to tired, and not total exhaustion of tired. The results of the measurement of output production, after the intervention of ergonomics, is as follows: an increase in production output of 49.01%. Novelity of the research on the environment ergonomics intervention is a workplace environment in particular relation to the craftsman batik. The hallmarks of this novelitas is the creation of collaboration and commitment of all parties, the employees craftsmen, business owners and management. Besides the above mentioned novelitas presented in this study, we note bargain (compromise) mutual benefit, which improved the quality of interaction between the business owners with craftsmen.

Keywords: fatigue, traditional batik artisan work

1. Introduction

The Asean economic community (MEA) proclaimed Indonesia at the end of year 2015 as having significant influence to arouse the public face of the challenge of competing in Indonesia's economy, and the challenges of competence in those times that will come. The spirit of the MEA was tempestuous at seuruh Indonesia, in order to not lose people in rural areas due to Free trade commitment for Asean member countries in Asia Asean Free Trade (AFTA). Generally, the industrial sector in Indonesia is able to absorb the workforce more than 60%, see Sajiyo [5]. Batik craft industry in particular became the object of earnings, especially the rural women, although such work is performed in the interruption after the work on the farm and carried out household. However, on the other hand, because the working environment is not ergonomic, the many batik craftsmen complain of fatigue at the time of work activities. The mother-homemaker who has her secondary work in batik is able to provide the economy of her family of more than 50% of everyday economic needs. In order for batik craftsmen to work more efficiently, it is necessary to make a work room designed off of the batik craftsmen ergonomic.

2. Literature and Methodology

2.1. Ergonomics

Ergonomics is the study of human, environmental, facilities, and systems of work that are interacting in cumulative for convenience in working, see Grandjean [2]. From the standpoint of ergonomics, between tasks and a capacity of work there should exist a balanced condition, see Manuaba [3].

2.2. Fatique

Is the loss of physiological fatigue temporarily of a psycho-physiological capacity due to excitation continuously? Psychological fatigue of man is a pseudo-fatigue, arising in the feelings and behavior seen in action, among his other non stringent opinion and his labile space is, see Adiputra [1].

2.3. Production Output

The production output is external beyond relation to work activities that have a direct impact on the productivity of work. The production output is heavily influenced by efficient work. The efficiency of the philosophically is construed as proportional value to the work activities, see Sajiyo [5].

2.4. Study of Location and Time

The location of the research is the batik craftsmen "cahaya sari" room in the hamlet Turisari, village Pungsari, district Plupuh, Kabupaten Sragen, Central Java. This research was conducted over 12 days, from September 7 to 19 March 2016.

2.5. Populations and Samples

The population in this research are 12 batik craftsmen, all of them women. They were taken as samples and created total sampling.

2.6. Research Variables

Research variables are as follows: (i) dependent variable that corrsponds to fatigue from work and to output production; (ii) independent variable relates to the conditions of the working place and to new work conditions.

2.7. Research Procedure

Research was conducted in the 5 following stages: (i) the preparatory stage included determination of the subject of the research, provide a description of the research process, and preparation of research instruments; (ii) the stage of implementation included the installation of the mounting position of equipment, exhaust lamp lighting, positioning work; (iii) the stage of environmental measurements included the measurement of light intensity, temperature, humidity, and wind speed; (iv) the stage of intervention impact ergonomics measurements includes the measurement of fatigue, and output production; and (v) the stage of data analysis and discussion research results. In the study conducted, it was performed intervention ergonomics, which included the following: (i) the installation of exhaust fan; (ii) the setting of position and measurement of light intensity; (iii) settings of working position, as described in the following.

2.7.1. Installation of Exhaust Fan

The installation of exhaust fan is based on wind direction either during the dry season or the rainy season. It aims so that air circulation in the room works in synergy with wind direction naturally outside of the work place. The batik craftsmen work on old conditions only installed in one window on the west, see Figure 1.

The batik craftsmen work in a space in a new condition that includes the mounted 5 (five) exhaust fans; 2 (two) fruit exhaust fans, mounted on the north side; the 2 (two) fruit exhaust fans, installed on the south side; one (1) fruit exhaust fan, installed on the west side; and 2 (two) windows, mounted on the north side and south side. The west side window is replaced by exhaust fan, see Figure 2.

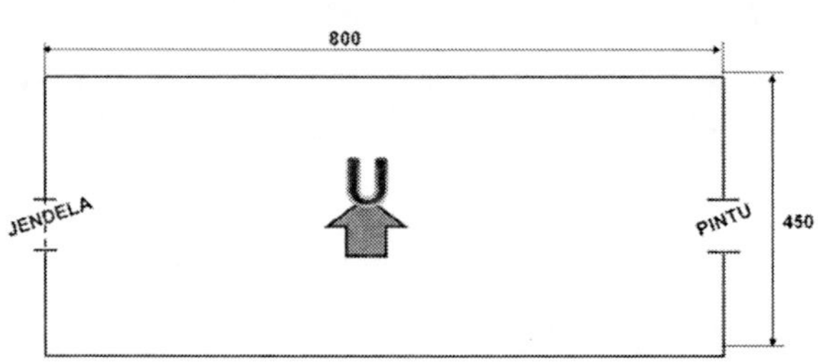

Figure 1. Work space before mounting exhaust fan.

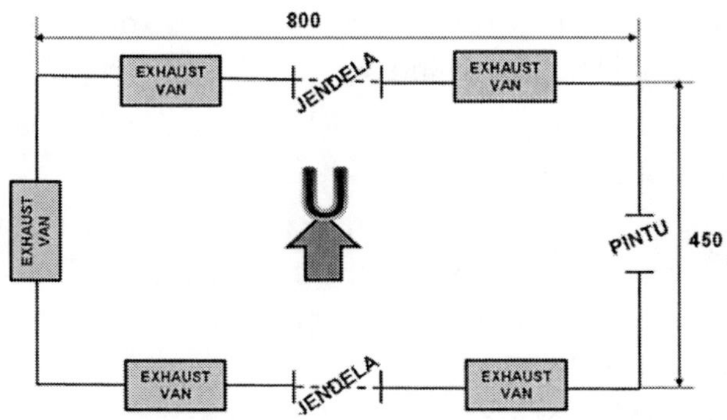

Figure 2. New work space (after installation of 5 exhaust fans and 2 windows).

The window is mounted on the side of the North and South and adapts to the wind direction naturally. Wind direction in Central Java during the dry season moves from the south – southwest towards the North-East of the sea, instead of the rainy season moving from the North-North-East towards the South-southwestern.

2.7.2. Setting the Lamps

Setting the position of the lamps to meet the needs in light and glare for more comfortable work for the workers. The old work space included only one (1) light source, placed in the middle of the room, see Figure 3.

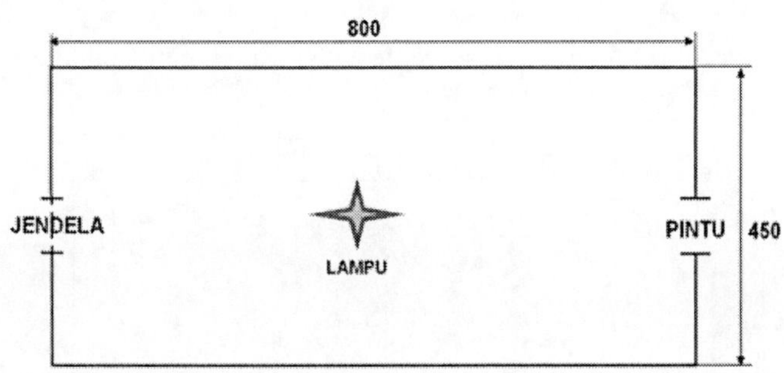

Figure 3. Old work space (only 1 light source is mounted).

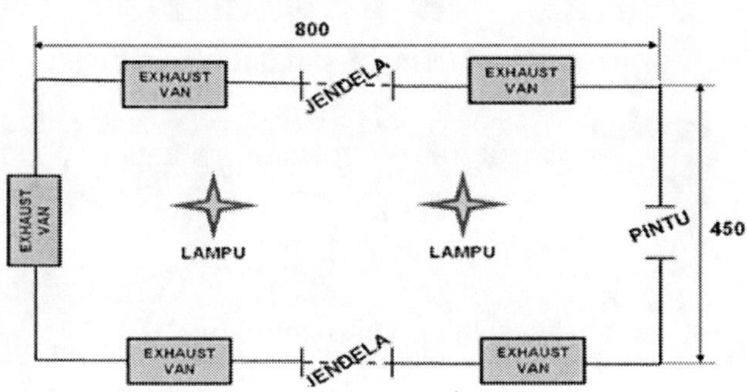

Figure 4. New work space (after installation of 2 light sources).

A new workspace included 2 (two) light sources, in order to meet the needs of working space and without glare, see Figure 4.

2.7.3. Setting the Working Position

Setting the working plays correspond to the circulation of wind and location of light sources, in order to be more comfortable without glare and hot. The position of the old work is composed of three (3) groups, each group of 4 (four), and working places located in the middle of the room, see Figure 5.

New working positions comprised from 4 (four) groups, each group consists of 3 (three) persons, working positions of 2 (two) groups are on the north side, and 2 (two) groups on the south side, see Figure 6.

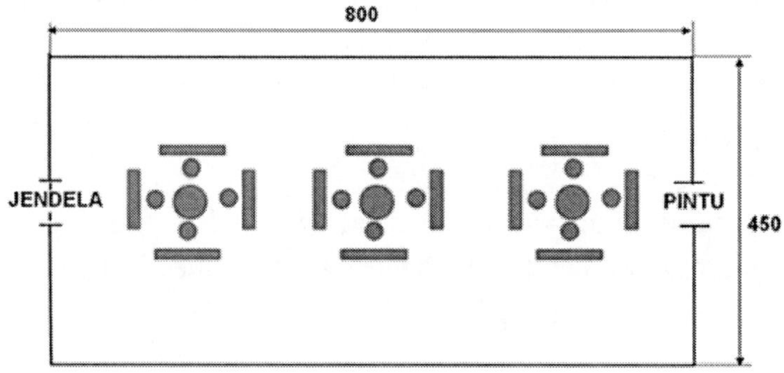

Figure 5. Old working places (each group of 4 people with position in the middle of room).

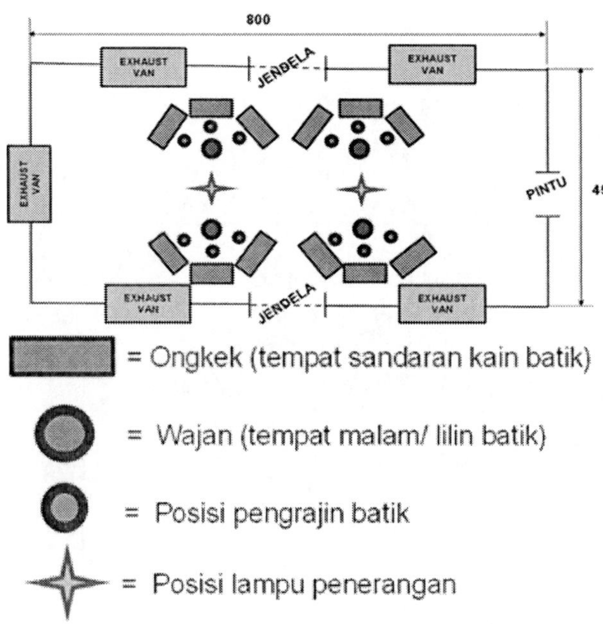

Figure 6. New work places (each group of 3 persons with position at the north and south sides from window).

3. Data Analysis

3.1. Data Analysis of Light Intensity

The light intensity data analysis obtained the following results: (i) the average light intensity at the old conditions of work (prior to intervention of ergonomics) was 167.44 Lux, and (ii) the average light intensity at the new working conditions (after the intervention of ergonomics) became 350.05 Lux.

3.2. Data Analysis of Air Humidity

The air humidity data analysis obtained the following results: (i) the average air humidity at the old conditions of work (prior to intervention of ergonomics) was 70.19%, and (ii) the average air humidity at the new working conditions (after ergonomics interventions done) became 58.36%.

3.3. Data Analysis of Air Temperature

The air temperature data analysis obtained the following results: (i) the average air humidity at the old conditions of work (prior to intervention of ergonomics) was 29.91°C, and (ii) the average air temperature at the new working conditions (after the intervention of ergonomics) became 26.07°C.

3.4. Data Analysis of Wind Speed

The wind speed data analysis obtained the following results: (i) the average air humidity at the old conditions of work (prior to intervention of ergonomics) was 0.10 m/s, and (ii) the average wind speed at the new working conditions (after the intervention of ergonomics) became 0.15 m/s.

3.5. Data Analysis of Subjective Fatigue

The subjective fatigue data analysis obtained the following results: (i) the average score of subjective fatigue at rest at the old work conditions (prior to intervention of ergonomics) was 1.21, and an average score of subjective fatigue was 3.20, and (ii) the average score of subjective fatigue during interruption at the conditions of work recently (after the intervention of ergonomics) became 1.14, and average score of subjective fatigue at the conditions of work recently (after the intervention of ergonomics) was 1.99.

3.6. Data Analysis of Pulse

The analysis of pulse data obtained the following results: (i) the average resting pulse rate (RPR) at the conditions of work prior to intervention of ergonomics was 77.60 beats/minute, and the average pulse rate at work (WPR) at the conditions old work (prior to intervention ergonomics) was 150.71 pulse/minute, (ii) the average resting pulse rate (RPR) at the conditions of work recently (after the intervention of ergonomics) was 77.40 beats/minute, and the average pulse rate at work (WPR) at the conditions of work recently (after the intervention of ergonomics) was 113.80 pulse/minute, and (iii) increase in average pulse at work (IWPR) at the conditions of work (prior to intervention of ergonomics) became 60.55%, and the average increase in the pulse of work (IWPR) at the conditions of work recently (after the intervention ergonomics) was 29.69%.

3.7 Data Analysis of Output Production

The subjective fatigue data analysis results obtained the following results: (i) the average production output at the conditions of work (prior to intervention of ergonomics) was 2.02 strands/person/week, and the average initial production output at room conditions of new work (after the intervention of ergonomics) became 3.30 strands/person/week, and (ii) the average increase in production output, after the intervention of ergonomics, was 63.37%.

CONCLUSION AND SUGGESTION

Conclusion

By redesigning the work environment, the ergonomics approach can reduce fatigue from work from the category "very tired" to "not tired," decrease the musculoceletal disorders from "deeply disturbed" to "not disturbed," and can increase the production output to 49.01%.

Suggestion

To further enhance the efficiency and productivity of the work of the craftsmen of batik, advanced research needs to be performed on the designing equipment, namely: batik design, batik place, pouring the wax, wax heaters, and wok burner.

REFERENCES

[1] Adiputra, *Macro Ergonomics*. Denpasar: Ergonomics Study Program of Work Physiology at the Postgraduate Program at Udayana University, 1998 (In Indonesian).

[2] Grandjean, *Fitting the Task to the Man*. A Textbook of Ocupational Ergonomics, Taylor & Francis, 2000.

[3] Manuaba, *Ergonomics Effect on Work Productivity*. Advanced Development Seminar on the Development of Kedoteran Science. Jakarta, 1992 (In Indonesian).

[4] Manuaba, *Ergonomics of Occupational Safety and Health*. Ergonomics National Seminar in Surabaya. Publisher Widya, 2000 (In Indonesian).

[5] Sajiyo, *Design Work with an Ergonomic Approach to Increase Work Productivity in the Cigarette Industry in East Java*. Denpasar: Dissertation Postgraduate Program of the Udayana University, 2008 (In Indonesian).

In: Proceedings of the 2018 International Conference … ISBN: 978-1-53615-862-5
Editors: Ivan A. Parinov et al.

Chapter 47

IS IT REALLY BRIGHT COLOR DOMINATING THE FACADE/EXTERIOR OF HOUSES IN SURABAYA, INDONESIA?

Uniek Praptiningrum, W*
Department of Architecture, Faculty of Engineering,
Universitas 17 Agustus 1945,Surabaya, Indonesia

ABSTRACT

The color of the house facade becomes an important part because, for the first time, people see the building as one of them also from the color. In addition, the color of the building can also be used as a guide and character of the inhabitants. To categorize building as a good architectural work, the building should be: (i) attractive in terms of appearance of the building, (ii) comfortable when occupied, (iii) beautiful views, (iv) safe in construction, (v) have local wisdom, and (vi) friendly to the nature and environment. There is a great variety of professions and people's tastes, who live within the Surabaya city and in very hot climatic conditions. Color selection for the facade building is quite influential. From the results of the interim survey, it was found that 80% of the 50 respondents live in Surabaya and choose bright colors for the facade of their homes. The remaining 20% of the respondents chose a combination of dark and bright colors. The research method used is the direct field survey, study of references, treatment in Internet, and collecting some of the data research that has been conducted by some competent sources with the following analysis.Preliminary, the following results were obtained: in Surabaya (Indonesia), many human professional sources impact and contribute to choosing, implemented in almost 80%.

Keywords: architectures, local elements, coloring facade properly, environmental care, secure

* Corresponding Author's E-mail: uniek@untag-sby.ac.id; uniekw99@yahoo.com.

1. INTRODUCTION

To implement an architectural work, such as dwellings, homes, etc. architecture elements, such as points, lines, spaces, shapes and colors, should be considered. The color element in architecture has a significant influence on architectural work. At a first glance, the human eye, color and shape, besides being a marker, (such as profession, social status, character of the homeowner), can also give a nuance and atmosphere to the surrounding environment [1].

Elements of facade are designated by the major providers to communicate the main function and significance of a building [2].

Building materials are enough to play a role in supporting the shape and color of the building. In addition, paints and building materials can also provide natural and artificial colors, which are caused by the material itself. For example, wood has a natural color of brown or blackish brown, brick gives a natural red color, and palm leaf roofs gives a dark brown color, as well as other ingredients. Wall-covering materials, such as ceramics, provide artificial colors from the factory. If they are examined more deeply, then basically almost all building materials that are commonly used will definitely has its advantages and disadvantages.

Likewise, the natural colors of building materials really have the advantages of artistic values. The development of technology makes weaknesses that can be minimized, and the advantages can be more effective. However, unfortunately, up to now, many advanced technologies have not answered a number of building materials that are very suitable for buildings located in humid tropical climates and friendly to climate conditions themselves [3].

For example, building materials from bamboo, such as woven bamboo or gedeg (Jawa), in terms of their natural color, and artistic value, is quite high. With high technology, this building material should be processed to become a qualitative building material, both in terms of durability and safety, of course, with prices that can be reached by all people. Likewise, it is attained with other local materials, such as roof coverings with raw materials of sago palm, shingles, and many others.

This condition is also, among other things, the cause of the lack of interest in the community to use or choose these materials

The results of the initial survey conducted on the owner/occupant of a residence in Surabaya (Indonesia) showed that 80% of the 50 respondents gave bright colors to the exterior/facade of their house, such as green, beige, blue, white, while 20% of the respondents varies. 80% of those who determine the color are housewives, 10% are children, and 10% are others. The reason for choosing colors is to want to differ from the others, trends, follow cartoons, and the like. Most responses with choosing the color are not based on climate considerations that follow the rules of the environment or are environmentally friendly. The material used on average is artificial coloring paint, such as paint and the like. The colors obtained from the base material itself are rarely used [4].

The survey was carried out by collecting data based on interviews and distributing questionnaires to respondents in several residences in Surabaya, surrounded in architecture.

The problems that are often caused are:

1. Most residents of residential houses, especially housewives, in determining color choices, are only based on their personal desires, and the minority of them have paid attention to the art of architecture.
2. Color selection is often done without the approval of the whole family, so that the husband and children are disappointed, if by chance not one's taste. The quality of the material used is often not environmentally friendly. Most of them have not used the color of the building material (not paint/coloring), which has the basic color of the building material itself, which actually has a high artistic value.

The purpose of this study is to find out how much the occupants/owners/developers apply the bright colors of their residential buildings and the extent to which it is relate to the architectural principles, building materials, and local climate. The assessment is only limited to the selection of colors on the facade/exterior of some houses in Surabaya and its surroundings relation to architecture.

2. Field Survey Results

Building houses in Indonesia today generally prefer practicality and security, especially in terms of their care. Basically, local ingredients with natural colors are very good and appropriate for buildings with a humid tropical climate. They have high aesthetic and artistic values. However, because there is no technology that develops such materials, like gedeg (woven bamboo. jw), bamboo grass, palm leaf, palm fiber and so on, especially in terms of durability and affordable prices, these building materials are abandoned. Most are not even almost touched by the urban community to build their homes. The purposes of conducting the assessment of color selection in houses are present in Table 1 and the stages of research process are shown in Figure 1.

Table 1.

Intent/Purpose To know how deeply the habitants understand about color selection in facade/exterior of residence and conjunction of the science of architecture and environment with local climate.
Objective To look in details the reasons for the selection of a color for the dwelling house, especially in Surabaya, Indonesia.
Subject Adding the knowledge of public common on color, consistent with architecture, the environment and local climate. Socialization to the community, the importance of the conjunction between color selection in facade/exterior houses, architecture, the environment and local climate. In architecture, the color selection is not because like, but there are other basic things, whichshouldbeconsidered fundamental.
Programming

The results of a survey of 50 respondents living in several residential or real estate environments in Surabaya and surrounding areas showed that 15% used a 10% brown coated ceramic tile cover with 80% concrete tile cover, ceramic tile 15%, while the rest consists of not concrete and asbestos. In the survey conducted in not one house, thatched roof coverings, shingles, palm fiber, or other local natural ingredients were used.

For the construction of walls and columns, nearly 100% of respondents use brick materials, and concrete is used for the columns. All respondents use ham pear finishing with se men coating painted in bright colors. None of the respondents used wall tiles or bamboo, or wood or anything else. For ceilings, 60% use gypsum, 20% use ether nit, the rest use wood, lam bisering or asbestos. Asbestos generally functions as a roof. None of them use gedeg (bambu) or other materials that are more like the traditional materials.

Several types of domestic building materials with rare natural colors are used in residential houses in residential areas in the Surabaya city and its surroundings.

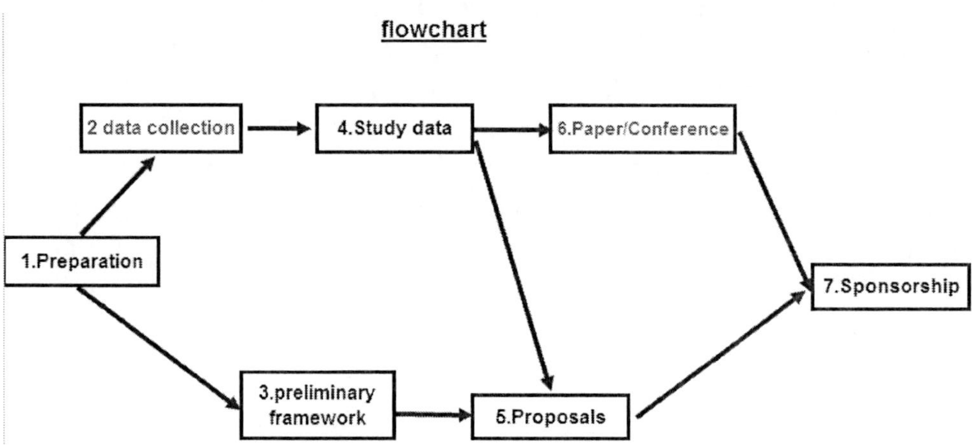

Figure 1. Stages of research process.

No	COLOUR OF FAÇADE/ EXTERIOR/responden	Ammount responden	NUMBER OF HOUSING/25 H WITH 3 KIND OF B M SAMPLING																								EXPL
			1	2	3	4	5	6	7	8	9	10	11	12	13	14	15	16	17	18	19	20	21	22	23	24	25	
A	**WALL (outside)**	**25**																										
1	red	2	v																								v	
2	green	6						v	v				v			v							v		v			
3	yellow	3						v							v										v			
4	orange	1																							v			
5	blue	4						v		v												v				v		
6	pink	2																v	v									
7	white	8		v		v	v				v			v			v	v	v									
8	cream	**8**		v	v		v					v	v						v	v	v							
9	other	**3**										v	v											v				
	Reason																											
1	economic/cheap																											
2	easy to do/maintenance/simple	1					v																					
3	easy to found	2					v				v																	
4	aestetic/like/attractif/elegant	19	v					v		v	v	v			v	v		v	v	v	v	v	v	v	v	v	v	
5	wide/leisure/enlargeclear away	4					v										v				v						v	
6	strong/serving/lasting	0																										
7	flexible	1			v																							
8	sun isolator/fresh/comfort	7					v		v				v	v			v	v								v		
9	trend	0																										
10	other	11		v	v	v			v		v	v	v									v	v	v			v	
B	**ROOF**	**25**																										
1	red	9				v		v						v	v	v	v	v			v						v	
2	green	0																										
3	blue																											
4	grey	10	v	v	v		v		v	v	v												v	v		v		
5	black	4																	v	v		v			v			
6	brown	2										v	v															
7	white																											
8	combination																											
9	other																											
	Reason (%)																											
1	economic/cheap	5			v			v						v									v			v		
2	easy to do/maintenance/simple	21	v	v	v			v		v	v	v	v	v	v	v	v	v	v	v	v		v	v	v	v	v	
3	easy to found	20	v	v		v	v	v		v	v	v	v	v	v	v	v	v	v		v		v	v		v	v	
4	aestetic/like/attractif/elegant	**9**							v			v	v			v		v		v		v		v	v			
5	wide/leisure/enlargeclear away	2																			v						v	
6	strong/serving/lasting	23	v	v	v	v	v	v	v	v	v	v	v	v	v	v	v	v	v	v	v	v	v	v	v			
7	flexible	0																										
8	sun isolator/fresh/comfort	4	v	v														v									v	
9	trend																											
10	other																											
C	**CHOOSY**	**25**																										
1	father	8	v		v	v			v				v						v		v	v						
2	mother	10								v	v			v	v			v		v				v	v	v	v	
3	father and mother	4		v			v	v								v												
4	son/daughter	1										v																
5	family	**1**																					v					
6	grandfather/grandmother	1															v											
7	other																											
	Total Sampling																											

The location houses of Responden (25 houses from 50 total houses) : Sources :Survey Result Student Department of Architecture Untag Surabaya 2018
I Perum Wiyung Brantas Permai
II:Perum Western Village, Rungkut Mananggal,Griya Amerta Regency Medokan Ayu and Others

edited by uniek praptiningrum,w 18

Figure 2. 25 sampling from 50 respondents.

Figure 3. Two colours for facades of houses in Surabaya.

3. Results and Discussion

Colors can be divided into two categories/properties: those that are seen in plain sight, and those that are felt by the soul.

Physical Objectives: Properties emitting; the light that appears before the eyes is one of the forms of energy emission, which is a narrow part of the electromagnetic wave.

Psychological Subject: Part of the understanding of the human senses; appearance of color, can be distinguished into hue (hue of color or color), value (dark, light) and chroma (pure - dirty, bright - gloomy, dim - light, and color intensity).

When building houses in Indonesia today, most people generally prefer practicality and security, especially in terms of their care. Basically, local ingredients with natural colors are very good and appropriate for buildings with humid tropical climate; they have high aesthetic and artistic values. However, because there is no technology that develops, such as gedeg (woven bamboo.jw), bamboo grass, palm leaf, palm fiber and so on, especially in terms of durability and affordable prices, this building material is abandoned, even almost not touched by the urban community to build their homes.

The results of a survey of 50 respondents living in residential areas in several residential houses or real estate environments in Surabaya and surrounding areas showed that 15% used a 10% brown coated ceramic tile cover with 80% concrete tile cover, ceramic tile 15%, while the rest consists of not concrete and asbestos. In the survey conducted, not one house used thatched roof coverings, shingles, palm fiber, or other local natural ingredients.

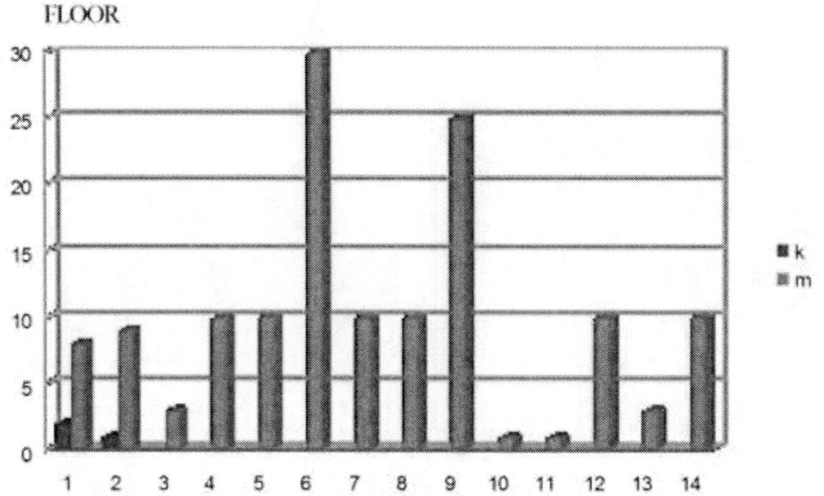

Figure 4. Color of Facade/Exterior: 50 residentskinds color material for: wall – fabricant; roof – fabricant; floor – fabricant; door/window – fabricant/natural.

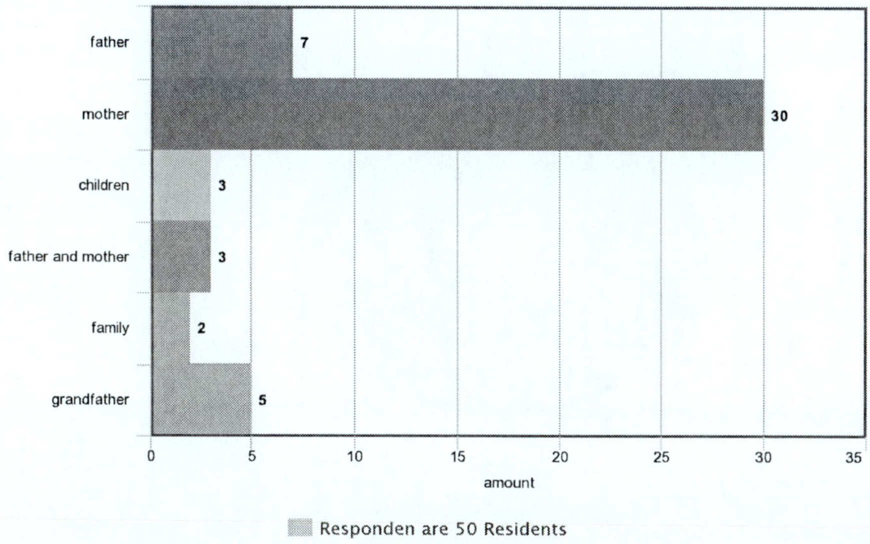

Figure 5. Respondents are Persons of respondents: (1) father;(2) mother; (3) children; (4) father andmother; (5) family; (6) grandfather.

For the construction of walls and columns, nearly 100% of respondents use brick materials, and concrete is used for the columns. Almost 100% finished with product coating painted in bright colors. None of the respondents used wall tiles or bamboo, or wood or anything else. For ceilings, 60% use gypsum, 20% use plasterboard, the rest use wood, lambrisering or asbestos. Asbestos generally functions as a roof. None of them use gedeg (bambu) or other materials that are more like traditional materials. The results of processing questionnaires are present in Figures 4 – 9.

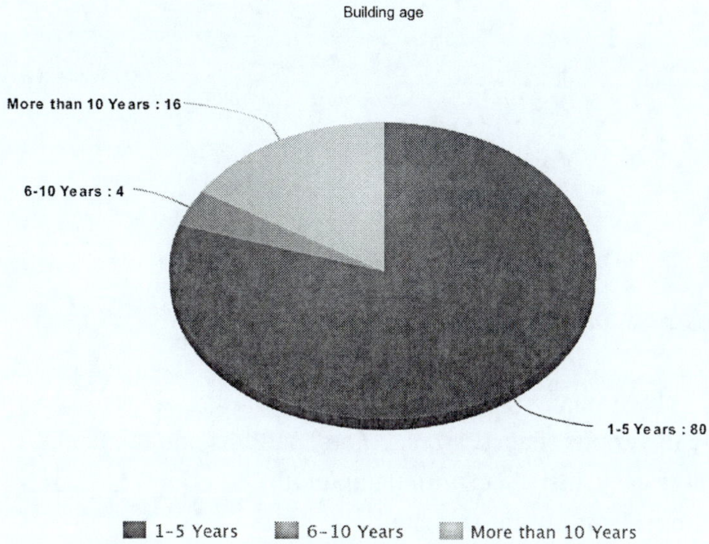

Figure 6. Sampling building age of 50 Respondents (–2018).

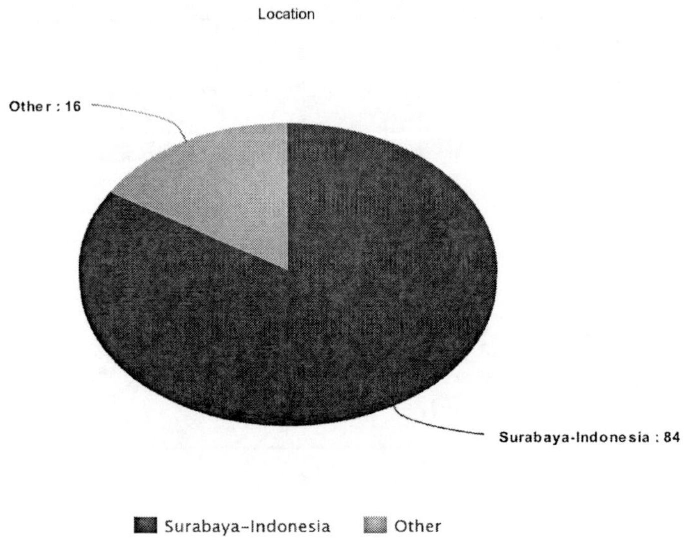

Figure 7. Location.

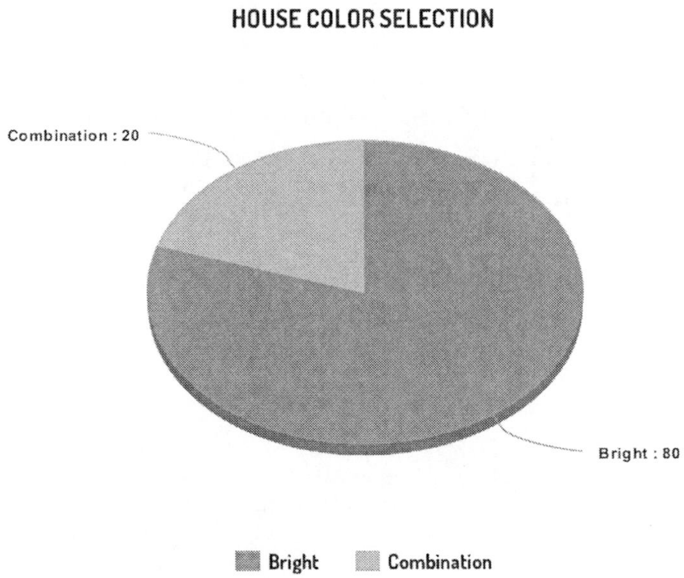

Figure 8. Sampling top cup of floor 50 Respondents (–2018).

Following reasons for their choices of color are:

(i) durable, able to survive for tens of years and more;
(ii) available in various shapes, sizes, colors, patterns and textures;
(iii) treatment is easy, just clean with damp cloth;
(iv) hold and not absorb water and sun light;
(v) prices, varied widely from cheap to expensive, depending on the quality and size;
(vi) trend and like.

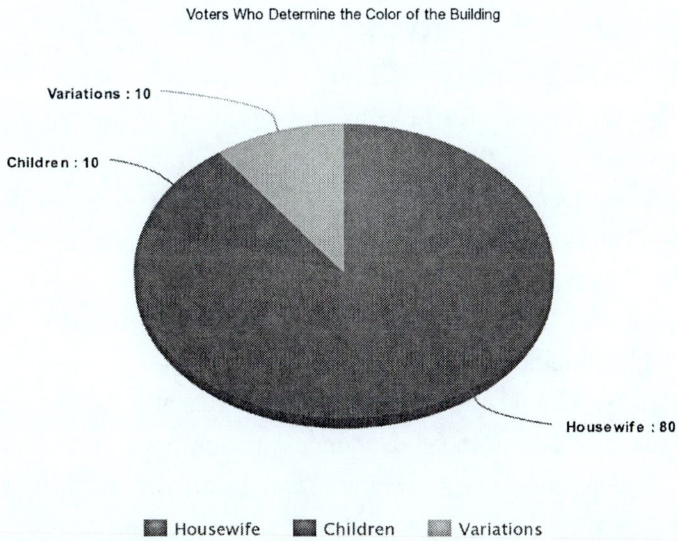

Figure 9. Reasons of color materials for choices of 50 respondents (combine brightcolor 80% + 20% others).

CONCLUSION

The colors, chosen and used on the facade/exterior of the residence in Surabaya, include 90% of the light colored walls with artificial color paint. 50% of frames, doors, and windows are painted with white color paint, and 50% is covered with polish paint. Modern advanced technology allows one to improve some local/traditional building materials, especially in terms of durability. The raw materials used do not damage the environment. Therefore, the application of natural colors, owned by the building material itself, can be used as the best alternative applied by the occupants/owners/operators to color the facade exterior of the house.

Technology is actually very potential and expected to be able to support the continuation of the use of local building materials, processed into building materials that are environmentally friendly, durable, safe, and still emphasize their artistic and aesthetic value and easy maintenance.

ACKNOWLEDGMENT

This work was financially supported by the university and the rest by private.

References

[1] Praptiningrum.W, Uniek. (2011). In the trend of using building materials in residential houses: Case Study of Houses in Surabaya and Yogyakarta Code. *Jurnal Faculty of Architecture Engineering Atmajaya University,* vol. *9*(1).

[2] Krier, Rob. (1988) *Architectural Composition*. New York, U.S.A: Rizzoli Intl Pubns.

[3] Mangunwijaya, Y. B. (1981). *Building physics conveyor articles*. Jakarta, Indonesia: PT Gramedia Pustaka Utama Press.

[4] Praptiningrum.W, Uniek. (2018). Study of the reasons for choosing bright colors for some facades / exteriors of residences in Surabaya and surrounding areas in Indonesia, *Architectural Engineering Student Survey Results 17 August 1945 University*.

In: Proceedings of the 2018 International Conference … ISBN: 978-1-53615-862-5
Editors: Ivan A. Parinov et al.

Chapter 48

NOISE LEVEL CONTROL DUE TO TRANSPORTATION ACTIVITIES IN EDUCATIONAL FACILITIES

Handy Febri Satoto** and Herlina
Industrial Engineering Department,
University of 17 Agustus 1945 Surabaya, Indonesia

ABSTRACT

One of the causes of noise is an increase in transportation activities in an area. This study area is an educational facility located on the roadside of Mulyorejo with quite high transportation activities. Our objective was to determine the L_{sm} equivalent noise level and determine how to control noise level due to transport activity in this study area. The sampling point in this study is in front of one of the educational facilities. The measurement of noise level is divided into daytime (L_s) and night (L_m). The Lsm results are the results of the calculation of L_s and L_m, which are then compared with KEP-48/MENLH/11/1996 concerning quality standards for noise levels for educational facilities. The measurement results of L_{sm} equivalent noise is 74.98 dBA. The value of the Lsm exceeds the quality standard of the noise level for educational facilities of 58 dBA. Efforts to control noise by using an artificial barrier is effective and can reduce noise level to 56.98 dBA that meets quality standards for educational facilities.

Keywords: noise, educational facilities, barriers

* Corresponding Author's E-mail: handyfebri@gmail.com.

1. INTRODUCTION

Noise is a term used to express an unwanted sound, including sound that is a side effect of activities, such as industrial activities and transportation. The sounds considered disturbing, including speech and music activities that are not desired by the listener, can also be referred to as noise [11]. On estimation of KepMen LH RI No.48/MENLH/PER/XI/1996, noise is an unwanted sound from a business or activity at a certain level and time that can cause human health and environmental discomfort.

The rate of growth of transportation in Indonesia has increased from year to year. The rate of motorcycle growth in Indonesia is the highest compared to ASEAN countries, which is 13.2% compared to other modes of transportation. The cause of the increasing growth rate of motorbikes is that motorbikes are cheap and affordable means of transportation [2]. The transportation sector potentially contributes in increasing noise. Increasing its number of vehicles crossing the transport activity in an area will be denser and accompanied by increasing noise.

Noise, caused by motorized vehicles as a mode of transportation, is included as a line source. The source of the line is the source of the point that moves following a straight line, for example, the motion of a car, train or aircraft during take off and landing. The source of the sound of the line is radiated in a cylindrical form [4] (see Figure 1).

Decibel (dB) is a unit used to express the intensity of sound in everyday life [1]. The decibel scale consists of three types, namely decibel A (dBA), decibel B (dBB) and decibels C (dBC). Various decibels basically refer to its frequency. Most of the votes are expressed in dBA noise level [4].

A land use in a certain area has a different noise level according to its designation, which is usually called the noise level quality standard. Some criteria for noise level quality standards, used for educational facilities are 55 dBA with a tolerance of + 3 dBA. The effect of noise is not only reducing our hearing capacity, but also other bodily functions. The impact of noise on the body is the same as the effect of stress on the human body.

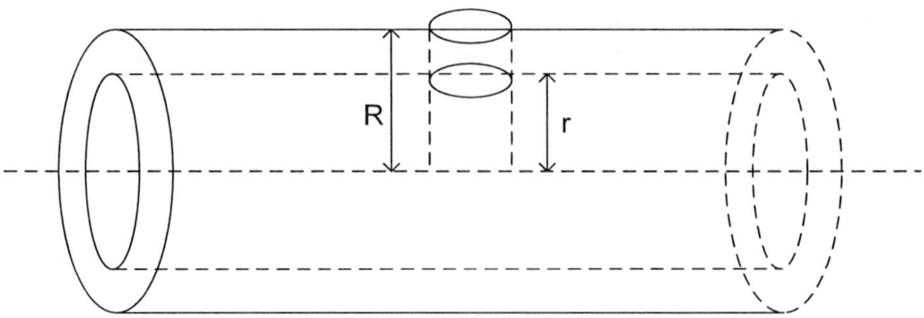

Figure 1. Distribution of line source sounds.

Installation of a barrier aims to prevent or block transmission of noise from the source to the receiver. Selection of barrier material for educational areas should pay attention to various things, among others: location, size of comfort, safety, strength and others. This can affect the psychology and health of the recipient [9].

Educational facilities are one place to get knowledge, which is the right of every human being. In educational facilities, it is expected to get enough peace so as not to interfere in the teaching and learning process. According to KepMen LH RI No.48/MENLH/PER/XI/1996, the noise level quality standard for the education area is 55 dBA. But, in this research area, it was found that transportation activities were quite high, so that the noise produced was also quite high. The object in this study is one of the educational facilities located on Mulyorejo's street Surabaya, which is affected by noise due to transportation activities in the area.

The purpose of this study is knowing the Lsm noise equivalent to the area of study when compared to the quality standard and how to control noise level due to transport activity in the study area.

2. Experimental Work

2.1. Tools Preparation

(1) *Sound Level Meter* (SLM) is used for detecting the noise of the study area. It must be calibrated first. It aims to obtain noise levels that reflect noise in the study area;
(2) GPS is used to determine the coordinates of the sampling point.

2.2. Sampling Point

The sampling point is determined in order to represent the actual situation in the study area. The sampling point in this study is in front of one of the educational facilities in Mulyorejo's street Surabaya with coordinates 7° 15.870' S and 112° 46.837' E.

2.3. Days for Sampling

In this study, the electoral considerations of sampling can show the actual situation in the study area:

(1) Monday, which can represent other working days, Tuesday, Wednesday, and Thursday;
(2) Friday, which is a work day, but at noon there is Friday prayer;
(3) Saturday, when not all people work but there are some who still work;
(4) Sunday, which is a holiday.

2.4. Noise Level Measurement

Noise measurements are according to KEP-48/MENLH/11/1996 that is done so by measuring for 10 minutes with readings every 5 seconds, so that obtained is 120 data (600 seconds/5 seconds). Noise level measurement in the study area is divided into several time intervals L_i, namely:

L_1 = 06.00 – 10.00 was taken at the peak range of 06.30 – 8.30

L_2 = 10.00 – 14.00 was taken at the peak range of 10.00 – 12.00

L_3 = 14.00 – 18.00 was taken at the peak range of 16.00 – 17.30

L_4 = 18.00 – 22.00 was taken at the peak range of 18.00 – 20.00

L_5 = 22.00 – 24.00

L_6 = 00.00 – 03.00

L_7 = 03.00 – 06.00

The consideration of the peak time selection is based on direct observation of the conditions in the study area for several weeks.

2.5. Noise Level Data Processing

120 data generated were then processed using the L_{eq} formula so that a noise level data was obtained representing the time range. The formula for L_{eq} is as follows [4]:

$$Leq = 10 Log \frac{1}{T} \sum \{T_1 10^{0.1Li}\},$$

where L_{eq} is the *equivalent continuous noise level* or level noise (dBA); T is the time period (seconds); T_1 is the reading period (seconds); L_i is the noise level at the time range (dBA).

Furthermore, the primary data will be measured by using the formula in accordance with KEP-48/MENLH/11/1996 concerning Noise Level Standards:

(i) Data from the time period of the interval $L_1 - L_4$ (daytime)

$$L_s = 10 \log 1/16 \{T_1 \times 10^{0.1 \times L1} + ... + T_4 \times 10^{0.1 \times L4}\} \text{ dBA}$$

(ii) Data from the time period of the interval $L_5 - L_7$ (evening)

$$L_m = 10 \log 1/8 \{T_5 \times 10^{0.1 \times L5} + T_6 \times 10^{0.1 \times L6} + T_7 \times 10^{0.1 \times L7}\} \text{ dBA}$$

then L_{sm} is calculated by the formula:

$$L_{sm} = 10 \log 1/24 \{16 \times 10^{0.1 \times LS} + 8 \times 10^{0.1 \times (LM + 5)}\} \text{ dBA}$$

the final step is carried out by evaluation by comparing the value of L_{sm} with the standard noise level which is set to +3 dBA tolerance.

2.6. Noise Level Control Efforts

Efforts to control noise levels in the study area are carried out with the assumption of using a barrier. The barrier used is an artificial barrier (pair of bricks) and natural barrier (tree). Furthermore, it is calculated whether after using the barrier, it can meet the noise quality standards for the educational facilities. Calculation of the level of reduction by an artificial barrier (pair of bricks) (see Figure 2) is calculated using the following equation [9]:

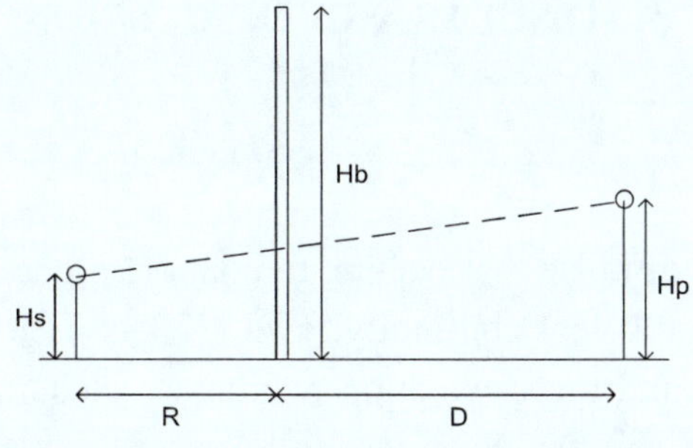

Figure 2. Relationship noise source receiver to barrier.

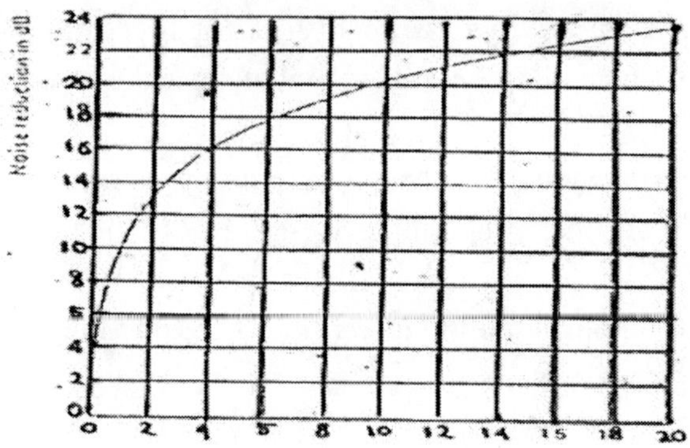

Figure 3. Sound reduction by barrier.

$$N = \frac{2}{\lambda} + (x + y - z)$$

where

$$x = \sqrt{R^2 + (Hb - Hs)^2}$$

$$y = \sqrt{D^2 + (Hb - Hp)^2}$$

$$z = \sqrt{(R + D)^2 + (Hp - Hs)^2}$$

N is a Fresnel number; *R* is the distance from the source of noise to barrier (m); *D* is the receiver distance to barrier (m); H_b is the barrier height (m); H_p is the receiver height (m); H_s is the high noise source (m).

By assuming that a Fresnel is known, the calculation is continued by using the graph in Figure 3 to determine the magnitude of attenuation of noise levels achieved [8].

3. Results and Discussion

3.1. Noise Level

To find out the level of noise that occurs, the data is processed using the L_{eq}, L_s, L_m, and L_{sm} formulas. The equivalent continuity level (L_{eq}) is used because the noise in the study area is continuous and the calculation results are close to L_{eq}. Using the formula above, the results are obtained for Monday as follows:

$L_1 = 77.26$; $L_2 = 76.29$; $L_3 = 74.01$; $L_4 = 73.84$; $L_5 = 68.71$; $L_6 = 66.34$; $L_7 = 71.75$.

So that the following values are obtained: $L_s = 75.60$; $L_m = 69.58$; $L_{sm} = 75.28$.

Then, by using the same method, the calculation of L_{eq}, L_s, L_m, and L_{sm} in each other day can be seen in Table 1.

Table 1 shows the maximum L_s occurring in Monday with noise levels of 75.60 dBA and minimum L_s occurring in Sundays with a noise level of 73.52 dBA. The daytime noise level of L_s in the study area is caused by the activities of the surrounding community and is more dominant due to transportation activities. The maximum value of L_m, occurring in Saturday has a noise level of 71.11 dBA and a minimum L_m, occurring in Monday and has a noise level of 69.58 dBA. Night noise level (L_m) is the background noise, which is the original noise study area with little transport activity, which could affect the noise level. The maximum value of L_{sm}, occurring in Saturday has a minimum noise level of 75.45 dBA and a minimum L_{sm}, occurring in Sunday that has a noise level of 74.11 dBA. This shows that the study area exceeds the quality standard for education, which is equal to 55 dBA, so it is not feasible for use as an educational facility.

The noise level of the study area (L_{sm}) in Monday, Friday, Saturday, and Sunday shows the values, which are almost the same. Therefore, it is necessary to have an equivalent noise level that can represent all days in the study area by using the following formula:

$$L_{sm\ eq} = 10\log 1/4 \times \{10^{0.1\times L_{sm\ Monday}} + 10^{0.1\times L_{sm\ Friday}} + 10^{0.1\times L_{sm\ saturday}} + 10^{0.1\times L_{sm\ Sunday}}\} = = 74.98 \text{ dBA}$$

Furthermore, an evaluation of the results of noise level (L_{sm}) equivalents was carried out in the study area with noise level quality standards with +3 dBA tolerance. The noise level in the study area uses quality standards for education equal to 55 dBA + 3 dBA = 58 dBA. From the results of equivalent L_{sm} calculations, the level of noise in the study area exceeds the quality level of noise levels for education (74.98 dBA > 58 dBA).

Table 1. Calculation Results of Ls, Lm, and Lsm (dBA)

Monday			Friday			Saturday			Sunday		
L_s	L_m	L_{sm}	L_s	L_m	L_{sm}	L_s	L_m	L_{sm}	L_s	L_m	L_{sm}
75.60	69.58	75.28	74.93	69.98	74.94	75.09	71.11	75.45	73.52	70.10	74.11

3.2. Noise Control

Efforts to reduce noise levels that occur in the study area can be done with the selection of several alternative barriers. The barriers used are two types, namely, an artificial barrier in the form of a couple of brick and natural barriers such as trees, shrubs, and bushes. The artificial glass barrier can reduce noise up to 30 dBA [5]. This glass barrier can be used as a door. The noise level can also be reduced by the presence of a natural barrier. A natural barrier can reduce noise by 4.6 dBA [7]. According to [10], several species are commonly used as a natural barrier, such as Adhatoda vasica, Adhatoda Zeylancia, Grasses, Terminalia Arjuna, and others. Natural barriers with a width of 15 m can reduce noise up to 9.9 dBA.

Calculation of the reduction noise level with artificial barrier (pair of bricks) is performed by using the below equations [9].

(1) The noise source height H_s = 50 cm, because the noise source lies in the motor vehicle exhaust, which averages 0.5 m in height.
(2) The height of the receiver (H_p) is the average Indonesian human height equal to ±1.6m.
(3) The noise source distance to the barrier R = 4 m.
(4) The distance from receiver to the barrier D = 4 m.
(5) The effective height of the barrier H_b = 2 m
(6) For wavelength, we have: $\lambda = c/f = 0.34$ m, because we select wave velocity in air c = 340 m/s the and frequency f = 1000 Hz.

According to [3], the value for wave velocity in air equal to 340 m/s should be used at a temperature of 27 °C with a pressure of 1 atm and the frequency used equal to 1000 Hz (this frequency is acceptable to the human ear). Calculation of the level of reduction by an artificial barrier (pair of bricks) is calculated using the following equation:

$$X = \sqrt{R^2 + (Hb - Hs)^2}$$

that is X = 4.27;

$$Y = \sqrt{D^2 + (Hb - Hp)^2}$$

that is Y = 4.02;

$$Z = \sqrt{(R + D)^2 + (Hp - Hs)^2}$$

that is $Z = 8.08$;

$$N = \left(\frac{2}{\lambda}\right) + (X + Y - Z)$$

that is $N = 6.092$.

At the Fresnel number known, the calculation is continued by using the graph in Figure 3 to find out the reduction in noise level achieved [8]. By using value of $N = 6.092$, we obtained a reduction in noise level equal to 18 dBA.

Artificial barriers can be used as efforts to control noise in the study area, so that the noise level (L_{sm}) in the study area can be reduced to L_{sm} after artificial barrier = 74.98 dBA – 18 dBA = 56.98 dBA.

The results of noise reduction with an artificial barrier become 56.98 dBA, so that it meets the noise quality standards for the education area with a quality standard of 55 + 3 = 58 dBA. Placement of this barrier can be placed in front of the entrance to the educational facility with a height of 2 m, which currently has a barrier/wall of less than 1m. The selection of the pair of bricks is based on the following things, which are more economical than concrete, durable and long lasting, and can be used as a fence for buildings inside.

Conclusion

Noise level measurement result in the study area was obtained: $L_{sm\ equivalent} = 74.98$ dBA. Values of L_{sm} generally exceeded the quality standard for the noise levels of 58 dBA for education area. To control noise level in the area study, we suggested the use of an artificial barrier. The noise level at the area study became 56.98 dBA by an artificial barrier so that it met quality standards for educational facilities.

References

[1] Basuki, H. *Designing, Planning the Field*. Alumni Publisher. Bandung. 1986 (In Indonesian).

[2] Beritatrans. *Motorcycle population growth in Indonesia is highest*. Jakarta. 2015. http://beritatrans.com (In Indonesian).

[3] Doelle, L. L. *Environmental Acoustics*. Erlangga Publisher. Jakarta. 1972 (In Indonesian).

[4] Harris, M. *Handbook of Acoustical Measurement and Noise Control.* 3rd Edn., Mc Graw Hill Book Company. New York. 1991.

[5] Hobbs, F. D. *Traffic Planning and Engineering*. 2nd Edn., Gajah Mada University Press. Yogyakarta. 1995 (In Indonesian).

[6] Kementerian Lingkungan Hidup, R. I. *Decree of the Minister of Environment Number KEP-48 / MENLH / 11/1996 concerning Noise Level Standards*. Jakarta. 1996 (In Indonesian).

[7] Krisindarto, A. *Noise Level Mapping Due to Transportation Activities and Alternatives to the Selection of Barriers in Central Surabaya Region*. Final Project of Environmental Engineering Department FTSP – ITS. Surabaya. 2006 (In Indonesian).

[8] Lord, P.; Templeton, D. *Acoustic Details*. 3rd Edn, Penerbit Erlangga, Jakarta. 2001 (In Indonesian).

[9] Miller, R. K.; Wayne. V. M. *Handbook of Acoustical Enclosure and Barrier*. The Fair Mont Press Inc. Atlanta. 1978.

[10] Tyagi, V., et al. *Applied Acoustics*. 2006, *vol. 67*, 926 – 935.

[11] Wilson, C. E. *Noise Control: Measurement, Analysis and Control of Sound and Vibration.* Harper and Row Publisher. Chambridge. 1989.

In: Proceedings of the 2018 International Conference … ISBN: 978-1-53615-862-5
Editors: Ivan A. Parinov et al.

Chapter 49

STUDY OF WATER RESOURCES UTILIZATION AS ALTERNATIVE ELECTRICITY IN JEMBER, EAST JAVA, INDONESIA

Zainal Arief[1,*], Siti Mundari[1] and Erma Yuliaty[1]
[1]Industrial Engineering Department, University of 17 Agustus 1945, Surabaya, East Java, Indonesia

ABSTRACT

Water resources are natural resources, which have renewable properties and its presence is quite abundant in Indonesia. Utilization of water resources in an appropriate way is an alternative among the fuel crisis and fuel prices increased simultaneously in the last decade. Moreover, to encourage an economic growth in rural areas that has not been reached by PLN, the natural resources utilization is the right policy of Indonesia Government. As the engineering developments in the hydropower electric field, the power of the turbines with the head height $H = 1 - 1.5$m can be achieved to 4 – 5kW. One of the reasons why Indonesia Government pays attention to the development of this small-power Micro Hydro Power Plants (MHPPs) is that world oil prices tend to increase. Another reason is a requirement for electrical energy increases to reach a countryside. The analysis performed consists of hydrological analysis, developed optimization analysis and technical feasibility analysis. The analysis begins with data processing to obtain the basic profile of population, physical area and regional transportation. The result obtained is a recommendation on turbine selection. The turbines with a higher specific speed is more economical, because the higher specific speed means the turbine unit is more compact. However, the speed of the turbine wheel depends on the construction and strength of the turbine material and generator load. Therefore, the speed of water turbine shaft usually ranges between 125 and 750rpm. At the same time, the

* Corresponding Author's E-mail: maula.nafi@untag-sby.ac.id.

synchronous speed of the generator depends on the frequency and number of pole pairs. The output power generated by MHPPs in several rivers in Jember is 16,720VA. If it is assumed that each family gets 200VA, then the people who can use MHPPs is 83 families.

Keywords: water resources, energy potential, electricity

1. Introduction

Human needs for energy are increasing, according to an increase in living standards. The energy resources crisis has confronted us with serioius problems, especially the problem of development continuity and process of industrialization. Therefore, it is necessary to study alternative energy resources, including the study of Micro Hydro Electric Stations (MEHSs) or Micro Hydro Power Plants (MHPPs) [4, 5]. Water resources are natural resources that have renewable properties and its presence is quite abundant in Indonesia. Utilization of water resources in an appropriate way is an alternative among the fuel crisis and fuel prices increased simultaneously in the last decade. Moreover, to encourage economic growth in rural areas that has not been reached by PLN, the utilization of natural resources is the right policy of Indonesia Government.

2. Experiment

The experiment begins with collecting data. The micro hydro power plant (MHPP) requires data such as map, hydrometeorology, etc. The other data are present in Table 1. The flowchart of this research experiment can be seen in Figure 1 [2, 4, 5].

The analysis performed consists of hydrological analysis, developed optimization analysis and technical feasibility analysis. The analysis begins with data processing to obtain the basic profile of population, physical area and regional transportation.

3. Results and Discussion

3.1. Micro Hydro Power Plant (MHPP)

Micro hydro is a small-scale hydroelectric power plant (can reach several hundred kW). The small amount of energy, produced by micro hydro (compared to large-scale hydropower), has implications for the simplicity of the equipment and the small area of land needed for the installation and micro hydro operation.

Table 1. Required data

No	Type of Data	Sub Data
1.	Map	Topographic map with scale of 1:100,000 Land use map Geological map
2.	Hydro-meteorology	Climate data includes air temperature, humidity, solar radiation, wind speed, and evaporation Rainfall intensity and river water debit
3.	Other Data	Conditions and location of water resources Environmental aspects of potential water resources Population conditions and potential customers Distance of location to the PLN electricity sources Estimated debit and head Socio-economic conditions Topographic and lithological conditions Road / transportation facilities that already exist or are building Other related information such as: existing sources of electrical energy, fluctuations in electricity prices at the consumer level and so on

This is one of micro hydro advantages, which does not cause environmental damage. Micro hydro is suitable to apply in rural areas that have not been reached by electricity from Complete Guide for PLN [4, 5]. Micro hydro gets energy from water streams that has certain height differences. This energy is used to turn turbines connected to electric generators. Micro hydro can utilize water levels that are not high, for example at 2.5m water level, 400W can be generated. The energy potential of using national micro hydro is estimated to reach 7,500MW, while the current quantity used is only around 600 MW. The energy potential should not be too large, but micro hydro should be considered to expand electricity coverage in all sites of Indonesia.

The distribution of MHPPs with small capacity is: (i) micro MHPPs < 100kW, (ii) mini MHPPs < 100 – 999kW, (iii) small MHPPs 1,000 – 10,000kW. With technical progress, height $H = 1 - 1.5$m of turbine can be used and its capacity can be reached to 4 – 5kW. One of the reasons for developed countries to pay attention to the development of small-capacity MHPPs is that oil price has continued to increase in the past, in addition to increasing electricity demand.

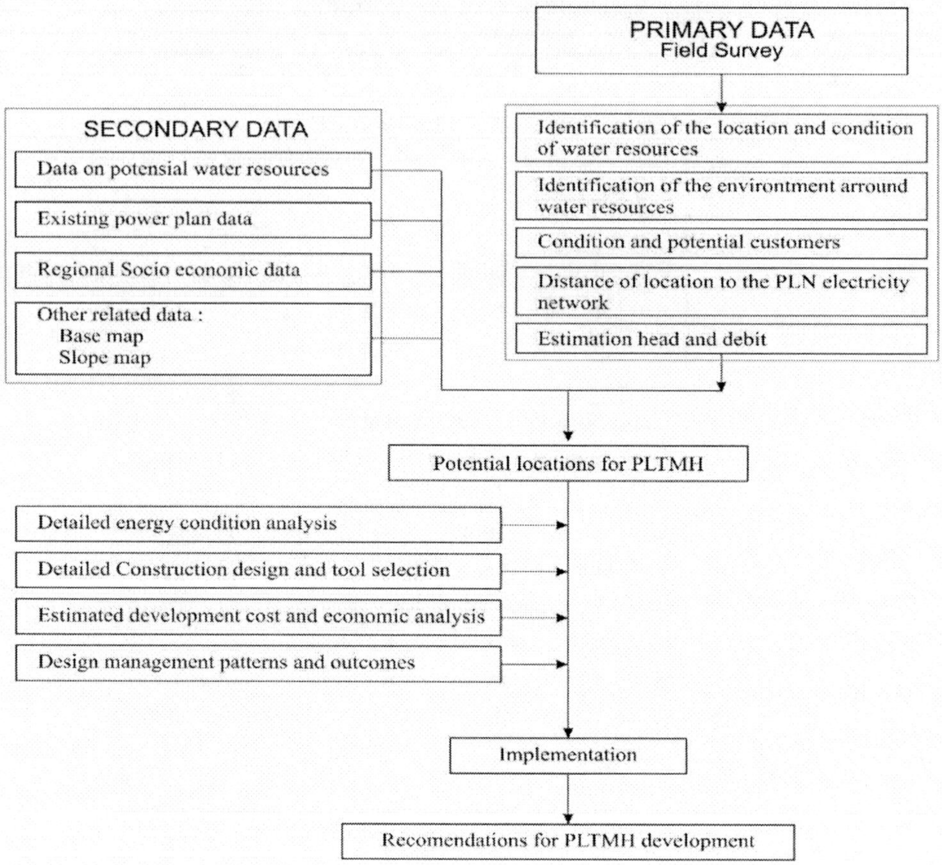

Figure 1. Research experiment flowchart.

In planning the development of MHPPs, Prof. Dr. E. Mosonyi, an expert on waterpower with an international reputation argues [3]:

(i) micro-words do not mean that for planning and implementation, the "microbrain" is needed; MHPPs require engineering graduates, besides being knowledgeable, they must have high knowledge;
(ii) it is true that all efforts need to be made to reduce the investment costs, some possibilities need to be examined regarding local conditions;
(iii) in developing countries, it is necessary to pay attention to social and demographic aspects, besides the economic and physical conditions, as well as people who will get electricity need to be studied;
(iv) MHPPs for rural areas need special attention according to the needs of irrigation and home industries.

Then the planning of a MHPP, especially for rural areas in developing countries, should be kept as small as possible. The generation cost per kWh is:

$$\text{Generation Cost per kWh} = \frac{\text{Cost per year}}{\text{kWh per year}}$$

$$\begin{aligned}&\text{Generation Cost per kWh}\\&= \frac{(\text{Capital Recovery Factor} \times \text{investment costs} + \text{Exploitation costs})\ \text{per year}}{\text{kWh per year}}\end{aligned}$$

To get the generation cost per kWh as small as possible it can be attempted:

(i) The smallest possible investment costs include:

- a simple MHPP plan,
- using local materials,
- using standard turbine/generator,
- using electronic load control (electronic load controller) instead of governor,
- using a turbine without a spiral house (open flume setting).

(ii) The smallest possible costs of exploitation include:

- expensive costs should not be used directly for exploitation and central supervision; as much as possible these expenses are used for the villagers by paying attention to their education;
- it is should not be used remote control or to send periodically personnel from the nearest PLN office to check MHPP (so that a salary is not included in the cost of MHPP exploitation);
- using a machine does not require maintenance and it should include only the factory cost;
- in order to clean the filter, to dispose the mud and so on, periodic personnel from the people of the nearest villages should be used.

(iii) Increasing the quantity of kWh to the larger amount means that the number of kWh used by the village must be larger. The efforts to utilize the generated quantity of kWh are necessary to expand by:

- expanding home industries that use electricity;
- using the electricity for joint needs as implemented by the Rupert Armstrong Evart Engineering in Nepal (1979) for supplying the hot water and cooling units;
- establishing new rural industries, such as sawmills, rice mills, or directed on reprocessing grated coconut, corn shells, and so on.

3.2. Calculation of Water Discharge for MHPP

Water debit calculation of flux discharge for MHPP is conducted by floating method. This measurement is performed by placing objects that cannot sink on the river surface at certain distance one of other and recording the time required by the floating object to move from one observation point to another certain point of observation. Floating objects used in this experiment could be any objects as long as they can float in the stream. The choice of a test place should be at relatively straight part of the river without many irregular streams. The width of the river is divided into three segments, and measurements are performed for each of segments. Then, the velocity of surface stream in each segment is calculated. The value of the speed of river stream (V) on surface in m/s is [1 – 3]: $V = L/t$, where L is the distance between two observation points (m), t is the time of floating (s).

To calculate the debit for each of segments, we used the formula: $Q = VA$, where Q is the internal discharge (m^3/s), V is the surface speed (m/s), medium value of A is the cross-section area of river segment (m^2).

3.3. Calculation of Initial Output Power (*P*)

The output power (P) is determined from the water discharge (Q) that enters the turbine and the head height difference. We used the following formula [1 – 3]: $P = 9.81\mu QH$, where $\mu = \eta_t\eta_{belt}\eta_g$, defining the MHPP efficiency equal to 60% (theoretically); Q is the debit, obtained from field measurements; H is the natural river elevation difference, being the river height difference at a distance of 10 m. So, the value of power obtained is the initial estimation of the power. Then $\eta_t = 0.74\%$ is the turbine efficiency for cross-flow turbines T-14 or equal to 0.75% for propeller open flume local; $\eta_{belt} = 0.98\%$ is the belt efficiency for flat belt or equal to 0.95% for V-shape belt; η_g is the generator efficiency (%).

3.4. Output Power, Generated by MHPP

Secondary data, obtained for several rivers in Jember area (from March 2017 to March 2018), consist of at least 290 liters/second and maximum debit is equal to 310 liters/second. Therefore, the average debit $Q = (290 + 310/2)$ liter/second $= 0.30\ m^3/s$. The average height is equal to 7.4 meters, then the output power, generated by the MHPP, is defined as

$$P = 9.81QH\eta_t\eta_{belt}\eta_g = 9.81 \cdot 0.3 \cdot 7.4 \cdot 0.74 \cdot 0.98 \cdot 0.9 = 14.21 \text{ kW}$$

or 14.21/0.85 kVA = 16.72 kVA

If it is assumed that each household gets 200 VA, the population, which can use the power from this MHPP includes 83 families. The waterwheel installed as one of the turbine types is shown in Figure 2.

The results were obtained with the aim of turbine optimization. The turbines with higher specific speed are more economical, because higher specific speed means that turbine unit is more compact.

Figure 2. Waterwheel turbine installed.

CONCLUSION

The output power, generated by MHPP in several rivers of Jember, is equal to 16720 VA. If it is assumed that each family gets 200 VA then the people who can use MHPP include 83 families.

REFERENCES

[1] Buyer, A. *Micro Hydro Power System,* Natural Resources Canada. 2008.

[2] Mandiri, Y, MHPP Planning. Padasuka, Yayasan Bina Desa Mandiri, Bandung. 2007 (In Indonesian).

[3] Masonyi, *Water Power Development-1.* Low Head Power Plants, Akademiai Kiado, Budapest. 2007.

[4] Nugroho, Sallata, MHPP (Micro Hydro Power Plant) *Complete Guide to Making Self-supporting Renewable Energy Sources,* BPPT. 2017. (In Indonesian).

[5] Notosudjono, D. MHPP Planning in Indonesia, BPPT, 2002, P. 68 (In Indonesian).

In: Proceedings of the 2018 International Conference … ISBN: 978-1-53615-862-5
Editors: Ivan A. Parinov et al.

Chapter 50

WEB-APPLICATION DEVELOPMENT FOR NETWORK ACCESS TO THE FEM MODULES OF THE ACELAN PACKAGE

***Dmitry N. Nadolin*[1], *Pavel A. Oganesyan*[1], *Arkady N. Soloviev*[1,2], *Artem E. Paterikin*[1] and *Sergey I. Kholostov*[1]**

[1]I. I. Vorovich Mathematics, Mechanics and Computer Sciences Institute, Southern Federal University, Rostov-on-Don

[2]Department of Theoretical and Applied Mechanics, Don State Technical University, Rostov-on-Don, Russia

ABSTRACT

The Finite element method (FEM) is a crucial tool for computer-aided engineering. Various packages are based on different implementation of FEM. This method involves solving complex computational problems. This paper considers the development of module architecture, web API and specific tools for ACELAN package to allow multiple users network access to computation modules. Client part of the project consists of user interfaces for FEM models description and visual postprocessing tool. The server side is capable of automated computations, scripts parsing, answering API requests, and storing model and materials information. The inner structure of the described model is present in this paper, along with results of load tests and performance analysis. Material storage module is based on the relational database build with object-relational mapping to allow usage of shared code base for different database engines. Visual postprocessing is developed with WebGL library. The scripting tool consists of script parser for Python in .NET framework called IronPython and custom DSL for describing FEM models and performing analyses of solution and automated computations. The developed system is scalable; each part can be used as a separate tool or integrated with extremal packages.

Keywords: finite element method (FEM), composites, client-server architecture, ACELAN

1. Introduction

Modern CAE packages rely on client server architecture, but in most cases both server and client applications are executed on the same machine. In case of problems with large computation time and high memory or CPU usage, it can be inefficient. Splitting computation and model description into different modules can be an advantage for CAE package, especially if web interface is present and server-side computing is used. ACELAN package use material properties identification [1, 2], which involves solving a series of direct problems for each representative volume. Most studies include series of numerical experiments with a described scenario. That makes automation of the process and computational resource management important. Visual modules of the package are obligatory for client application, but most computational tasks can be performed on the remote server. In this study, we compare efficiency of some approaches to software architecture and implementation, including both client and server-side applications.

2. Application Development

Development of client-server architecture for FEM package includes safe scripting tool, visual postprocessing, material properties database, and network API to exchange data with computational modules and allow shared user access to FEM problems in terms of material properties identification problems.

2.1. Server-Side Development

The server part is implemented using the cross-platform technology .NET Core and the .NET Standard platform, which allows using the received application both on servers running various operating systems and in application of solutions for PCs.

Material identification for composites with two components and coherence type, described in [3], starts with setting material properties for components. Each component can be isotropic or have any class of anisotropy. Effective properties of a mixture can produce another class of anisotropy. In most common cases, material in ACELAN COMPOS can have next properties: (i) scalar density and (ii) stiffness, electric permittivity, permeability, and coupling properties in matrix form, which is used in computational FEM modules of ACELAN [5, 6]. In most cases, a significant part of the material properties is zeros. A method for storing material properties for all anisotropy classes has been developed and implemented, which makes it possible to save space by effectively locating data in a relational database. Based on basic idea of sparse matrices storing as triplets, this method relies on object-relational mapping between custom classes for different materials and data about location and value of each non-zero element in property matrices.

The separation of materials by users was implemented, which allows us to limit the scope of data. Thus, a single library can be replenished in a multi-user mode, which simplifies the filling of the database with information and allows one to work with the application to replenish both the individual library and the general library of materials. For isotropic materials, there is a possibility to transform the properties into different types: the Young's modulus and Poisson's ratio, the modulus of all-round compression, and the Lamé coefficients. In the case of anisotropic materials with polarization properties, it can be useful to rotate the axes of the local coordinate system, in which material properties are described. Such a possibility was also implemented. Consider the formation of a rotation matrix for the elastic properties of anisotropic material. For the tensors of the first, second, third, and fourth ranks, there are known transformation equations from the orthogonal system to new orthogonal system. A model of three turns was considered: (i) the first rotation through an angle φ around axis; (ii) the second rotation through an angle ω around axis, (iii) the third rotation through an angle θ around axis. In this case, the components of the rotation matrix for elastic properties are expressed, as shown in [6].

The application also analyzes the material properties of composites obtained by numerical methods. The symmetry of the property matrices can be violated because of the computational error, as well as the zero matrix elements that can be replaced by values close to tolerance. The module offers an automated tool for correcting such inaccuracies and anisotropy class analysis for the resulting matrix.

The following requirements for the runtime environment of user programs were considered during the development: fault tolerance, isolation, multitasking, and scalability. The most obvious requirement for third-party programs execution environment is fault tolerance and the ability to recover from failures. Thus, specific features were implemented to ensure that the completion of user scripts does not lead to the loss of ACELAN package performance and the computer running the package, and that crashing one user script does not lead to disruption of other user scripts. Since the ACELAN package is available as a client-server application in the web-interface, it is necessary to isolate the execution environment of user scripts from the server environment. An important characteristic of software products is interactivity. The ACELAN package can be used by many users simultaneously by using the web-interface.

Considering the fact that to solve the identification problem, it is necessary to solve from 3 to 6 direct problems [3] with different boundary conditions, the waiting time for the completion of the work of the user scripts can reach unacceptably high values. Therefore, it is necessary to provide parallel operation of several independent scripts. In addition, simultaneous execution of multiple tasks more efficiently uses the resources of multiprocessor systems.

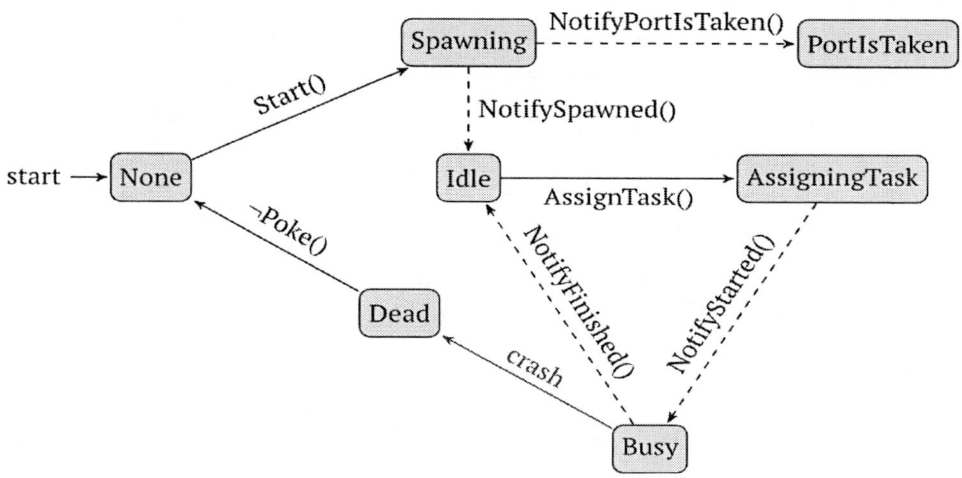

Figure 1. State machine of script interpreter.

Using an architecture, based on running each script in an isolated process, it is relatively easy to meet the requirements for fault tolerance and multitasking. The implementation of interaction between server processes and interpreter environments with the help of a network allows one to run the latter on different computers. It is possible that IN conditions, which establish a connection in both directions and limit the access of the interpreter processes to the file and operating systems, were achieved by running these processes on behalf of OS users with a limited set of rights. A protocol was developed for the interaction of the server and scripting environments, ensuring uninterrupted and fault-tolerant behavior of the system. When the system is initialized, N instances of the class responsible for executing the scripts are created, where N is the number of dedicated ports during configuration. These objects represent a kind of abstraction over the interpreter processes in server memory. They are responsible for interacting with these processes. Schemes of work for the cases $N = 1$ and $N > 1$ are no different, since the environments themselves do not exert any influence on each other. Thus, the entire system, consisting of a server and a single environment, can be represented as a finite state machine, represented in Figure 1, where solid lines indicate transitions initiated by the server side of the system, and dotted-in response to the server receive the corresponding event from the interpreter. Lines of red color indicate unexpected transitions, which, as a rule, occur due to failures in the execution of user scripts. The states of the yellow color have corresponding representations in the program itself, while the blue states are added to the circuit of the automaton for clarity.

2.2. Client-Side Development

Client side is a web application built with common front-end tools in .NET Core technology stack. Model creation is available both in interactive UI mode and in scripting mode. REST API is used to perform data transfer between client and server side, which is a common approach in a modern web development. Specific FEM features were developed in a WebGL visual postprocessing module.

Figure 2. Heat map visualization for whole representative volume and for single component.

Core possibilities of that module are: (i) visualization of models of different shapes with the possibility to provide optimization for the ACELAN data format; (ii) separation of different materials, from which the model is composed; (iii) visualization of heat maps for the analysis of results (Figure 2), with the presence of an accurate determination of the displayed value and averaging the data for the elements of the element; (iv) demonstration of various components of solution, including vector fields, visualization of the deformed state with the possibility of indicating the scale of the deformation, interacting with the model, rotating and scaling; (v) mapping the results of the solution to the problem.

3. Results and Discussion

3.1. Scripting Tool Performance Analysis

A series of experiments was conducted to evaluate additional time costs and the presence of which is due to the addition of the communication factor for the network and the need to compile user scripts before they are executed. As an instrument for solving direct problems, the ACELAN modules were used, based on the iterative methods of solving systems of linear equations [7]. The transition from Python user scripts to executable C# computational code was based on IronPython technology [8]. The purpose of the numerical experiment was to compare the performance of the two system configurations with the simultaneous operation of many users. In the first case, one thread was used to perform tasks. The second number of threads was equal to the number of available virtual computer cores. Since a computer with a four-core Intel Core i5-6400 processor was used for the experiments, the results presented below were obtained with four executing processes. The system was transferred to 50 copies of the tasks at a time, and the processing time of the entire set was measured, as was the time spent on each task individually.

The schedules of the average, maximum, and minimum waiting time for an individual task in this experiment are shown in Figs. 3 and 4. Figure 3 confirms the results of the experiment while waiting for execution of the entire package of tasks: execution of scripts in separate processes. So, it is calculated on the server by the time of waiting for an individual task at a depth of partitioning the model greater than 3. However, this effect can only take place if the server's load is optimal. In a real situation, while simultaneously receiving multiple requests, the server without deliberately lowering the load on the server by placing incoming requests in a queue or stack starts processing all these requests in different threads simultaneously. The parallel execution of tasks in the number of more processors on the computer inevitably leads to frequent switching of the context by the operating system, which increases the average time for waiting for tasks performed.

Thus, based on the results of experiments, it can be concluded that the performance of tasks in individual processes is objectively much slower than in the server process. Nevertheless, when the system is under high load due to the limitation of the number of simultaneously executed tasks, the time for performing custom tasks can be less by an amount comparable to the amount of additional costs for network interaction, the support of the necessary infrastructure, compilation, and execution in the DLR environment. During this study, an environment was created to execute user-side scripts on the server side, which has the properties of isolation, scalability, fault tolerance, and multitasking. This environment was implemented as a separate process-host of the interpreter of the developed object-oriented language. Scalability and multitasking are achieved with the help of running multiple copies of such processes. The implementation of the interaction between the server and the environment based on the network also allows us to run the processes of executing on separate (virtual or physical) computers. Isolation is achieved by means of the operating system, namely, launching executing processes on behalf of users with limited rights. A series of experiments was conducted to evaluate the effectiveness of the implemented system. The results of these experiments showed that there is a noticeable slowdown in work, but when the load increases, the implemented system shows the best result. The developed language and interpreter were successfully implemented in the finite-element package ACELAN.

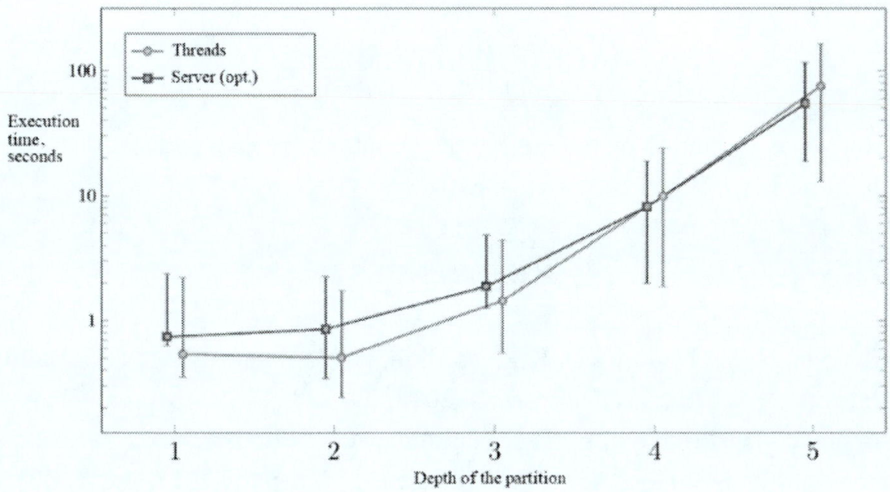

Figure 3. Execution time for the solution of one task for a simultaneous arrival of a set of tasks with optimal server usage case.

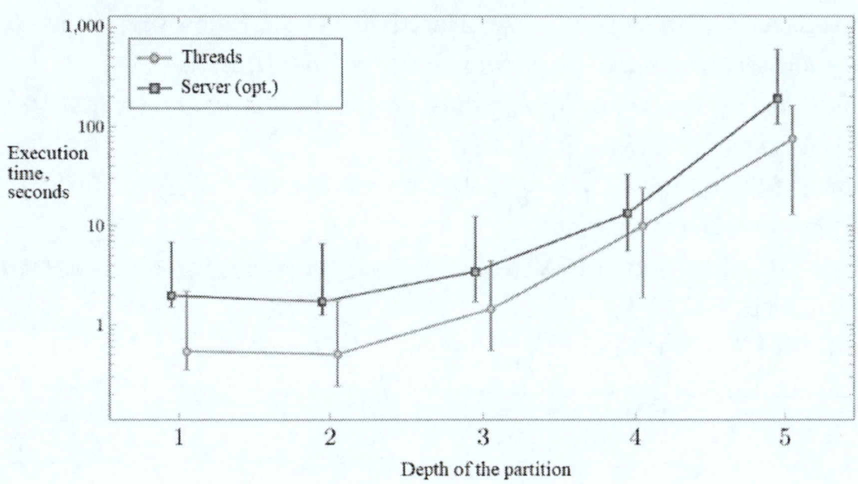

Figure 4. Execution time for the solution of one task for a simultaneous arrival of a set of tasks with non-optimal server usage scenario.

Conclusion

Development of the described modules allowed us to create universal architecture based on single responsibility ideas. Each of the modules can be integrated into other package or is replaced with external software. Most modules come with replacement ones, such as UI and scripting for model creation or desktop and web visualization tools. Computational modules can be executed on server, cloud platforms on a single client machine. These advantages allow ACELAN COMPOS to be an effective modeling tool for specific problems of material science.

Acknowledgment

The study was supported by the government contract (project part) 9.1001.2017/PCh.

References

[1] Kurbatova, N. V.; Nadolin, D. K.; Nasedkin, A. V.; Oganesyan, P. A.; Soloviev, A. N. In: *Advanced Structured Materials*. 2018, 69 – 88.

[2] Kudimova, I.; Mikhayluts, D.; Nadolin, A.; Nasedkin, A.; Nasedkina, A.; Oganesyan, P.; Soloviev, A. *Procedia Structural Integrity*. 2017, *vol. 6*, 301 – 308.

[3] Kurbatova, N. V.; Nadolin, D. K.; Nasedkin, A. V.; Nasedkina, A. A.; Oganesyan, P.A.; Skaliukh, A. S.; Soloviev, A. N. In: *Wave Dynamics and Composite Mechanics for Microstructured Materials and Metamaterials*, 2017.

[4] Belokon, A. V.; Eremeyev, V. A.; Nasedkin, A. V.; Solov'yev, A. N. *Journal of Applied Mathematics and Mechanics*. 2000, *vol. 64(3)*, 367.

[5] Belokon, A. V.; Nasedkin, A. V.; Solov'yev, A. N. *Journal of Applied Mathematics and Mechanics*. 2002, *vol. 66(3)*, 481.

[6] Lehnitsky, S. G. *Elastic Theory of Anisotropic Bodies*. 2nd Edn. Moscow. Science, 1977, pp. 1 – 416 (In Russian).

[7] Paige, C. C.; Saunders, M. A. *SIAM Journal on Numerical Analysis*, *vol. 12 (4)*, 617 – 629 (1975).

[8] Foord, M. J.; Muirhead, C.; Hugunin, J. *IronPython in Action*. Manning, Greenwich, 2009.

About the Editors

Ivan A. Parinov, DrSc,

I. I. Vorovich Mathematics, Mechanics and Computer Sciences Institute,
Southern Federal University, Rostov-on-Don, Russia, Chief Research Fellow

DrSc Ivan A. Parinov received the MSc in Mechanics from the Rostov State University, Mechanics and Mathematics Department in 1978 (Rostov-on-Don, Russia), the PhD in Physics and Mathematics from the Rostov State University in 1990 and the DrSc in Techniques from the South-Russian State Technical University in 2008. He is Corresponding Member of Russian Engineering Academy (2014) He is working in the I. I. Vorovich Mathematics, Mechanics and Computer Sciences Institute of the Southern Federal University (former Rostov State University) from 1978. Today, he is Chief Research Fellow. He has head in 1993-2018 the Research Grants and Programs from Soros Foundation, Collaboration for Basic Science and Education (COBASE, USA), Russian Foundation for Basic Research (13 grants), Russian Ministry for Science and Education (8 grants and scientific proposals). He has published about 300 scientific technical publications among them 28 monographs (in particular, 8 books published with Springer and 13 books with Nova Science Publishers). He has 14 Russian and Soviet patents and 4 certificates on Russian state registration of programs for computer. His developments were awarded 30 gold medals and special awards of European exhibitions. DrSc Ivan A. Parinov is reviewer of Zentralblatt fur Mathematik (Germany) and Mathematical Reviews (USA) from 1994 (above 450 reviews), review papers of journal SOP Transactions on Theoretical Physics (Scientific Online Publishing) from 2013. He was INTAS expert (2004), FP-7 expert (2010), ERA.NET RUS expert (2011, 2015, 2017) and he is the Russian Academy of Science expert (from 2016), the Russian Foundation for Basic Research expert (from 2017), the Russian Science Foundation (from 2017) and Russian Federal expert of scientific-technical area (2012-2021). He is winner of competition of the Russian Department of Education and Science (2014-2016 and 2017-2019) on direction “Organization of Fulfillment of the Scientific Researches”. DrSc Ivan A. Parinov is Active Member of the New York Academy of Sciences (1994), member of the American Mathematical Society (from 1996), the European Mathematical Society (from 1998) and The Indian Structural Integrity Society (from 2016). He has honorary title "Honorary Member of All-Russian Society of Inventors and Rationalizers" (2008). His research interests include R&D of novel materials and composites (in particular, high-temperature superconductors and ferropiezoelectrics), fracture mechanics and strength physics, acoustic emission, optics, mathematical modeling, different applications of advanced materials and composites. More common information is located into INTERNET-page: http://www.math.rsu.ru/niimpm/strl/welcome.en.html.

Shun-Hsyung Chang, PhD

National Kaohsiung University of Science and Technology,
Department of Microelectronics Engineering, Kaohsiung, Taiwan, Professor

Dr. Shun-Hsyung Chang received the Ph.D. degree in electrical engineering from National Sun-Yat Sen University, Taiwan, in 1990. Afterwards, he served as an associate professor and has been a professor since 1997 in the Department of Electrical Engineering at National Taiwan Ocean University, Keelung, Taiwan, where he was the chairman 1998~2003. He also had served as the Vice President of National Kaohsiung Marine University, Kaohsiung, Taiwan, during 2003~2006. He is a professor in the Department of Microelectronic Engineering at this university presently. He was head of 3 grants in 2010-2017 on joint program of the Ministry of Science and Technology of Taiwan and the Russian Foundation for Basic Research. He has published over 250 scientific-technical publication, among them 13 monographs (4 books with Springer and 5 books with Nova Science Publishers). His outstanding performance in research brought him great honor. In 1984, he was granted The Creative Youth prize by the Ministry of Education. He received the Long-Term Paper Award, organized by Acer Incorporated, in 1994, 1996 and 1998. He and his graduate students won the Third Prize of Graduate Team in TI-Taiwan 1994 DSP Design Championship. He also obtained the Distinguished Award in the contest of Special Topic of Communication System in graduate division held by the Ministry of Education in 2000. He contributes greatly in editorship. He was the editor-in-chief of the Journal of Marine Science and Technology 1998~2001, of which he is an editor at present. He is currently the editor-in-chief of the Journal of Ocean and Underwater Technology 2008-2017. His research interests include electrical engineering; electronics and communication engineering; signal, image, video and array processing; underwater and acoustic communication; underwater acoustics; Kalman filtering and detection. More common information is located into INTERNET-page: https://www.researchgate.net/profile/Shun-Hsyung_Chang

Yun Hae Kim, PhD

ex-Acting President of Korea Maritime and Ocean University,
Director of Idea Factory Center, Busan, South Korea, Professor

Professor KIM Yun-Hae graduated from Department of Marine Engineering, Korea Maritime and Ocean University in 1983 (B.E.) and in 1987 (M.E.). Moreover, he graduated from Department of Mechanical Engineering, The University of Tokushima, Japan (M.E.) in 1990. He received PhD degree from Kyushu University in Mechanical Engineering, Japan by 1993. He worked at NIST, USA, as a visiting professor in 1997-1998, and his studies focused on the fabrication and evaluation of composite materials. He worked as research assistant (1985-1987) at Korea Maritime and Ocean University. He published over 150 papers in international and national journals. He joined over 60 conferences and symposia in international and national level as general speaker, invited speaker or keynote speaker. He founded two international conferences, namely, Asian Conference on Engineering Education (ACEE, 2009) and Asian Conference on Campus Sustainability (ACCS, 2015). He was a president of Korea Association of Green Campus Initiatives (KAGCI). He was the ex-Acting president of Korea Maritime and Ocean University, ex-Dean of Student Affairs and ex-Dean of Academic Affairs from 2014 to 2016. He was a conference chair of the Advanced Materials Development and Performance (AMDP2014). At that time, over 400 delegates participated in this conference. He was the editor of SCI journal, International Journal of Modern Physics B. Now, he is working as the Editor-in-Chiefs of the Journal of Ocean Engineering and Technology for 9 years and Nano Hybrids Composites. In particular, he joined in last five years (2013 - 2017) over 20 conferences as Keynote Speaker and Conference Co-Chairs. He is working as Director of Idea Factory Center at KMOU. He received many awards: Ministry of Science and Technology (Jang Young Sil Best Award, Korea), Ministry of Science and Technology, Korea, Distinguished Professor Award, The Korean Society of Engineering Education, Distinguished Service Medal, The Korean Society of Engineering Education, Ministry of Education, Science and Technology (Ministry Award, Engineering Education Innovation Field), Ministry of Education, Science and Technology, Korea, Best Paper Award, ICPNS, Seattle, USA. His research interests include R&D of novel materials and composites based on polymer materials (in particular, environmental characteristics), fracture mechanics and strength evaluations, ocean applications of advanced materials and composites. More common information is located into INTERNET-page: https://www.researchgate.net/scientific-contributions/2011562234_Yun-Hae_Kim.

INDEX

D

E

F

G

H

I

L

M

N

T

U

V

W

Z

Related Nova Publications

Carbon Steel: Microstructure, Mechanical Properties and Applications

Editor: Clyde Phelps

Series: Materials Science and Technologies

Book Description: This compilation opens with a section discussing the key interactions required to successfully friction stir weld carbon steels. Friction stir welding is a relatively recent development, predominantly technologically established using aluminum, where the lower melting point of the material greatly aided the process.

Hardcover ISBN: 978-1-53615-435-1
Retail Price: $160

Dielectric Materials and Applications

Editor: Pankaj K. Choudhury, PhD

Series: Materials Science and Technologies

Book Description: The book *Dielectric Materials and Applications* focuses on the recent research advancements in the area of dielectrics that can be utilized in a variety of technology-oriented applications.

Hardcover ISBN: 978-1-53615-316-3
Retail Price: $230

To see a complete list of Nova publications, please visit our website at www.novapublishers.com

PHYSICAL CHEMISTRY RESEARCH FOR ENGINEERING AND APPLIED SCIENCES

VOLUME 1

Principles and Technological Implications